**Statistical Quality Control**

# Statistical Quality Control
Using Minitab, R, JMP, and Python

*Bhisham C. Gupta*
*Professor Emeritus of Statistics*
*University of Southern Maine*
*Portland, ME*

This edition first published 2021

© 2021 John Wiley & Sons, Inc.

The right of Bhisham C. Gupta to be identified as the author of this work has been asserted in accordance with law.

*Registered Office*
John Wiley & Sons, Inc., 111 River Street, Hoboken, NJ 07030, USA

*Editorial Office*
111 River Street, Hoboken, NJ 07030, USA

For details of our global editorial offices, customer services, and more information about Wiley products visit us at www.wiley.com.

Wiley also publishes its books in a variety of electronic formats and by print-on-demand. Some content that appears in standard print versions of this book may not be available in other formats.

*Library of Congress Cataloging-in-Publication Data*
ISBN 978-1-119-67163-3 (hardback)
ISBN 978-1-119-67170-1 (ePDF)
ISBN 978-1-119-67172-5 (ePub)
ISBN 978-1-119-67171-8 (oBook)

Cover image:
Cover design by

Set in 9.5/12.5pt STIXTwoText by SPi Global, Pondicherry, India

SKY10025840_032521

*In loving memory of my parents, Roshan Lal and Sodhan Devi*

# Contents

# Preface

This is an introductory textbook on statistical quality control (SQC), an important part of applied statistics that is used regularly to improve the quality of products throughout manufacturing, service, transportation, and other industries. The objective of this book is to provide a basic but sound understanding of how to apply the various techniques of SQC to improve the quality of products in various sectors. Knowledge of statistics is not a prerequisite for using this text. Also, this book does not assume any knowledge of calculus. However, a basic knowledge of high school algebra is both necessary and sufficient to grasp the material presented in this book.

Various concepts of the Six Sigma methodology are also discussed. Six Sigma methodology calls for many more statistical tools than are reasonable to address in one book; accordingly, the intent here is to provide Six Sigma team members with some basic statistical concepts, along with an extensive discussion of SQC tools in a way that addresses both the underlying statistical concepts and their applications.

## Audience

This book is written for second-year or third-year college/university students who are enrolled in engineering or majoring in statistics, management studies, or any other related field, and are interested in learning various SPC techniques. The book can also serve as a desk reference for practitioners who work to improve quality in sectors such as manufacturing, service, transportation, medical, oil, and financial institutions, or for those who are using Six Sigma techniques to improve the quality of products in such areas. This book will also serve those who are preparing for certification in Six Sigma training, such as the Six Sigma Green Belt. In the Western world, these kinds of certifications are very common, particularly in the manufacturing, service, transportation, medical, and oil industries.

## Topics Covered in This Book

In Chapter 1, we introduce the basic concept of quality, as discussed by various pioneer statisticians, engineers, and practitioners. A few important pioneers in this area are W. Edwards Deming, Joseph M. Juran, Philip B. Crosby, and Armand V. Feigenbaum. Concepts of quality control and quality improvements are discussed. Finally, we discuss how management can help improve product quality.

In Chapter 2, we introduce the Six Sigma methodology. In the current era, Six Sigma methodology is considered the statistical standard of quality. During the past three decades, Six Sigma methodology has become an integral part of many manufacturing and service companies throughout the world. The benefits of Six Sigma methodology are also discussed.

In Chapter 3, we introduce the different types of data that we encounter in various fields of statistical applications, and we define some terminology, such as *population, sample,* and different measures. We then introduce various graphical and numerical tools needed to conduct preliminary analyses of data. We introduce the statistical software packages Minitab version 19, Rx64 3.6.0. Finally, some discrete and continuous probability distributions commonly encountered in the application of SQC are discussed.

In Chapter 4, we study different types of sampling methods – simple random sampling, stratified random sampling, cluster random sampling, and systematic random sampling – that are encountered whenever we use sampling schemes to study specific populations. Then we discuss how estimates of various population parameters are derived under different sampling methods.

In Chapter 5, we discuss Phase I control charts for variables. Phase I control charts are used to detect large shifts in a given process in any manufacturing or service industry. Large shifts usually occur when a process is either brand-new or has undergone drastic changes: for example, a new industry is started and/or new machines are installed in an established industry. This also occurs when a significant new technology is introduced. These kinds of changes can also usually occur whenever a high level of administration is changed: for example, a new CEO or CTO (Chief Technical Officer) is hired.

In Chapter 6, we discuss Phase I control charts for attributes. Control charts for attributes are used whenever we collect a set of counted data. Again, Phase I control charts are used to detect significant shifts in a given process established in any manufacturing or service industry.

In Chapter 7, we discuss Phase II control charts to detect small shifts. Small shifts usually occur when a well-established process has just experienced small changes. Some examples include workers having gone through some kind of training or specific machines having had their annual checkups, calibrations, or renovations, etc.

In Chapter 8, we discuss various types of process capability indices (PCIs). PCIs are used when a process is well established and the manufacturer or a service provider wants to ensure that they can deliver a product that meets the consumer's requirements. In this chapter, we discuss how to use PCIs to help any given industry avoid the severe consequences of heavy losses. Then we discuss various aspects of pre-control, which is an important tool in SQC. As we discuss later, in Chapter 9, different sampling schemes are used to sort out the defective parts from shipments. Pre-control is a mechanism to reduce the amount of sampling and inspection required to validate that a process is producing products consistent with customer's expectations. In the last part of this chapter, we discuss aspects of measurement system analysis (MSA). Sometimes a measurement system is responsible for the bad quality of a product; proper MSA can avoid a measurement system that may become the root cause of a product's failure.

In Chapter 9, we discuss various kind of acceptance sampling schemes that are still in use in some places in the world. These schemes were used frequently in the United States during World War II, when armaments and ammunitions were being produced in huge quantities. However, these sampling schemes are currently not very popular since they are sometimes very expensive to use and are often not very effective. The methods discussed in the previous chapters are more effective in achieving processes that produce better-quality products. Finally, we discuss some sampling standards and plans: ANSI/ASQ Z1.4-2003, ANSI/ASQ Z1.9-2003, Dodge's Continuous Sampling Plans and MIL-STD-1235B.

Chapter 10 is available for download on the book's website: www.wiley.com/college/gupta/SQC. This chapter presents more detail about the statistical software packages used in this text. In Chapter 3, we introduce the statistical software packages Minitab and R; but due to lack of space, certain basic concepts of these packages are not discussed. In Chapter 10, we discuss these additional aspects of Minitab and R. We also discuss two other statistical packages: JMP and Python. Then, using JMP and Python, we study SQC techniques using examples handled earlier in the book with Minitab and R. After going through the material presented in this chapter, you will be able to analyze different data situations using Minitab, R, JMP, and/or Python; cover all of the SQC techniques discussed in this book; and implement them in various sectors whenever and wherever high-quality products are desired.

## Approach

In this text, we emphasize both theory and application of SQC techniques. Each theoretical concept is supported by a large number of examples using data encountered in real-life situations. Further, we illustrate how the statistical packages Minitab® version 19, R® version 3.6.0, JMP® Pro-15, and Python® version 3.7 are used to solve problems encountered when using SQC techniques.

## Hallmark Features

As indicated, we incorporate the statistical packages Minitab and R throughout the text and discuss JMP and Python examples in Chapter 10. Our step-by-step approach with these statistical packages means that no prior knowledge of their use is required. After completing a course that includes this book, you will be able to use these statistical packages to analyze statistical data in quality control. Familiarity with the quality control features of these packages will further aid you as you learn additional features pertaining to other related fields of interest.

## Student Resources

- Data sets with 20 or more data points in examples and exercises are saved as Minitab, CSV (ANSI), and JMP files and available on the book's website: www.wiley.com/college/gupta/SQC. The CSV (ANSI) files can easily be imported into the R and Python software discussed in this book.
- Solutions to all of the odd-numbered review practice problems presented in this text are available to students on the book's website: www.wiley.com/college/gupta/SQC.

## Instructor Resources

- Data sets with 20 or more data points in examples and exercises are saved as Minitab, CSV (ANSI), and JMP files and made available on the book's website: www.wiley.com/college/gupta/SQC. The CSV (ANSI) files can easily be imported into the R and Python software discussed in this book.

- Solutions to all review practice problems presented in this text are available to instructors on the book's website: www.wiley.com/college/gupta/SQC.
- PowerPoint slides to aid instructors in the preparations of lectures are available on the book's website: www.wiley.com/college/gupta/SQC.

## Errata

I have thoroughly reviewed the text to make sure it is as error-free as possible. However, any errors discovered will be listed on the book's website: www.wiley.com/college/gupta/SQC.

If you encounter any errors as you are using the book, please send them to me at bcgupta@maine.edu so that they can be corrected in a timely manner on the website and in future editions. I also welcome any suggestions for improvement you may have, and I thank you in advance for helping me improve the book for future readers.

## Acknowledgments

I am grateful to the following reviewers and colleagues whose comments and suggestions were invaluable in improving the text:

Dr. Bill Bailey, Kennesaw State University
Dr. Raj Chikkara, Professor Emeritus, University of Houston
Dr. Muhammad A. El-Taha, Professor of Mathematics and Statistics, University of Southern Maine
Dr. Sandy Furterer, University of Dayton, Ohio
Dr. Irwin Guttman, Professor Emeritus of Statistics, SUNY at Buffalo and Univ. of Toronto
Dr. Ramesh C. Gupta, Professor of Statistics, University of Maine
Dr. Kalanka P. Jayalath, University of Houston
Dr. Jamison Kovach, University of Houston
Dr. Eric Laflamme, Associate Professor of Statistics, Plymouth State University
Dr. Mary McShane-Vaughn, Principal, Partner-University Training Partners
Dr. Daniel Zalewski, University of Dayton, Ohio
Dr. Weston Viles, Assistant professor of Mathematics and Statistics, University of Southern Maine

I would like to thank George Bernier (M.S. Mathematics, M.S. Statistics), who is a lecturer in mathematics and statistics at the University of Southern Maine. He provided assistance in the development of material pertaining to R and also helped by proofreading two of the chapters.

I would also like to express my thanks and appreciation to Dr. Eric Laflamme, Associate Professor of Mathematics and Statistics at Plymouth State University of New Hampshire, for helping by proofreading five of the chapters. Last but not least, I would also like to thank Mark W. Thoren (M.S. Electrical Engineering, staff scientist at Analog Devices) for providing assistance in the development of material pertaining to Python. Finally, I would like to thank Dr. Mary McShane-Vaughn, Principal at University Training Partners, for providing Chapter 2 on Lean Six Sigma. Her concise explanation of the subject has helped give context to why we must monitor the variability of processes to achieve and sustain improvements.

I acknowledge Minitab® for giving me free access to Minitab version 19 and allowing me to incorporate the Minitab commands and screenshots in this book. Minitab® and the Minitab logo are registered trademark of Minitab. I also thank the SAS Institute for giving me free access to JMP Pro-15 and allowing me to incorporate the JMP commands and screenshots in this book. JMP® and SAS® are registered trademarks of the SAS institute in the United States and other countries. I would also like to thank all the contributors to the libraries of R version 3.6.0 and Python version 3.7.

I would like to gratefully thank my family and acknowledge the patience and support of my wife, Swarn; daughters, Anita and Anjali; son, Shiva; sons-in-law, Prajay and Mark; daughter-in-law, Aditi; and wonderful grandchildren, Priya, Kaviya, Ayush, Amari, Sanvi, Avni, and Dylan.

Bhisham C. Gupta

# About the Companion Website

This book is accompanied by a companion website:

**www.wiley.com/go/college/gupta/SQC**

The companion websites include:

- Chapter 10 (instructor and student sites)
- SQC Data Folder (instructor and student sites)
- PowerPoint Files (instructor site only)
- Instructors' Solution Manual (instructor site only)
- Students' Solution Manual (student site only)

# 1

# Quality Improvement and Management

## 1.1 Introduction

Readers of this book have most likely used or heard the word *quality*. The concept of quality is centuries old. Many authors have defined *quality*, and some of these definitions are as follows:

- Joseph M. Juran defined *quality* as "fitness for intended use." This definition implies that *quality* means meeting or exceeding customer expectations.
- W. Edwards Deming stated that the customer's definition of *quality* is the one that really matters. He said, "A product is of good quality if it meets the needs of a customer and the customer is glad that he or she bought that product." Deming also gave an alternative definition of *quality*: "A predictable degree of uniformity and dependability with a quality standard suited to the customer."
- Philip B. Crosby defined *quality* as "conformance to requirements, not as 'goodness' or 'elegance.'" By *conformance*, he meant that the performance standard must be zero defects and not "close enough." He is known for his concept of "Do it right the first time."

The underlying concept in all these definitions is much the same: consistency of performance and conformance with the specifications while keeping the customer's interests in mind.

## 1.2 Statistical Quality Control

*Statistical quality control* (SQC) refers to a set of statistical tools used to monitor, measure, and improve *process* performance in real time.

> **Definition 1.1** A *process* may be defined as a series of actions or operations that change the form, fit, or function of one or more input(s) as required by a customer. A process may also be defined as a combination of workforce, equipment, raw material, methods, and environment that work together to produce a product. Figure 1.1 shows various steps that usually take place in any process, whether in a manufacturing or non-manufacturing environment.

The quality of the final product depends on how the process to be used is designed and executed.

The concept of SQC is less than a century old. Dr. Walter Shewhart (1931), working at the Westinghouse Hawthorne plant in Cicero, Illinois, drew the first statistical process control (SPC) chart

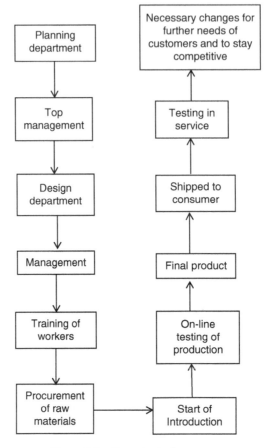

**Figure 1.1** Flow chart of a process.

in 1924. While working at Hawthorne, Shewhart met and influenced W. Edward Deming and Joseph Juran; later, they went on to champion Shewhart's methods. Shewhart, Deming, and Juran are often considered the three founders of the quality improvement movement.

As mentioned above, SQC is a set of statistical tools used to monitor, control, and improve process performance. These essential tools are (i) SPC, (ii) acceptance sampling plans, and (iii) design of experiments (DOE). DOE is used to improve the process and find important control factors, whereas SPC monitors these factors so that the process remains in a steady state. SPC is one of the important tools that makes up SQC. However, the term *statistical process control* is often used interchangeably with *statistical quality control*.

### 1.2.1 Quality and the Customer

The customer or consumer plays a very important role in achieving quality, for it is the customer who defines the quality of the product. If the customer likes the product and is satisfied with it the way it functions, then the probability is high that they will be willing to buy the product again in the future, indicating that you have a quality product. However, quality can also be achieved through innovation. Quality is not static; rather, it is an ongoing process. For example, a given product may be of great quality today – but if no further innovative improvements are made, it may become obsolete in the future and consequently lose its market share. It should be obvious that the required innovation can only be defined by the producer.

The customer is not in a position to tell how a product should look like 5 or 10 years from now. For example, five decades ago, a customer could not imagine electric cars or self-driven cars, or small computers replacing the huge computers that used to occupy entire rooms. The customer is only the judge of the product in its current form. In other words, a concern about quality begins with the customer, but the producer must carry it into the future. The producer or their team has to incorporate their innovative ideas at the design stage. This is called *continuous improvement* or *quality forever*. We will have a brief look at this concept later in this chapter.

It is important to note that a customer can be *internal* or *external*. For example, a paper mill could be an internal or external customer of a pulp mill. If both the paper and the pulp mill are owned by the same organization, then the paper mill is an *internal* customer; otherwise, it is an *external* customer. Similarly, various departments are internal customers of the Human Resources department. Another example is that students from various departments of a university taking a course from another department are internal customers, whereas a part-time student from outside the university is an external customer. In such cases, the company or organization should not assume that if its internal customers are satisfied, external customers are also automatically satisfied. The needs of external customers may be entirely different from those of internal customers, and the company must strive to meet both sets of needs. Furthermore, the goal of a company or an organization should be that all customers are satisfied not only for the moment but forever.

In summary, to achieve quality and competitiveness, you must first achieve quality today and then continue to improve the product for the future by introducing innovative ideas. To do this, an organization must take the following steps:

1) Make the customer its top priority. In other words, it should be a customer-focused organization.
2) Make sure the customer is fully satisfied and, as a result, becomes a loyal customer. A *loyal* customer is the one who will always give reliable feedback.
3) Create an environment that provides the most innovative products and has as its focus quality improvement as an ongoing process.
4) Take data-driven action to achieve quality and innovation. That is, the organization must collect information systematically, following appropriate sampling techniques to obtain data from internal as well as external customers about their needs and analyzing it to make necessary improvements. This process should be repeated continuously.

## 1.2.2 Quality Improvement

Different authors have taken different steps to achieve *quality improvement*. In this chapter, we quote the steps suggested by four prominent advocates of quality who revolutionized the field of SQC: Philip B. Crosby, W. Edwards Deming, Joseph M. Juran, and Armand V. Feigenbaum. We first discuss ideas suggested by Crosby, Feigenbaum, and Juran; later, we will look those from W. Edwards Deming.

Following are Juran's 10 steps to achieve quality improvement (Uselac 1993, p. 37; Goetsch and Davis 2006):

1) Build awareness of both the need for improvement and opportunities. Identify gaps.
2) Set goals for improvement.
3) Organize to meet the goals that have been set. They should align with the company's goal.
4) Provide training.

5) Implement projects aimed at solving problems.
6) Report progress.
7) Give recognition.
8) Communicate results.
9) Keep scores. Sustain these and continue to perfection.
10) Maintain momentum by building improvement into the company's regular system.

Next, we summarize Armand V. Feigenbaum's philosophy for total management (Tripathi 2016; Watson 2005):

- Quality of products and services is directly influenced by nine Ms: Markets, Money, Management, Men, Motivation, Material, Machines and Mechanization, Modern information methods, and Mounting product requirements.
- Three steps to quality: (i) management should take the lead in enforcing quality efforts and should be based on sound planning; (ii) traditional quality programs should be replaced by the latest quality technology to satisfy future customers; (iii) motivation and continuous training of the entire workforce gives insights about organizational commitment to the continuous quality improvement of products and services.
- Elements of total quality to enable a *total customer focus* are as follows:
  - Quality is the customer's perception.
  - Quality and the cost are the same, not different.
  - Quality is an individual and team commitment.
  - Quality and innovation are interrelated and mutually beneficial.
  - Managing quality is managing the business.
  - Quality is a principle.
  - Quality is not a temporary or quick fix.
  - Productivity is gained by cost-effective, demonstrably beneficial quality investment.
  - Implement quality by encompassing suppliers and customers in the system.

Feigenbaum was the first to define a system engineering approach to quality. He believed that total quality control combines management methods and economic theory with organizational principles, resulting in commercial leadership. He also taught that widespread quality improvement performance in a nation's leading businesses is directly related to quality's long-term economic impact.

Philip B. Crosby is well known for his "Quality Vaccine" and 14 steps to quality improvement. The Quality Vaccine consists of the following three ingredients:

- Determination
- Education
- Implementation

Crosby's suggested set of 14 steps to quality improvement are as follows (Goetsch and Davis 2006):

1) Make it clear that management is committed to quality for the long term.
2) Form cross-departmental quality teams.
3) Identify where current and potential problems exist.
4) Assess the cost of quality and explain how it is used as a management tool.
5) Increase the quality awareness and personal commitment of all employees.
6) Take immediate action to correct problems that have been identified.

7) Establish a zero-defects program.
8) Train supervisor to carry out their responsibilities in the quality program.
9) Periodically hold "zero defects days" to ensure that all employees are made aware there is a new direction.
10) Encourage individuals and teams to establish both personal and team improvement goals.
11) Encourage employees to tell management about obstacles they face in trying to meet quality goals.
12) Recognize employees who participate.
13) Implement quality councils to promote continual communication.
14) Repeat everything to illustrate that quality improvement is a never-ending process.

Note that many of these steps are covered if the projects in the Six Sigma methodology are well executed.

### 1.2.3 Quality and Productivity

During and after World War II, America was under a lot of pressure to increase productivity. The managers of manufacturing companies in America believed that productivity and quality were not compatible, and their way to increase productivity was to hire more workers and put "quality" on the back burner. Japan and Germany were also coming out of the ashes of World War II. So, until 1960, America dominated the world with its productivity – but in 1948, Japanese companies started to follow the work that many pioneers such as Shewhart, Juran, and Deming practiced at Westinghouse. The managers of Japanese companies observed that improving quality not only make their products more attractive but also increased productivity. However, this observation did not sink into the minds of the managers of American companies: they continued working with the assumption that improving quality cost more and inhibited productivity and consequently would mean lower profits.

Deming's famous visit to Japan in (1950) brought about a quality revolution in Japan, and the country became a very dominant power of quality throughout the world. During his visit, he gave a seminar that was attended not only by engineers but also by all the top managers. He told the Japanese managers that "they had an obligation to the world to uphold the finest of management techniques." He warned them against mistakenly allowing into Japanese companies the use of certain Western management practices, such as management by objective and performance standards, saying that "these practices are largely responsible for the failure of Western industry to remain competitive." As Deming noted in his book *Out of the Crisis*, which resulted from his visit to Japan, the chain reaction shown in Figure 1.2 became engraved in Japan as the way of industrial life. This chain reaction was on the blackboard during every meeting he held with top management in Japan.

Furthermore, Deming noted that "Once management in Japan adopted the chain reaction, everyone there from 1950 onward had one common aim, namely, quality." But as remarked earlier, this idea was not adopted by American management until at least the late 1980s. In the 1960s and 1970s, American companies continued to dominate in productivity, mainly by increasing their workforce. However, as a result of ignoring the quality scenario, America started to lose its dominance in terms of competitiveness and thus productivity. During this period, Germany and South Korea also became competitors with America. Ultimately, in the 1990s, American management started to work on quality; and as a result, America began to reemerge as a world-class competitor.

The gurus and advocates for quality – Deming, Feigenbaum, Juran, and Crosby – were the most influential people in making the move from production and consumption to total quality control

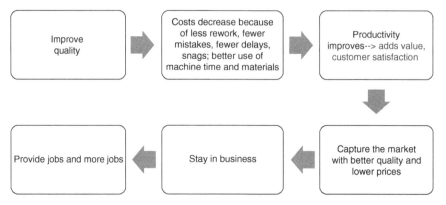

**Figure 1.2** A chain reaction chart used by the Japanese companies in their top management meetings.

and management. According to Joseph A. DeFeo, president and CEO of the Juran Institute, "the costs of poor-quality account for 15 to 30% of a company's overall costs." When a company takes appropriate steps to improve its performance by reducing deficiencies in key areas (cycle time, warranty costs, scrap and rework, on-time delivery, billing, and others), it reduces overall costs without eliminating essential services, functions, product features, and personnel increases as outlined by Goetsch and Davis (2006). Feigenbaum also said that up to 40% of the capacity of a plant is wasted through *not getting it right the first time*.

Furthermore, we note that often, *flexibility* in manufacturing can increase productivity without affecting quality. For example, the best Japanese automaker plants can send a minivan, pickup truck, or SUV down the same assembly line one after another without stopping the line to retool or reset. One Nissan plant can assemble five different models on one line. This flexibility obviously translates into productivity (Bloomberg Businessweek 2003).

## 1.3   Implementing Quality Improvement

Earlier in this chapter, we noted that the characteristic of quality improvement is not static; rather, it is an ongoing process. It becomes the responsibility of all management that all appropriate steps are taken to implement quality improvement. The first step by management, of course, should be to transform "business as usual" into an improved business by instilling quality into it. Deming's 14-point philosophy is very helpful to achieve this goal:

1) Create constancy of purpose for improving products and services.
2) Adopt the new philosophy. That is, management must learn about the new economic age and challenges such as competitiveness, and take responsibility for informing and leading their business.
3) Cease dependence on inspections to achieve quality.
4) End the practice of awarding business based on price alone; instead, minimize total costs by working with a single supplier.
5) Constantly improve every process for planning, production, and service.
6) Institute training on the job.
7) Adopt and institute leadership.

8) Drive out fear.
9) Break down barriers between staff areas.
10) Eliminate slogans, exhortations, and targets for the workforce.
11) Eliminate numerical quotas for the workforce and numerical goals for management.
12) Remove barriers that rob people of pride of workmanship, and eliminate the annual rating or merit system.
13) Institute a vigorous program of education and self-improvement for everyone.
14) Put everybody in the company to work to accomplish the transformation.

This philosophy can be used in any organization to implement total quality management (TQM). For more details and examples, we refer you to *Out of the Crisis* (Deming 1986).

### 1.3.1 Outcomes of Quality Control

The outcomes of quality control are obvious. Some of these outcomes are the following:

- The quality of the product or service will improve, which will make it more attractive and durable. Better quality will result in a higher percentage of the product meeting the specifications of the customer. Consequently, only a small percentage (or none) of the products will be rejected.
- Since few or no products are rejected, fewer need rework, and consequently there are fewer delays in delivery. This makes the customer happy, and they are bound to buy the product again. All this adds up to more savings, and that results in a lower price for the product – which makes it more competitive.
- Consequently, there will be better use of resources, such as manpower, raw material, machine hours, etc. All of these outcomes result in lower costs, better quality, higher productivity, and hence a larger market share.

### 1.3.2 Quality Control and Quality Improvement

Quality control helps an organization to create products that, simply put, are of better quality. Continuous quality improvement makes operators, engineers, and supervisors more focused on customer requirements, and consequently, they are less likely to make any "mistakes."

#### 1.3.2.1 Acceptance Sampling Plans

Quality control may use a technique called *acceptance sampling* to improve quality. An *acceptance sampling plan* is a method for inspecting a product. Acceptance sampling may inspect only a small portion of a lot or 100% of the lot. In some cases, inspecting 100% of the lot means all products in that lot will be destroyed. For example, if we are testing the life of a new kind of bulbs for a particular type of projector, then inspecting 100% of the lot means all the bulbs in that lot will be destroyed.

But acceptance sampling plans increase quality only of the end product or service, not of what is still being manufactured or of services that are still being performed, which means any defects or errors that occurred during the production process will still exist. In certain service industries, nothing can be done until the service has been fully provided or after it has been provided. For example, if a patient is receiving treatment, then nothing can be done during or after the treatment if the treatment was bad. Similarly, if a dentist has pulled out the wrong tooth, then nothing can be done after the dentist has completed the service. Thus quality improvement is extremely important in such situations. In manufacturing, acceptance sampling very often requires rework on defective units; after rework, these units may turn out to be acceptable or not, depending on what kind of defects these units had in the first place. All of this implies that acceptance sampling is not a very effective method for quality improvement. We will study acceptance sampling plans in more detail in Chapter 9.

### 1.3.2.2 Process Control

We turn now to process control. In *process control or statistical process control*, steps are taken to remove any defects or errors before they occur by applying statistical methods or techniques of five kinds: define, measure, analyze, improve, and control. We discuss these techniques in Chapter 2. Deming describes *statistical quality* as follows: "A state of statistical control is not a natural state for the manufacturing process. It is instead an achievement, arrived at by eliminating one by one, by determined effort, the *special causes* of excessive variation." Another way of describing statistical quality is as an act of taking action on the process based on the result obtained from monitoring the process under consideration. Once the process-monitoring tools (discussed in detail in Chapters 5–8) have detected any cause for excessive variation (excessive variation implies poor quality), the workers responsible for the process take action to eliminate the cause(s) and bring the process back into control. If a process uses statistical control, there is less variation; consequently, quality is better and is continuously improved. If the process is under control, then it is more likely to meet the specifications of the customer or management, which helps to eliminate or significantly reduce any costs related to inspection.

Quality improvement is judged by the customer. Very often, when a customer is not satisfied with quality improvement, they do not bother to file a complaint or demand compensation if the product is not functioning as it is expected to. On the other hand, if there is significant quality improvement, the customer is bound to buy the product repeatedly. These customers we may define as *loyal customers*. So, quality improvement is best judged by loyal customers, and loyal customers are the biggest source of profit. If there is no significant improvement in quality, then not only do we lose dissatisfied customers but we also lose some of the loyal customers. The loss due to dissatisfied customers or losing loyal customers usually is not measurable – but such a loss is usually enormous, and sometimes it is not recoverable and can cause the collapse of the organization. Thus, quality control and quality improvement are the best sources of good health for any company or organization.

### 1.3.2.3 Removing Obstacles to Quality

Deming's 14-point philosophy helped Western management transform old-fashioned "business as usual" to modern business, where concern for quality is part of the various problems that face any business. Note, however, that there is a danger that these concerns may spread like wildfire, to the detriment of the business as a whole. Further, some problems are what Deming calls "deadly diseases" and become hurdles on the way to fully implement the transformation (Deming 1986, Chapter 3). Deming describes the deadly diseases as follows:

1) Lack of constancy of purpose to plan products and services that have a market sufficient to keep the company in business and provide jobs.
2) Emphasis on data analysis, a data-based decision approach, and short-term profits. Short-term thinking that is driven by a fear of an unfriendly takeover, and pressure from bankers and shareholders to produce dividends.
3) Performance evaluations, merit ratings, or annual reviews without giving sufficient resources to accomplish desired goals.
4) Job hopping by managers for higher ranks and compensation.

5) Using only visible data or data at hand in making decisions, with little or no consideration of what is unknown or unknowable.
6) Excessive medical costs.
7) Excessive liability cost that is jacked up by lawyers who work on contingency fees and unfair rewards given by juries.

Deadly diseases 1, 3, 4, and 5 can usually be taken care by using a total quality approach to quality management, but this topic is beyond the scope of this book. However, deadly diseases 2, 6, and 7, add major costs to the organization without contributing to the health of the business. They are more cultural problems, but they pressure companies to implement quality improvement and consequently compete globally.

#### 1.3.2.4 Eliminating Productivity Quotas

Numerical quotas for hourly workers to measure work standards have been a common practice in America. This is done by the Human Resources (HR) department to estimate the workforce that the company needs to manufacture $X$ number of parts. While doing these estimates, it could be that nobody bothers to check how many of the manufactured parts that have been produced are defective, or how many of them meet the specifications set by customers or will be rejected/returned. HR normally does not take into account the cost of such events – which, of course, the company has to bear because of rework on defective or nonconforming parts or rejected and trashed parts. All of this adds to the final cost.

In setting up numerical quotas, the average number of parts produced by each worker is often set as a work standard. When we take an average, some workers produce a smaller number of parts than the set standard, and some produce more than the set standard. No consideration is given, while setting the standard, to who produced a small or large number of parts that meet customer specifications. Thus, in this process, workers who produce more parts than the set standard – regardless of whether the parts meet the specifications – are rewarded, while other workers are punished (no raises, no overtime, etc.). This creates differences between workers and, consequently, chaos and dissatisfaction among workers. The result is bad workmanship and more turnover, and workers are unable to take the pride in their workmanship to which they are entitled.

### 1.3.3 Implementing Quality Improvement

It is the responsibility of top management to lead the quality improvement program and, by removing any barriers, to implement the quality improvement program. Then the process owners have the responsibility to implement quality improvement in their company or organization. To do this, they first must understand that quality improvement takes place by reducing variation. The next step for them to understand is what factors are responsible for causing variation. To control such factors, the best approach is if management collaborates with the workers who work day in and day out to solve problems, who know their jobs, and who know what challenges they are facing. By collaborating with workers, management can come to understand everything about quality improvement and what is essential to achieve it. This, of course, can be achieved if management has better communication with those workers who do work for quality improvement.

The next step after the implementation of quality improvement is to focus on customers and their needs. By focusing on customers, loyal customers are created, and they can be relied on for the future of the company. Note that when there is focus on the needs of customers, the use of SPC tools becomes essential to achieve the company's goals, which in turn helps improve quality on a continuous basis. According to Crosby, quality improvement is based on four "absolutes of quality management":

1) Quality is defined as *conformance to requirements* (a product that meets the specifications), not as "goodness" or "elegance."
2) The system for causing quality is *prevention* (eliminating both *special and common causes* by using SPC tools), not appraisal.
3) The performance standard must be *zero defects,* not "close enough."
4) The measurement of quality is the price of *nonconformance* (producing defective parts, or parts that do not meet specifications), not indices.

Thus, to implement quality improvement, top management should follow these absolutes. Management must also implement training for workers so that they understand how to use all the tools of SPC and thus can avoid any issues that infiltrate the system. Management must make sure that supplier(s) also understand that quality improvement is an ongoing part of the company's policy and get assurance from suppliers that they will also use SPC tools to maintain the quality of everything they supply.

## 1.4 Managing Quality Improvement

Managing quality improvement requires accountability, daily defect analysis, preventive measures, data-driven quality, teamwork, training, and optimizing process variables.

Managing quality improvement is as important as achieving improved quality. Managing quality improvement becomes part of implementing quality improvement. Quality management plays a key role in the success of achieving total quality.

Hiring a consulting firm that "specializes in quality improvement" and not taking control into their own hands is the biggest mistake made by top management of any organization. Hiring a consulting firm means top management is relinquishing their responsibility for sending the important message to their employees that quality improvement is a never-ending policy of the company.

Top management should play an important role in team building and, as much as possible, should be part of the teams. Top management and supervisors should understand the entire process of quality improvement so that they can guide their employees appropriately on how to be the team players. Teamwork can succeed only if management supports it and makes teamwork part of their new policy. Forming a team helps to achieve quality improvement, and establishing plans is essential: it is part of the job or process of managing quality improvement.

Another part of quality improvement is that management must provide the necessary resources and practical tools to employees who are participating in any project to achieve quality improvement. Arranging 5- or 10-day seminars on quality improvement for employees without giving them any practical training and other resources is not good management practice.

### 1.4.1 Management and Their Responsibilities

The management of a company must create an environment of good communication so that everyone in the company knows their duties and goals or purpose. Management must make sure the purpose remains steadfast and does not change if a top manager leaves the company. Management must keep the communication lines with employees open so that they know which direction the business is moving and what their responsibilities are to take the business where it would like to be in 5 or 10 years. In discussing his 14 points, Deming says, "Adopt the new philosophy. We are in a new economic age created by Japan. Western management must awaken to the challenge, must learn their responsibility, and take on leadership for change."

It is the responsibility of management to have a continuous dialogue with customers and suppliers. Both customers and suppliers play an important role in achieving quality improvement. Management must make a commitment to sustained quality improvement by providing resources for education and practical training on the job and showing their leadership. This can be done if they are willing to increase their understanding and knowledge about every process that is taking place to improve quality. Leadership that just passes on orders but doesn't understand anything about the processes that are in the works or under consideration for the future will be disappointed and will also disappoint their customers and investors.

### 1.4.2 Management and Quality

In this modern era, management and quality go hand in hand. Global customers are not only looking at the quality of the product they buy but also are looking at who manufactured it and how much they are committed to backing it up. Customers are also interested in determining the reliability of that commitment. Global competition puts so much pressure on management that they must make quality their top priority for the company. Managers can no longer satisfy employees, customers, or investors with just encouraging words: they must show solid outcomes, such as how much sales have gone up, how many new customers have been attracted, and the retention rate of old customers. All of this will fall in line only if management has made quality an important ongoing process.

Furthermore, management must understand that the customer defines quality, so management must make commitments to customers about the increased quality and value of the company's products. In addition, management should understand that it is the employees who build the quality into products. Thus, management must make commitments to employees and give them the resources and tools they need. Management must also obtain the same kind of commitments from suppliers. Any sloppiness on the part of suppliers can ruin all the plans for ongoing process improvement or quality improvement. In the modern economic age, only companies and managements that make such commitments and follow through on them can assure themselves a bright future, job guarantees, and better compensation for their employees.

Management and quality are a two-way street. Any company with good management delivers better quality, and having better quality means there is good management.

### 1.4.3 Risks Associated with Making Bad Decisions

It is important to note that whenever decisions are made based on samples, you risk making bad decisions. Bad decisions in practice lead to difficulties and problems for producers as well as consumers. These bad decisions in statistical terms are referred to as *type I and type II errors* as well as *alpha* ($\alpha$) and *beta* ($\beta$) risks, respectively. It is important to know the following key points about these risks:

- Sooner or later, a bad decision will be made.
- The risks associated with making bad decisions are quantified in probabilistic terms.
- $\alpha$ and $\beta$ risks added together do not equal 1.
- Even though $\alpha$ and $\beta$ go in the opposite direction (that is, if $\alpha$ increases, $\beta$ decreases), there is no direct relationship between $\alpha$ and $\beta$.
- The values of $\alpha$ and $\beta$ can be kept as low as you want by increasing the sample size.

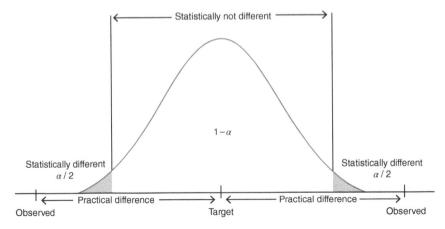

**Figure 1.3** Detecting practical and statistical differences.

**Definition 1.2** *Producer risk* is the risk of failing to pass a product or service delivery transaction on to a customer when, in fact, the product or service delivery transaction meets customer quality expectations. The probability of making a producer risk error is quantified in terms of $\alpha$.

**Definition 1.3** *Consumer risk* is the risk of passing a product or service delivery transaction on to a customer under the assumption that the product or service delivery transaction meets customer quality expectations when, in fact, the product or service delivery is defective or unsatisfactory. The probability of making a consumer risk error is quantified in terms of $\beta$.

In Figure 1.3, our comparison points change from the shaded region under the distribution tails to the center of the distribution. A practical decision then requires that we consider how far off the intended target the observed process behavior is as compared with the statistical difference identified in Figure 1.3. Note that differentiating between a practical and statistical difference is a business or financial decision. When making a practical versus a statistical decision, we may well be able to detect a statistical difference; however, it may not be cost-effective or financially worth making the process improvement being considered.

## 1.5 Conclusion

In this chapter, we have given a general overview of quality improvement and its management. For more details on these topics, we refer you to the works of Philip B. Cosby, W. Edwards Deming, Joseph M. Juran, and Armand V. Feigenbaum (see the Bibliography). Also, the Six Sigma methodology, which we introduce in the next chapter, is a step forward to help achieve and manage quality improvement, since understanding the idea of Six Sigma means customer requirements must be met. In the remainder of this book, we discuss statistical techniques and SPC tools that are essential to implementing the Six Sigma methodology for improving process performance.

# 2

# Basic Concepts of the Six Sigma Methodology

## 2.1 Introduction

Far from being just another quality fad, Six Sigma has continued to grow in popularity and influence since its creation at Motorola in 1986. Six Sigma techniques have been adopted by a wide range of manufacturing firms and have also translated successfully into other sectors, including retail, hospitality, financial services, high tech, transportation, government, and healthcare. According to the American Society for Quality, as of 2009, 82 of the largest 100 companies in the US had deployed Six Sigma. Fifty-three percent of Fortune 500 companies have adopted Six Sigma practices, saving an estimated $427 billion over the past 20 years [1]. In a broad sense, it can be said that if a company has customers and a process, it can benefit from implementing Six Sigma.

## 2.2 What Is Six Sigma?

In statistics, the lowercase Greek letter sigma ($\sigma$) represents the population standard deviation, which is a measure of variation or spread. In the quality field, we aim to reduce the process standard deviation to achieve more consistent outcomes. Whether we are measuring the dimension of a metal flange, inner diameter of a pipe, burst strength of a package, monthly labor costs for a division, or repair time of a subassembly, a repeatable process is the desired state.

If we were limited to a single-sentence elevator speech to describe Six Sigma, a reasonable definition might be "Six Sigma is a quality approach that strives to reduce variation and decrease defects, which results in increased customer satisfaction and improved bottom-line results." If we dig a little deeper, however, it becomes clear that there are at least three different definitions of the term *Six Sigma*. Six Sigma can be correctly classified as a management philosophy; it is also defined as a systematic approach to problem-solving. Last, *Six Sigma* is a term used for a statistical standard of quality. Let's explore each of these definitions in more detail.

### 2.2.1 Six Sigma as a Management Philosophy

Six Sigma is a management philosophy that emphasizes reducing variation, driving down defect rates, and improving customer satisfaction. Specifically, the tenets of the philosophy include:

- Enterprise-wide deployment of Six Sigma to improve processes
- Implementation driven from the top down
- Process improvements achieved through projects completed by teams

*Statistical Quality Control: Using Minitab, R, JMP, and Python*, First Edition. Bhisham C. Gupta.
© 2021 John Wiley & Sons, Inc. Published 2021 by John Wiley & Sons, Inc.
Companion website: www.wiley.com/go/college/gupta/SQC

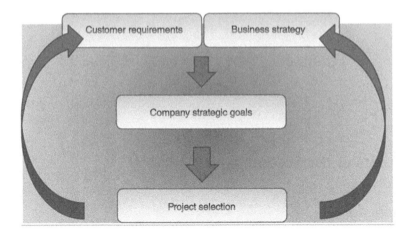

**Figure 2.1** Six Sigma project selection.

- Project benefits linked directly to the organization's bottom line
- Rigorous application of data analysis and statistical tools
- An extremely high standard for quality

In this management framework, quality is no longer relegated to the quality department, nor is it reserved only for the shop floor. Each facility, each department, and each employee plays a part in improving quality.

In larger companies with an established Six Sigma program, a high-level steering committee chooses Six Sigma projects that align with and help advance the company's strategic goals. These strategic goals are derived from customer requirements and upper management's overall business strategy. The relationship among these elements is illustrated in Figure 2.1. These projects are completed by cross-functional teams, and the benefits are reported in terms of defect reduction and, especially, dollar savings.

The Six Sigma philosophy also places an emphasis on measurement. Decisions are based on data, not on company folk wisdom or gut feel. In addition, there is a relentless emphasis on reducing variation, driven in large part by an extremely high standard of quality in which virtually no defects are produced.

### 2.2.2 Six Sigma as a Systemic Approach to Problem Solving

The second definition of *Six Sigma* refers to a systematic approach to problem-solving. The emphasis in Six Sigma is on solving process problems. A *process* is a series of steps designed to produce products and/or services.

As shown in Figure 2.2, the inputs to a process may include materials, people, or information. These inputs flow through the steps of the process, which produces a product or service as a final output. The concept can be applied to any industry or function. A manufacturing process may take raw materials and transform them into a finished product. In a front-office process, invoices may go through process steps that then create vendor payments. A physician may order a series of tests, which finally leads to a diagnosis and treatment.

The Six Sigma methodology is driven by team projects that have a clearly defined purpose and specific goals. The projects concentrate on improving processes and have relatively short durations,

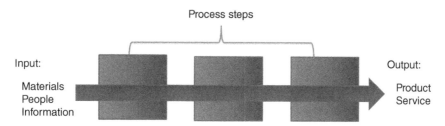

**Figure 2.2** Flow chart of a process.

with the majority completed within two to nine months. Project goals generally focus on reducing variation in a process and consequently reducing defects. Teams work together to accomplish project goals by following problem-solving approach or five defined phases: Define, Measure, Analyze, Improve, and Control, (DMAIC, pronounced "duh may' ik"), as shown in Figure 2.3.

**Figure 2.3** The DMAIC cycle.

In the *Define phase*, a team sifts through customer feedback and product performance metrics to craft project problem and goal statements and establish baseline measures. Labor, material, and capital resources required by the project are identified, and a rough timeline for project milestones is created. This information is collected into a project charter that is approved by upper management.

During the *Measure phase*, the team identifies important metrics and collects additional data to help describe the nature of the problem.

In the *Analyze phase,* the team uses statistical and graphical techniques to identify the variables that are major drivers of the problem. In this stage, root causes of the problem are identified.

In the *Improve phase*, the team identifies ways to address the root cause(s) and prioritizes the potential solutions. Then, changes to the process are designed, tested, and then implemented.

In the final *Control phase*, the team works to create a plan to prevent the newly improved process from backsliding to previous levels of defects. Mechanisms for ongoing monitoring and control of key variables are also established at this point. During the project wrap-up, the team's success is celebrated, and the lessons that the team learned during the project are shared throughout the company so that other teams can benefit from their discoveries.

### 2.2.3 Six Sigma as a Statistical Standard of Quality

The third definition of *Six Sigma* refers to a level of quality that produces very few defects in the long term. With this definition, Six Sigma is often written using the numeral 6 and the Greek letter sigma: 6σ.

Bill Smith devised this standard of quality at Motorola in 1986. According to Smith's definition, over the long run, a process running at a 6σ level will produce 3.4 defects per million

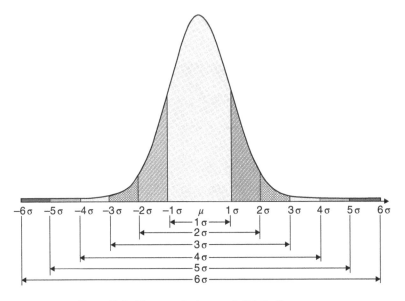

**Figure 2.4**  The standard normal distribution curve.

opportunities (DPMO). To achieve this extremely high level of quality, a process must produce outputs that consistently meet the target specifications. Producing outputs with very small variability around the target is the key to achieving Six Sigma quality. With such tight control over variability, even if the process drifts from the target over time, the output will still meet specifications and achieve the 3.4 DPMO standard.

As might be expected given this extremely high standard for quality, most processes are not running at a 6σ quality level. In fact, in most organizations, processes are running at the 3σ level, which translates into about 67,000 DPMO.

### 2.2.3.1  Statistical Basis for Six Sigma

The Six Sigma standard of quality has its basis in normal distribution theory (see Figure 2.4). We assume that the quality characteristic of interest – say, the diameter of a metal shaft – is normally distributed. The corresponding engineering specifications are assumed to be symmetric, and the process is centered at the specification target value.

For a centered process, 6σ quality is achieved when the lower and upper specifications map to ± six standard deviations from the mean, as shown in Figure 2.5. The centered process would produce approximately 2.0 defects per billion opportunities.

We then assume that no matter how well our process currently meets specifications, in the long run, the output will drift based on factors such as machine wear, incoming material variation, supplier changes, and operator variability. The assumption in Six Sigma is that even a well-behaved process will drift from the target, perhaps by as much as 1.5 standard deviations. By applying this *1.5 sigma shift* to a centered process, the rationale is that companies can gain a more realistic view of the quality level their customers will experience over the long term.

After adding a 1.5 sigma shift to a centered process, as shown in Figure 2.6, we then use a normal distribution table to calculate the fraction of parts that fall outside the specification limits. Thus the 6σ defect rate is equal to 3.4 DPMO *after the shift in the mean*. Table 2.1 lists the DPMO rates for various sigma levels for centered and shifted processes.

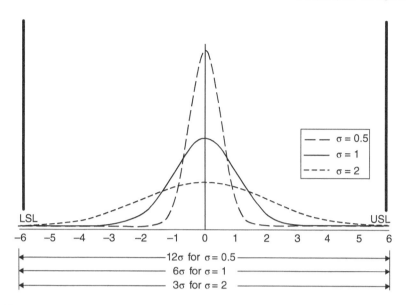

**Figure 2.5** For a normally distributed characteristic, centered at specification target, 6σ quality is achieved when the lower and upper specifications map to ±6σ.

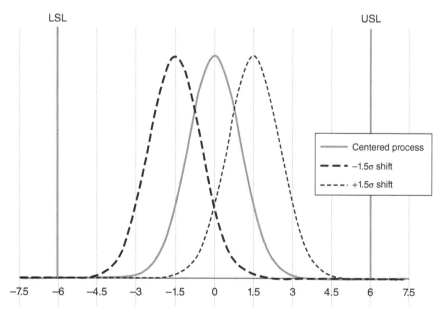

**Figure 2.6** Applying the 1.5σ shift to a centered 6σ process.

**Table 2.1** Defects per million opportunities (DPMO) for various sigma levels; centered and 1.5σ shifted processes.

| | DPMO | |
|---|---|---|
| **Sigma** | **Centered Process** | **1.5 sigma shift** |
| 6.0 | 0.00 | 3.4 |
| 5.0 | 0.57 | 232.6 |
| 4.0 | 63.34 | 6209.7 |
| 3.0 | 2699.80 | 66,811 |
| 2.0 | 45,500.26 | 308,770 |
| 1.0 | 317,310.51 | 697,672 |

### 2.2.4 Six Sigma Roles

As an enterprise-wide system, it follows that Six Sigma deployment requires the efforts of a cross-section of people filling various roles. These roles include Executive, Champion or Sponsor, Process Owner, and the Belt-holders.

An *Executive* is vital to the success of a Six Sigma implementation. Since the program is driven from the top down, executives must show support through their communications and actions. Without key support, the program will fade.

A *Champion* or *Sponsor* is a top-level manager familiar with Six Sigma principles who works with a project team. The champion's role is to support the team by providing necessary resources and removing roadblocks when they occur. A champion does not attend every team meeting but will check in at major project phases and serve as a go-to when the team needs top management support.

A *Process Owner* is a manager who is responsible for all aspects of a process and who also has the authority to make changes to the process. The process owner is a key member of a Six Sigma project team and often serves as a team leader.

If you are new to Six Sigma, you might be a bit confused by *quality* being described using a karate metaphor. It is admittedly a bit strange. Mikel Harry at Motorola is credited for coining the "belt" term. As the story goes, a plant manager told Harry that the Six Sigma tools he was applying were "kicking the hell out of variation" [4]. In Harry's mind, the comment conjured an image of a ninja who could expertly wield tools to make the data reveal what it knows about a process. Hence the "belt" nomenclature was born.

Within the Six Sigma hierarchy, there are several levels of *Belt-holders*. Depending on the size and complexity of the organization, there may be White, Yellow, Green, and Black Belts, as well as Master Black Belts.

*White Belts*, who often are also Executives in the organization, are not directly involved with improvement projects but have an awareness of Six Sigma concepts and help support enterprise-wide deployment.

*Yellow Belts* serve as members of Six Sigma project teams. They are familiar with Six Sigma principles and can apply basic tools.

*Green Belts* tend to work on projects on a part-time basis while maintaining their current positions in the company. A Green Belt may lead a small project team or assist a Black Belt on a larger project. Green Belts know statistical techniques and can apply a large array of non-statistical problem-solving tools.

At larger companies, *Black Belts* work full-time in their Six Sigma roles. They lead Six Sigma projects and train and mentor Green Belts. A Black Belt possesses sophisticated knowledge of statistics and data analysis tools, as well as team dynamics.

Finally, *Master Black Belts* work full time managing, mentoring, and training Black Belts. A Master Black Belt may also work with upper management and the steering committee to plan Six Sigma deployment efforts and to select and prioritize Six Sigma projects.

## 2.3   Is Six Sigma New?

The origin of quality standards can be traced back to the medieval guild system in which craftsman placed their marks on only those products that met their standards. Quality tools such as statistical process control (SPC) charts date back to Walter Shewhart in 1924, and both W. Edwards Deming and Joseph M. Juran introduced quality thinking and methods in Japan after World War II. Japanese industry continued to innovate and develop quality systems and tools, with major contributions from Ishikawa, Shingo, and Taguchi, among others. In the last 70 years, there has been no shortage of quality programs touted as definitive solutions, including Total Quality Management, Quality Circles, and Zero Defects. Given the long history of quality, then, is Six Sigma new?

On the one hand, it can be argued that Six Sigma is simply a repackaging of existing quality know-how. It is true that Six Sigma uses existing quality tools such as the Basic seven tools: cause and effect diagrams, check sheets, control charts, histograms, Pareto charts, scatter diagrams, and stratification; and the New seven tools: affinity diagrams, arrow diagrams, interrelationship diagrams, matrix diagrams, prioritization matrices, process decision program charts, and tree diagrams. (See Section 2.4.1.) Six Sigma's focus on team problem-solving is taken from Japanese practices such as quality circles. Both Deming and Juran advocated data-based decision-making and emphasized the vital role of management in successful quality implementation.

Several features of the Six Sigma approach are unique, however. Because Six Sigma requires data-driven decisions, it has led to the widespread use of sophisticated statistical techniques. Analysis tools once only known to high-level industrial statisticians, such as linear regression, chi-squared tests, and designed experiments, are now routinely employed in Six Sigma projects by non-statisticians. The project approach, in which upper management identifies project areas based on customer requirements and the overall business strategy, is unique to Six Sigma. In addition, the emphasis on reducing the variation in a process is a relatively new concept. Earlier quality efforts may have emphasized aligning processes to desired target levels but did not necessarily reduce the variation around those targets. Taguchi's work in the 1980s introduced the importance of reducing variability to improve quality and reduce quality costs.

Finally, the requirement that quality improvement be tied directly to the organization's bottom line is a distinct hallmark of Six Sigma. Six Sigma projects saved GE a reported $12 billion over five years (5). Results such as these have earned Six Sigma enduring popularity with executives.

## 2.4 Quality Tools Used in Six Sigma

The quality tools available to Six Sigma practitioners are vast. For example, *The Quality Toolbox* by Nancy Tague details over 100 separate tools that can be used in project planning, idea creation, process analysis, data collection and analysis, cause analysis, and decision making. The statistical tools used in Six Sigma projects include:

- Hypothesis testing
- Analysis of variance (ANOVA)
- Gauge repeatability and reproducibility (R&R)
- Statistical control charts
- Process capability
- Correlation and regression analysis
- Design of experiments
- Multivariate analysis
- Time-series analysis

### 2.4.1 The Basic Seven Tools and the New Seven Tools

Six Sigma practitioners also leverage the Basic Seven and New Seven Tools. The Basic Seven tools were first brought into focus by Kaoru Ishikawa, a professor of engineering at Tokyo University and the father of "Quality Circles." These tools can be used to collect and analyze data in the Measure, Analyze, Improve, and Control phases of a Six Sigma project:

1) *Cause-and-effect diagram* (also known as Ishikawa or fishbone diagram): A visual tool that identifies potential causes of a problem and sorts ideas into useful categories.
2) *Check sheet*: A structured, prepared form for collecting and analyzing the frequency of occurrences of various events.
3) *Control chart*: A graph used to study how a process changes over time. Comparing current data to historical control limits leads to conclusions about whether the process variation is consistent (in control) or unpredictable (out of control due to some special causes of variation).
4) *Histogram*: A bar chart that shows the shape of a data set using its frequency distribution, or how often each value occurs in a data set.
5) *Pareto chart*: A bar graph that shows the frequency of occurrence of events in descending order. The chart helps the team focus on the main drivers of a problem.
6) *Scatter diagram*: An X-Y plot that shows the relationship between two quantitative variables that are measured in pairs.
7) *Stratification*: A technique that separates data gathered from a variety of sources so that patterns can be seen.

All of these tools are discussed at length in the next three chapters.

In 1976, the Union of Japanese Scientists and Engineers (JUSE) saw the need for tools to promote innovation, communicate information, and successfully plan major projects. A team researched and developed the New 7 Tools, often called the Seven Management Tools:

1) *Affinity diagram*: A visual tool that allows a team to organize a large number of ideas, opinions, or issues into their natural relationship groupings.
2) *Arrow diagram*: A graphical tool that shows the required order of tasks in a project or process, the best schedule for the entire project, and potential scheduling and resource problems and their solutions.

3) *Interrelationship diagram*: A visual tool that shows cause-and-effect relationships and helps analyze the natural links between various aspects of a complex situation.
4) *Matrix diagram*: A tool that shows the relationship between two, three, or four groups and can give information about the relationship, such as its strength and the roles played by various individuals or measurements.
5) *Prioritization matrix*: A decision tool that rates various options based on predetermined criteria.
6) *Process decision program chart*: A risk analysis tool that systematically identifies what might go wrong in a plan under development.
7) *Tree diagram*: A visual tool that breaks broad categories down into finer and finer levels of detail, helping to move step-by-step thinking from generalities to specifics.

## 2.4.2 Lean Tools

Recently, the Six Sigma methodology has been enhanced by incorporating the tenets of *Lean Thinking*. The Lean philosophy places importance on mapping the value stream and stresses the elimination of waste and continuously improving processes. Lean tools such as Eight Wastes, visual management, the 5S method, value stream mapping, mistake-proofing, and quick changeover have become powerful tools used in the Improve and Control phases of Six Sigma projects. Together, Six Sigma and Lean deliver higher quality, increased efficiency, and flexibility at a lower cost.

The Lean philosophy follows five principles. First, we specify value, as perceived from the end user's point of view. Next, we identify the value stream, which includes all the activities performed from the time of customer order to delivery. Every task we perform can be classified as value-added or non-value-added. Third, we create flow so that our product – parts or paperwork or information – moves through the system without any hiccups. Fourth is the idea of pulling from the customer, meaning that we will make a product when a customer orders it, as opposed to making products that sit in inventory. Finally, we seek 100% quality and perfection in our operations. Contrast this to the Six Sigma goal of 3.4 DPMO, which is the result after the $1.5\sigma$ shift. The Lean goal of perfection lends itself to the idea of continuous improvement: we can always do better; we are never done; we will never quite reach the goal.

Non-value-added activities or policies are classified as *waste*. In Lean, there are Eight Wastes, and the goal is to identify and eliminate this waste. This contrasts with Six Sigma, in which we try to reduce variability.

### 2.4.2.1 Eight Wastes

As we have noted, the emphasis of Lean is on reducing waste, and there are eight named wastes: *t*ransportation, *i*nventory, *m*otion, *w*aiting, *o*verproduction, *o*ver-processing, *d*efects, and *s*kills (TIM WOODS). These wastes do not add value to the product or service and should be eliminated. We'll now define each of them.

*Transportation* waste is the movement of things, whether actual materials, paperwork, or electronic information. Moving things from place to place does not add value to the product or service. In fact, there is a higher probability of parts getting lost or damaged, the more we move them. There might also be delays in availability because objects are in route. This may lead to another waste: waiting. One of the causes of this type of waste may be poor workplace layout.

The next waste is *inventory*. This waste describes an excess of things, whether it is parts, supplies, equipment, paperwork, or data. Accumulating and storing inventory costs money, so a major effect of this waste is reduced cash flow. We may lose production because we are looking for things in our disorganized storage areas, and inventory may get damaged or become obsolete before we can use it. We know we have excess inventory if there are stockpiles of materials or messy storage areas. The root cause of this type of waste is a just-in-case mentality, which might also be driven by an unreliable supply chain.

Next is the waste of *motion*, which is the movement of people, in contrast to transportation, which is the movement of things. Often these two wastes occur together, since it is likely that people are moving the materials around. Motion waste could stem from a poorly designed process, where operators must use excessive motion to get their jobs done. Motion waste may result in worker injury.

*Waiting* waste occurs when people wait for other people. This waste is incurred when meetings start late, when people must wait for information before they can move on to the next step, and when people wait for products or machines to be available. We can also think of this waste another way, in which information, products, or machines are waiting for the people to act upon them. Waiting increases cycle times and may increase overtime hours. It can be due to unbalanced workloads or a push environment, where products are sent downstream even if the next process is not ready for them.

Next we have *overproduction* waste. A direct consequence of overproducing is inventory waste. It can also trigger waiting waste for work-in-process parts.

*Over-processing* waste is doing more than the customer is willing to pay for. This waste includes performing inspections. The customer is not willing to pay more for inspection; they expect the product to be made right in the first place. Over-processing waste could also be incurred by adding features to a product or service that are not valued by the customer. This type of waste may result in longer lead times for delivery and a frustrated workforce that is asked to do tasks that are not adding value.

Of course, a major contributor to waste is *defects*. When we have defects, we increase internal and external failure costs and create dissatisfied customers.

The final waste is the waste of *skills*. This is the waste of an organization not using employees' aptitudes to their fullest extent. This waste can result in frustrated workers, absenteeism, and turnover.

There are many interconnections among the various forms of waste. When there is transportation waste, chances are there is also motion waste. If these two wastes exist, there might also be waiting waste. Overproduction leads to inventory waste, and so on. Many of the root causes of these wastes also overlap. For example, lack of training is a root cause for transportation, motion, waiting, defects, and skills wastes. Root causes for each of the wastes are summarized in Table 2.2.

### 2.4.2.2 Visual Management

A major tenet of Lean is *visual management*, a technique that makes the current status of inputs, outputs, or the process readily apparent at a glance. In general, visual-management techniques are inexpensive, simple, unambiguous, and immediate. By using visual management, problems are easily detected and so can be corrected quickly.

Visual management tools are often low-tech, such as using colored tape on the factory floor to show where inventory belongs, where someone shouldn't stand, or where fork trucks will be traveling. Racks and bins can be color-coded by part type. Indicator lights on machines can be used to signal run status, and pictographs of work instructions allow operators to easily reference process steps. Maintenance charts can show the latest service performed, and statistical process-control charts show the current process performance. Visual management can certainly be used in a manufacturing plant, but it also can be used in an office, hospital, restaurant, or any type of workplace.

**Table 2.2** Root causes of the Eight Wastes.

| Waste | Root Cause |
|---|---|
| Transportation | Poor layout<br>Lack of cross-training |
| Inventory | Just-in-case mentality<br>Unreliable supply chain |
| Motion | Poor layout<br>Lack of cross-training<br>Insufficient equipment |
| Waiting | Too many handoffs<br>Push environment<br>Unbalanced workloads<br>Lack of cross-training |
| Overproduction | Lack of systems thinking<br>Push environment<br>Individuals valued over teams |
| Over-processing | Lack of trust<br>Unclear customer requirements |
| Defects | Poor training<br>Non-standard work<br>Lack of job aids<br>Poor communication |
| Skills | Lack of trust<br>Lack of training<br>Silo thinking |

### 2.4.2.3 The 5S Method

The tool known as *5S* is defined as a physical methodology that leads to a workplace that is clean, uncluttered, safe, and well organized, resulting in reduced waste and increased productivity. The tool helps create a quality work environment, both physical and mentally.

The 5S philosophy applies in any work area suited for visual control and Lean production. 5S is derived from the following Japanese terms that refer to creating such a workplace:

- *Seiri*: To separate needed tools, parts, and instructions from unneeded material and to remove the unneeded ones.
- *Seiton*: To neatly arrange and identify parts and tools for ease of use.
- *Seiso*: To conduct a cleanup campaign.
- *Seiketsu*: To conduct *seiri*, *seiton*, and *seiso* daily to maintain a workplace in perfect condition.
- *Shitsuke*: To form the habit of always following the first four S's.

Table 2.3 lists the Japanese terms and the English translations of the 5S methods. Benefits to be derived from implementing a 5S program include:

- Improved safety
- Higher equipment availability
- Lower defect rates
- Reduced costs
- Increase production agility and flexibility
- Improved employee morale

**Table 2.3** The 5S methods.

| Japanese | Translated | English | Definition |
|---|---|---|---|
| *Seiri* | Organize | Sort | Eliminate whatever is not needed by separating needed tools, parts, and instructions from unneeded material. |
| *Seiton* | Orderliness | Set in order | Organize whatever remains by neatly arranging and identifying parts and tools for ease of use. |
| *Seiso* | Cleanliness | Shine | Clean the work area by conducting a cleanup campaign |
| *Seiketsu* | Standardize | Standardize | Schedule regular cleaning and maintenance by conducting *seiri*, *seiton*, and *seiso* daily. |
| *Shitsuke* | Discipline | Sustain | Make 5S a way of life by forming the habit of always following the first four S's. |

- Better asset utilization
- Enhanced enterprise image to customers, suppliers, employees, and management

#### 2.4.2.4 Value-Stream Mapping

A *value-stream map* is a high-level process map that captures the steps of a process from the time a customer orders a product to the time it is delivered. The map also shows how information, work, and materials flow through the process. The team starts with a current state map to document the process as it exists now. This map is used to identify and quantify waste and to highlight areas that can be improved.

After the current state is well understood, the team builds a future-state map that eliminates unnecessary steps or other sources of waste that have been identified. This future-state map gives the team a blueprint for improvement.

#### 2.4.2.5 Mistake-Proofing

Mistake-proofing is also known by its Japanese name, *pokayoke*. The technique is always handy, but it is especially useful in preventing defects when they are rare or when errors are random and non-systematic.

There is a classic definition of *mistake-proofing*: the complete elimination of errors. But over the years, mistake-proofing has been expanded to include controls or warnings. With this version, having a machine shut down after a mistake is made, or a light or alarm sound when an error occurs, is considered mistake-proofing. Checklists also fall into the control category. All of these are effective if the warning is heeded or the checklist is adhered to. But the gold standard, of course, is to have a design in which the mistake does not occur in the first place.

A classic example of mistake-proofing is diesel and unleaded fuel nozzles at gas stations. You can't put diesel fuel in your unleaded gas tank; the nozzles just won't fit. This is an example of the elimination of a defect by design.

#### 2.4.2.6 Quick Changeover

The quick-changeover tool can be used to reduce or eliminate waiting waste and increase machine availability. In manufacturing, maintenance must change over a line when it is switched from making one product to the next. If a manufacturing plant can achieve shorter changeover times, it can run its production line with smaller batches. This reduces inventory waste and allows the plant to match its production output to customer demand. If a plant spends more time producing and less time changing over, it can increase its capacity.

Dr. Shigeo Shingo developed quick-changeover techniques while working with an auto manufacturer. He defined changeover time as the total time from the *last unit of production* to the *first unit of good production at full speed*. To dramatically reduce changeover time, teams convert *internal work* to *external work*.

Internal work consists of all the tasks performed while the line is shut down. External tasks are performed either before the line is shut down or after the line is up and running again. By performing a detailed work breakdown, the team can convert internal tasks to external task: doing things ahead of time, or after the fact, so that the actual downtime and internal work are minimal.

Unlike other tools used in Lean, quick changeover may require some significant capital investment, such as specialty tools and fasteners designed to minimize downtime. The goal is to re-design, re-engineer, and re-tool.

## 2.5 Six Sigma Benefits and Criticism

Motorola, along with the early adopters of the Six Sigma methodology (General Electric, Allied Signal, and Honeywell), have all reported remarkable financial benefits. From its inception in 1986 until 2001, Six Sigma saved Motorola $16.1 billion. GE, after investing $1.6 billion in the program, delivered a $4.4 billion savings in a three-year period. Allied Signal and Honeywell each attributed $500 million in savings to Six Sigma efforts in 1998 alone. In all, Six Sigma delivered savings on the order of 1.2 to 4.5% of company revenue [2].

The power of Six Sigma was also recognized by the US government. After implementing Six Sigma methods in 2006, the US Army claimed savings of $19.1 billion as of April 2011 [3]. In 2012, the White House praised the contributions of Six Sigma along with Lean methodology in eliminating unnecessary review steps for air and water permitting [4].

Despite these well-publicized benefits, not everyone has a favorable view of Six Sigma practices, and some doubt its efficacy. In 2007, a study conducted by Qual Pro, a consulting firm and proponent of a competing quality improvement system, reported that 53 of 58 companies that use Six Sigma have trailed the S&P 500 ever since they implemented it [5]. Of course, there is no proof of a causal relationship between Six Sigma implementation and poor financial results. In fact, it could be argued that companies adopting Six Sigma had a compelling quality or financial reason to do so and would have fared even worse if they hadn't adopted the program.

Others have argued that the rigor of the Six Sigma process can stymie innovation. The Six Sigma approach that is so successful in tackling problems in existing processes may have the opposite effect in research and development organizations. In these cases, the systematic, highly ordered approach may thwart the creativity necessary for cutting-edge advances.

Another perceived drawback of Six Sigma is its heavy training burden. In an enterprise-wide deployment, employees at all levels of the organization must be trained to some degree, with Green and Black Belts undergoing many additional hours of statistical training. This training is expensive and time-consuming and may not yield an immediate return.

### 2.5.1 Why Do Some Six Sigma Initiatives Fail?

Like many business initiatives, Six Sigma implementation can fail because of poor management execution and lack of support. Organizations new to Six Sigma should be aware of the dangers that poor leadership, cultural integration, and project selection can have on a Six Sigma implementation [5]:

- *Lack of leadership strategy and commitment:*
  - Following the "more is more" approach, companies implement so many projects that too many resources are pulled away from everyday operations and new product development. Customer satisfaction and the R&D pipeline both suffer.
  - Leaders expect superior results too quickly.
  - After initial success, management decides to move on to the next "new thing."

- *Poor cultural integration:*
  - Leaders fail to communicate progress throughout the training period.
  - The Black Belt role is not fully supported by management.
  - Six Sigma is treated as an "extra assignment" and perceived as a burden.

- *Poor project selection and assignment:*
  - Projects are not tied to customer requirements or the business strategy.
  - Projects concentrate on improving local processes and ignore the big picture.
  - Projects are assigned with leadership's suggestions on fixes and root causes.

For references [2], [3], [4] and [5] see the End Notes given on the last page of the Bibliography Section.

## Review Practice Problems

1   In which phase of the Six Sigma project does the team identify root causes of the problem?

2   A well-behaved process is currently running at a $5\sigma$ level. What is the expected number of defects per million opportunities (DPMO) if the mean of the process shifts by 1.5 standard deviations?

3   Using the Six Sigma methodology, a team improves a process, bringing the sigma level from up $1\sigma$ to $3\sigma$. What is the resulting change in the long-term defect rate?

4   In what ways is Six Sigma similar to other quality initiatives?

5   In what ways is Six Sigma different from other quality initiatives?

6   List and explain the differences between the goals of Six Sigma and the goals of Lean.

7   What is the main difference between the Basic 7 tools and the New 7 Tools?

8   Think of the last time you bought lunch at a fast-food restaurant. Use the Eight Wastes as a guide to identify wastes that you observed.

9   In what way is the Control phase of the DMAIC cycle similar to the last stage of the 5S method?

10   Give three examples of mistake-proofing devices that can be found in everyday life. For each, indicate whether the mistake-proofing feature is a control or warning, or if it completely eliminates the mistake.

# 3

# Describing Quantitative and Qualitative Data

## 3.1  Introduction

In all statistical applications, we collect sets of data. These sets could be small, containing just a few data points, or very large and messy, with millions of points or more. Not all data are of the same kind, and therefore not all data sets can be treated in the same manner. In this chapter, we introduce the classification of data into various categories, such as qualitative and quantitative, and apply appropriate graphical and numerical techniques to analyze them.

Sometimes data sets contain data points that are erroneous and thus may not belong to the data set. These data points are commonly known as *outliers*. In this chapter, we apply a method that can detect such outliers so that they may be either corrected and included in the analysis or eliminated from the analysis. In certain applications, we are studying two characteristics of an individual or item, and we may be interested in finding any association between these characteristics; a measure of such association is introduced in this chapter. The concept of descriptive and inferential statistics is also introduced. Finally, we study some important probability distributions that are quite frequently used in the study of statistical quality control.

## 3.2  Classification of Various Types of Data

In practice, it is common to collect a large amount of non-numerical and/or numerical data on a daily basis. For example, we may collect data concerning customer satisfaction comments of employees, perceptions of suppliers, etc. Or we may track the number of employees in various departments of a company, check weekly production volume in units produced or sales in dollars per unit of time, etc. However, not all of the data collected can be treated the same way, as there are different types of data. Accordingly, statistical data are typically divided into two major categories:

- Qualitative
- Quantitative

Each of these categories can be further subdivided into two subcategories. The two subcategories of qualitative data are called *nominal* and *ordinal,* and the two subcategories of quantitative data are called *interval* and *ratio*. Figure 3.1 summarizes this classification of statistical data.

In the figure, the data classifications – nominal, ordinal, interval, and ratio – are arranged from left to right in the order of the amount of information they provide. Nominal data provide minimum information, whereas ratio data provide maximum information:

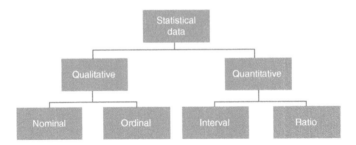

**Figure 3.1** Classifications of statistical data.

- *Nominal data:* As mentioned, nominal data contain the least amount of information. Only symbols are used to label categories of a population. For example, production part numbers with a 2003 prefix are nominal data wherein the prefix indicates only that the parts were produced in 2003 (in this case, the year 2003 serves as the category). No arithmetic operations, such as addition, subtraction, multiplication, or division, can be performed on numbers representing nominal data. As another example, the jersey numbers of baseball, football, and soccer players are nominal: that is, adding any two jersey numbers and comparing the result with another number makes no sense. Some other examples of nominal data are ID numbers of workers, account numbers used by a financial institution, ZIP codes, telephone numbers, genders, and colors.
- *Ordinal data*: Ordinal data are more informative than nominal data. When categorical data have a natural, logical order, the data collected are called ordinal. Examples include companies ranked according to the quality of their products, companies ranked based on their annual revenues, the severity of burn types, academic years of students, and stages of cancer among cancer-afflicted patients. Like nominal data, no addition, subtraction, multiplication, or division can be used on ordinal-type data.
- *Interval data:* Interval data are numerical data that are more informative than nominal and ordinal data but less informative than ratio data. A typical example of interval data is temperatures (in Celsius or Fahrenheit). Arithmetic operations of addition and subtraction are applicable, but multiplication and division are not. For example, the temperature of three consecutive parts A, B, and C during a selected step in a manufacturing process are 20 °F, 60 °F, and 30 °F, respectively. Then we can say the temperature difference between parts A and B is larger than the difference between parts B and C. Also, we can say that part B is warmer than part A, part C is warmer than part A, and part C is cooler than part B. However, it is physically meaningless to say that part B is three times as warm as part A or twice as warm as part C. Moreover, in interval data, zero does not have the conventional meaning of nothingness; it is just an arbitrary point on the scale of measurement. For instance, 0 °F and 0 °C (= 32 °F) have different values, and in fact they are arbitrary points on different scales of measurements. Other examples of interval data include the year in which a part is produced, students' numeric grades on a test, and date of birth.
- *Ratio data:* Ratio data are numerical data that have the potential to produce the most meaningful information of all data types. All arithmetic operations are applicable to this type of data. Numerous examples exist, such as height, weight, length of a rod, diameter of a ball bearing, RPM of a motor, number of employees in a company, hourly wages, annual growth rate of a company, etc. In ratio data, the number zero equates to nothingness or corresponds to the absence of the characteristics of interest.

Having defined statistical data, we now define statistics, descriptive statistics and inferential statistics. The term *statistics* is commonly used in two senses. First, colloquially, we use the word *statistics* to refer to a collection of numbers or facts. The following are some examples of statistics:

- In 2015, total compensation of CEOs from 10 selected companies ranged from $5 million to $10 million.
- On average, the yearly income of medical doctors in various specialties varies from $150 000 to $1 million dollars.
- In 2010, almost 10% of the population of the United States did not have health insurance.
- In 2016, the average tuition at private colleges was over $50,000 per academic year.
- In the United States, seniors spend a significant portion of their income on health care.
- The R&D budget of pharmaceutical division of a company is higher than the R&D budget of its biomedical division.
- In December 2018, almost all the states in US reported declining jobless rates.

Second, scientifically, the word *statistics* means the study of collecting, organizing, summarizing, analyzing, and interpreting data to make an informed decision. This broad field of statistical science can be further divided into two parts: *descriptive statistics* and *inferential statistics*:

- *Descriptive statistics* uses techniques to organize, summarize, analyze, and interpret the information contained in a data set to draw conclusions that do not go beyond the boundaries of the data set.
- *Inferential statistics* uses techniques that allow us to draw conclusions about a large body of data based on the information obtained by analyzing a small portion of these data.

## 3.3  Analyzing Data Using Graphical Tools

In Section 3.2, we briefly introduced various types of data, descriptive statistics, and inferential statistics. In this and the next chapter, we focus on taking a detailed look at both inferential and descriptive methods applicable to the various types of data.

Practitioners applying statistics in a professional environment often become overwhelmed by the large data sets they have collected. Occasionally, practitioners even have difficulty understanding data sets because either too many or too few factors influence a response variable of interest. In other cases, practitioners may doubt whether the proper statistical techniques were used to collect data. Consequently, the information in a data set may be biased or incomplete.

To avoid these situations, it is important to stay focused on the purpose or need for collecting the data, thus facilitating the appropriate utilization of data-collection techniques and the selection of appropriate factors. Descriptive statistics are very commonly used in applied statistics to help us understand the information contained in large, complex data sets. Next, we continue our discussion of descriptive statistics by considering an important tool called the frequency distribution table.

### 3.3.1  Frequency Distribution Tables for Qualitative and Quantitative Data

In statistical applications, often we encounter large quantities of messy data. To gain insight into the nature of the data, we frequently organize and summarize the data by constructing a table called a *frequency distribution table*. In any statistical application (as noted in Section 3.2), we can have data that are either qualitative (categorical) or quantitative (numerical). In this section, we discuss the construction of a *frequency distribution table* when the data is qualitative or quantitative.

### 3.3.1.1 Qualitative Data

A frequency distribution table for qualitative data requires two or more categories along with the number of data that belong to each category. The number of data belonging to any particular category is called the *frequency* of that category. We illustrate the construction of a frequency distribution table for qualitative data with the following example.

**Example 3.1 Industrial Revenue Data with the Following Example**

Consider a random sample of 100 small to mid-size companies located in the midwestern region of the United States, classified according to annual revenue (in millions of dollars). Construct a frequency distribution table for the data obtained by this classification.

**Solution**

We classify the annual revenues into five categories, as follows:

1) Under 250
2) 250 to less than 500
3) 500 to less than 750
4) 750 to less than 1000
5) 1000 or more

   Then the data collected can be represented as shown in Table 3.1, where we use the labels 1, 2, 3, 4, and 5 for the categories.

   After tallying the data, we obtain a frequency distribution table for these data, as shown in Table 3.2.

**Table 3.1** Annual revenues of 100 small to mid-size companies located in the midwestern region of the United States.

| | | | | | | | | | | | | | | | | | | | |
|---|---|---|---|---|---|---|---|---|---|---|---|---|---|---|---|---|---|---|---|
| 4 | 3 | 5 | 3 | 4 | 1 | 2 | 3 | 4 | 3 | 1 | 5 | 3 | 4 | 2 | 1 | 1 | 4 | 5 | 3 | 2 | 5 | 2 | 5 | 2 |
| 1 | 2 | 3 | 3 | 2 | 1 | 5 | 3 | 2 | 1 | 1 | 2 | 1 | 2 | 4 | 5 | 3 | 5 | 1 | 3 | 1 | 2 | 1 | 4 | 1 |
| 4 | 5 | 4 | 1 | 1 | 2 | 4 | 1 | 4 | 1 | 2 | 4 | 3 | 4 | 1 | 4 | 1 | 4 | 1 | 2 | 1 | 5 | 3 | 1 | 5 |
| 2 | 1 | 2 | 3 | 1 | 2 | 2 | 1 | 1 | 2 | 1 | 5 | 3 | 2 | 5 | 5 | 2 | 5 | 2 | 2 | 2 | 2 | 2 | 1 | 4 |

**Table 3.2** Frequency distribution for data in Table 3.1.

| Categories | Tally | Frequency or count | Cumulative frequency |
|---|---|---|---|
| 1 | ///// ///// ///// ///// ///// /// | 28 | 28 |
| 2 | ///// ///// ///// ///// ///// / | 26 | 54 |
| 3 | ///// ///// ///// | 15 | 69 |
| 4 | ///// ///// ///// / | 16 | 85 |
| 5 | ///// ///// ///// | 15 | 100 |
| Total | | 100 | |

Note that sometimes a quantitative data set is such that it consists of only a few distinct observations, which occur repeatedly. This kind of data is usually summarized in the same manner as the categorical data, but the categories represent distinct observations. We illustrate this scenario with the following example.

**Example 3.2  Manufacturing Data**

The following data show the number of defective parts produced by manufacturing per day over a period of 40 days:

| 1 | 2 | 1 | 5 | 4 | 2 | 3 | 1 | 5 | 4 | 3 | 4 | 6 | 2 | 3 | 3 | 2 | 2 | 3 | 5 |
|---|---|---|---|---|---|---|---|---|---|---|---|---|---|---|---|---|---|---|---|
| 1 | 3 | 2 | 2 | 4 | 2 | 6 | 1 | 2 | 6 | 6 | 1 | 4 | 5 | 4 | 1 | 4 | 2 | 1 | 2 |

Construct a frequency distribution table for these data.

**Solution**

In this example, the variable of interest is the number of defective parts produced on each day over a period of 40 days. Tallying the data, we obtain a frequency distribution table (Table 3.3).

A frequency distribution table such as Table 3.3 in which each category is defined using a single numerical value is usually called a *single-valued frequency distribution* table.

**Table 3.3**  Frequency distribution table for the data in Example 3.2.

| Categories | Tally | Frequency | Cumulative frequency |
|---|---|---|---|
| 1 | ///// /// | 8 | 8 |
| 2 | ///// ///// / | 11 | 19 |
| 3 | ///// / | 6 | 25 |
| 4 | ///// // | 7 | 32 |
| 5 | //// | 4 | 36 |
| 6 | //// | 4 | 40 |
| Total | | 40 | |

### 3.3.1.2  Quantitative Data

So far, we have discussed frequency distribution tables for qualitative data and quantitative data that can be treated as qualitative data. In this section, we will discuss frequency distribution tables for quantitative data.

Let $X_1, ..., X_n$ be a set of quantitative data values. To construct a frequency distribution table for this data set, we follow these steps:

1) Find the range R of the data, which is defined as

$$R = \text{largest data point} - \text{smallest data point} \qquad (3.1)$$

2) Divide the data set into an appropriate number of *classes*. The classes are also sometimes called *categories, cells, or bins*. There are no hard-and-fast rules to determine the number of classes, but as a guideline, the number of classes, say *m*, should be somewhere between 5 and 20. However, *Sturges'* formula is often used, given by

$$m = 1 + 3.3 \log n \qquad (3.2)$$

or

$$m = \sqrt{n} \qquad (3.3)$$

where *n* is the total number of data points in a given data set and log denotes log base 10. The result often gives a good estimate for an appropriate number of intervals. Note that since *m*, the number of classes, should always be a whole number, you may have to round *m* up or down when using either Eq. (3.2) or (3.3).

3) Determine the width of classes as follows:

$$\text{Class width} = R/m \qquad (3.4)$$

The class width should always be a number that is easy to work with, preferably a whole number. Furthermore, this number should always be obtained only by rounding up (never by rounding down) the value obtained when using Eq. (3.4).

Finally, preparing the frequency distribution table is achieved by assigning each data point to an appropriate class. While assigning these data points to a class, we must be particularly careful to ensure that each data point is assigned to one, and only one, class; and that the whole set of data is included in the table. Another important point is that the class at the lowest end of the scale must begin at a number that is less than or equal to the smallest data point, and the class at the highest end of the scale must end with a number that is greater than or equal to the largest data point in the data set.

**Example 3.3   Rod Manufacturing Data**
The following data give the lengths (in cm) of 40 randomly selected rods manufactured by a company:

| | | | | | | | | | |
|---|---|---|---|---|---|---|---|---|---|
| 145 | 140 | 120 | 110 | 135 | 150 | 130 | 132 | 137 | 115 |
| 142 | 115 | 130 | 124 | 139 | 133 | 118 | 127 | 144 | 143 |
| 131 | 120 | 117 | 129 | 148 | 130 | 121 | 136 | 133 | 147 |
| 147 | 128 | 142 | 147 | 152 | 122 | 120 | 145 | 126 | 151 |

Prepare a frequency distribution table for these data.

**Solution**

Following the steps described previously, we have

1) Range $(R) = 152 - 110 = 42$.
2) Number of classes $= 1 + 3.3 \log 40 = 6.29$, which, by rounding down, becomes 6.
3) Class width $= R/m = 42/6 = 7$.

The six classes we use to prepare the frequency distribution table are therefore as follows:

110 to less than 117
117 to less than 124
124 to less than 131
131 to less than 138
138 to less than 145
145 to 152

Note that in the case of quantitative data, each class is defined by two numbers. The smaller of the two numbers is called the *lower limit*, and the larger is called the *upper limit*. Also note that, except for the last class, the upper limit does not belong to the class. For example, data point 117 will be assigned to class 2, not class 1. Thus, no two classes have any common point, which ensures that each data point belongs to only one class. For simplification, we use the following mathematical notation to denote the classes:

$$[110 - 117), [117 - 124), [124 - 131), [131 - 138), [138 - 145), [145 - 152]$$

Here, a square bracket symbol "[" implies that the beginning point belongs to the class and a parenthesis ")" implies that the endpoint does not belong to the class. Then the frequency distribution table for the data in this example is shown in Table 3.4

**Table 3.4** Frequency table for the data in Example 3.3.

| Classes | Tally | Frequency or count | Relative frequency |
| --- | --- | --- | --- |
| [110–117) | /// | 3 | 3/40 |
| [117–124) | ///// // | 7 | 7/40 |
| [124–131) | ///// /// | 8 | 8/40 |
| [131–138) | ///// // | 7 | 7/40 |
| [138–145) | ///// / | 6 | 6/40 |
| [145–152] | ///// //// | 9 | 9/40 |
| Total | | 40 | 1 |

## 3.4 Describing Data Graphically

### 3.4.1 Dot Plots

A dot plot is one of the easiest graphs to construct. In a dot plot, each observation is plotted on a real line. For illustration, we consider the following example.

**Example 3.4  Shipment Data**

The following data gives the number of defective parts received in 24 different shipments.

| 14 | 12 | 10 | 16 | 11 | 25 | 21 | 15 | 17 | 5 | 32 | 28 |
|----|----|----|----|----|----|----|----|----|----|----|----|
| 26 | 21 | 29 | 12 | 10 | 21 | 10 | 17 | 15 | 13 | 34 | 26 |

Construct a dot plot for these data.

**Solution**

To construct a dot plot, first draw a horizontal line, the scale of which begins at the smallest observation (5 in this example shown in Figure 3.2) or smaller and ends with the largest observation (34 in this case) or larger.

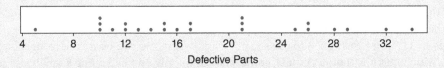

**Figure 3.2**  Dot plot for the data on defective parts received in 24 shipments.

Dot plots are more useful when the sample size is small. A dot plot gives us, for example, information about how far the data are scattered and where most of the observations are concentrated. For instance, in the previous example, we see that the minimum number of defective parts and the maximum number of defective parts received in any shipment were 5 and 34, respectively. Also, we can see that 67% of the time, the number of defective parts received in a shipment was between 10 and 21 (inclusive) for the shipment, and so on.

### 3.4.2  Pie Charts

Pie charts are commonly used to represent different categories of a population that are based on a characteristic of interest of that population: for example, allocation of the federal budget by sector, revenues of a large manufacturing company by region or by plant, technicians in a large corporation who are classified according to their basic qualification (i.e. high school diploma, associate degree, undergraduate degree, or graduate degree), and so on. A pie chart helps to better understand at a glance the composition of the population with respect to the characteristic of interest.

To construct a pie chart, divide a circle into slices such that each slice represents a proportional size of that category. Remember, the total angle of a circle is 360 degrees. The angle of a slice corresponding to a given category is determined as follows:

$$\text{Angle of a slice} = (\text{Relative frequency of the given category}) \times 360$$

We illustrate the construction of a pie chart for the data in Example 3.5 using technology.

**Example 3.5  Manufacturing Process**

In a manufacturing operation, we are interested in better understanding defect rates as a function of various process steps. The inspection points in the process are initial cutoff, turning, drilling, and assembly. These are qualitative data and are shown in Table 3.5. Now, using Minitab and R, we construct a pie chart for these data.

**Minitab**

The pie chart is constructed by following these steps in Minitab:

1) Enter the category in column C1. Enter the frequencies of the categories in column C2.
2) From the menu bar, select **Graph > Pie Chart** to open the **Pie Chart** dialog box onscreen.
3) Check the circle next to **Chart values from a table**. Enter **C1** under **Categorical values** and **C2** under **Summary variables**. Note that if you have the raw data without having the frequencies for different categories, then check the circle next to Chart counts of unique values. In that case, the previous dialog box won't have a **Summary variables** box.
4) Click **Pie Option**; in the new dialog box that appears, select any number of options you like and click **OK.** The pie chart will appear, as shown in Figure 3.2. Check the **Labels** option, check any options, and click **OK.**

Figure 3.3 shows the pie graph we obtained by using the previous steps in Minitab. The pie chart for these data may also be constructed manually by dividing the circle into four slices. The angle of each slice is given in the last column of Table 3.5.

Clearly, the pie chart in Figure 3.3 gives us a better understanding at a glance about the rate of defects occurring at different process steps.

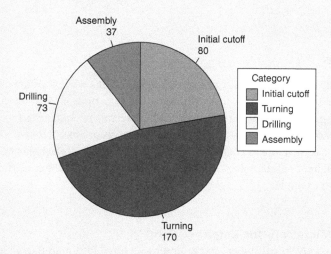

**Figure 3.3**   Pie chart for defects associated with manufacturing process steps.

**Table 3.5**   Understanding defect rates as a function of various process steps.

| Process steps | Frequency | Relative frequency | Angle size |
|---|---|---|---|
| Initial cutoff | 80 | 80/360 | 80 |
| Turning | 170 | 170/360 | 170 |
| Drilling | 73 | 73/360 | 73 |
| Assembly | 37 | 37/360 | 37 |
| Total | 360 | 1.00 | 360.00 |

**R**

To construct a pie chart using R, we use the built-in "pie()" function in R by running the following R code in the R **Console Window**:

```
Freq = c(80, 170, 73, 37)
#To label categories
Process =c('Initial Cutoff', 'Turning', 'Drilling', 'Assembly')
#To calculate percentages
Percents =round(Freq/sum(Freq)*100,1)
label =paste(Percents, '%', sep='') # add % to labels
#pie chart with percentages
pie(Freq, labels =label, col=c(2,3,4,5),main='pie chart of Process Steps')
```

Note: All statements preceded by "#" are ignored by R; they are included in the script to clarify the commands. The pie chart constructed by using R is identical to the one constructed by using Minitab.

### 3.4.3  Bar Charts

Bar charts are commonly used to study one or more populations when they are classified in various categories, such as by sector, by region, over different time periods, and so on. A bar chart is constructed by creating categories that are represented by intervals of equal length on a horizontal axis or *x*-axis. Within each category, we use the number of observations as a frequency of the corresponding category. Then, at each interval marked on the x-axis, we draw a rectangle or a bar of length proportional to the frequency of the corresponding category. We illustrate the construction of a bar chart with the help of an example.

**Example 3.6  Manufacturing Employee Data**
The following data give the number of new employees hired by a manufacturing company over six years (2014–2019):

| 128 | 245 | 130 | 154 | 152 | 165 |
|-----|-----|-----|-----|-----|-----|

Construct a bar chart for these data.

**Solution**

We can easily construct a bar chart manually by following the previous discussion. However, here we construct the bar chart using Minitab and R.

**Minitab**

1) Enter the category in column C1.
2) Enter frequencies of the categories in C2.

3) From the menu bar, select **Graph** > **Bar Chart** to open the **Bar Chart** dialog box. In this dialog box, select one of the three options under **Bars represent:** i.e. **Counts of unique values**, **A function of variables**, or **Values from a table**, depending on whether the data are sample values; functions of sample values, such as means of various samples; or categories and their frequencies. In this example, we are given categories and their frequencies, where the categories are the years in which the new employees were hired. Moreover, we are dealing with a sample from a single population, we select **Simple** and click **OK**. This prompts another dialog box to appear, as shown here.

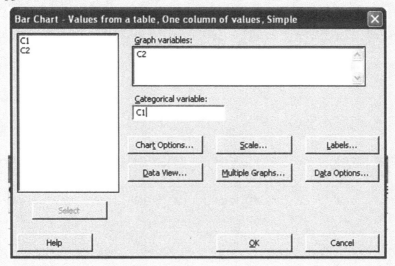

4) Enter **C2** in the box under **Graph variables**.
5) Enter **C1** in the box under **Categorical variable**.
6) There are several other options, such as **Chart Options, Scale,** etc. Click them and use them as needed. When you are finished, click **OK**.

The resulting bar chart will look like the one in Figure 3.4.

**R**

To construct bar chart using R, we use the built-in "barplot()" function in R by running the following R code in the R **Console Window**:

```
Freq = c(128, 245, 130, 154, 152, 165)
year =c(2014,2015,2016,2017,2018,2019)
# To obtain the bar chart
barplot(Freq, xlab='year', ylab='Employees Hired')
```

If you have raw data – that is, data that are not summarized as in the previous example, then the R code will be slightly different. For example, suppose you have the following data representing types of parts:

| | | | | | | | | | |
|----|----|----|----|----|----|----|----|----|----|
| 10 | 6  | 6  | 6  | 10 | 7  | 6  | 6  | 10 | 5  |
| 5  | 9  | 5  | 7  | 7  | 6  | 7  | 5  | 8  | 5  |

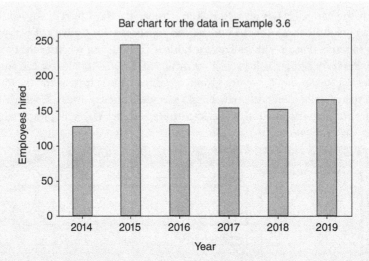

**Figure 3.4**  Bar chart for the data in Example 3.6.

Then use the following R code:

```
PartTypes <- c(10, 6, 6, 6, 10, 7, 6, 6, 10, 5, 5, 9, 5, 7, 7, 6, 7, 5, 8, 5)
# To obtain the frequencies
counts = table (PartTypes)
# To get the bar chart
barplot(counts, xlab='PartTypes', ylab='frequency')
```

### Example 3.7    Auto Parts Data

A company that manufactures auto parts is interested in studying the types of defects that occur in parts produced at a particular plant. The following data shows the types of defects that occurred over a certain period:

| | | | | | | | | | | | | | | | | | | | |
|---|---|---|---|---|---|---|---|---|---|---|---|---|---|---|---|---|---|---|---|
| B | A | A | A | B | A | E | D | C | A | B | C | D | C | A | E | A | B | A | B |
| C | E | D | C | A | E | A | D | B | C | B | A | B | E | D | B | D | B | E | A |
| C | A | B | C | E | C | B | A | B | D | | | | | | | | | | |

Construct a bar chart for the types of defects found in the auto parts.

### Solution

To construct a bar chart for the data in this example, we first need to prepare a frequency distribution table. The data in this example are qualitative, and the categories are the types of defects: A, B, C, D, and E. The frequency distribution table is shown in Table 3.6.

Now, to construct the bar chart, we can use Minitab by following the same steps as described in Example 3.6, since we have categories and frequencies. If you did not prepare a frequency distribution table, then from the first three options choose **Counts of unique values**; and the remainder of the procedure is exactly the same. Figure 3.5 shows the resulting bar chart.

**Table 3.6** Frequency distribution table for the data in Example 3.7.

| Type of defect | Tally | Frequency | Relative frequency | Cumulative frequency |
|---|---|---|---|---|
| A | ///// ///// //// | 14 | 14/50 | 14 |
| B | ///// ///// /// | 13 | 13/50 | 27 |
| C | ///// //// | 9 | 9/50 | 36 |
| D | ///// // | 7 | 7/50 | 43 |
| E | ///// // | 7 | 7/50 | 50 |
| Total | | 50 | 1.00 | |

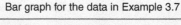

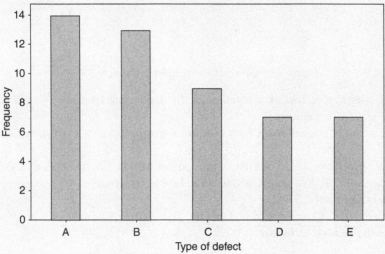

**Figure 3.5** Bar chart for the data in Example 3.7.

**Example 3.8   Auto Parts Data**

The following data give the frequency of defect types for auto parts manufactured over the same period of time in two different plants that have the same manufacturing capacity. Construct a side-by-side bar chart comparing the types of defects occurring in the auto parts that are manufactured at the two plants.

| Defect type | A | B | C | D | E | Total |
|---|---|---|---|---|---|---|
| Plant I | 14 | 13 | 9 | 7 | 7 | 50 |
| Plant II | 12 | 18 | 12 | 5 | 8 | 55 |

**Solution**

This graph can easily be constructed by using one of the statistical packages discussed in this book. For example, to use Minitab, take the following steps:

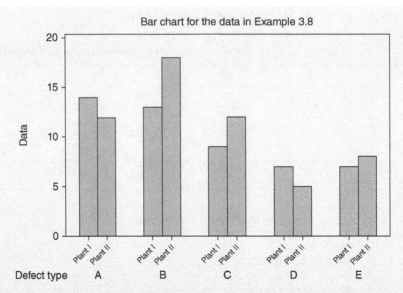

**Figure 3.6** Bar chart for the data in Example 3.8.

1) Enter the categories in column C1, and enter the frequencies for Plant I and Plant II in columns C2 and C3, respectively.
2) From the menu bar, select **Bar Chart > Values from a table > Two- way table cluster > OK**.
3) A new dialog box appears. In this dialog box, enter **C2** and **C3** in the box below **Graph variables**, and enter **C1** in the box below **Row labels**. Then select the second option from **Table Arrangements**.
4) Select any of the other options, and click **OK**. The Minitab output shown in Figure 3.6 will appear in the **Session Window**.

Such bar charts are very commonly used to compare two or more populations. From the example chart, we can see that defect types B, C, and E occur less frequently at Plant I than at Plant II, whereas defect types A and D occur more frequently at Plant I. From the data, we can also see that the total number of defects at Plant I is less than at Plant II.

**R**

Here the "matrix()" function takes a vector of data, "c(...)", and converts it into a matrix using various options: in this case, by stating that there are five columns ("ncol = 5") and that the vector values will fill the resulting matrix row by row ("byrow = TRUE"). Because we want the "barplot()" function to use default labeling, we assign the rows and columns names ("dimnames") via a list of two string vectors, where the first list element is the vector of row names and the second list element is a vector of column names. R has a special constant, "LETTERS", which is a vector of the letters $A$ through $Z$; here, we use the first five elements of "LETTERS". The "barplot()" function plots the given matrix, using different rows to plot separate side-by-side bars ("beside = TRUE") and displaying a legend that gives the matrix row names ("legend. text = TRUE"):

```
Plant = matrix(c(14, 13, 9, 7, 7, 12, 18, 12, 5, 8), ncol =5,
byrow = TRUE, dimnames = list (c("Plant I", "Plat II"), LETTERS [1:5]))
barplot(Plant, beside = TRUE, legend.text = TRUE)
```

### 3.4.4  Histograms

Histograms are very popular and are used to represent quantitative data graphically. They generally provide useful information about a data set such as insight about trends, patterns, location of the center of the data, dispersion of the data, and so on – that is not readily apparent from raw data.
   Constructing a histogram involves the following two major steps:

1) Prepare a frequency distribution table for the given data.
2) Use the frequency distribution table prepared in step 1 to construct the histogram. From here on, the steps are exactly the same as those for constructing a bar chart, except that in a histogram, there is no gap between the intervals marked on the *x*-axis. We illustrate the construction of a histogram in Example 3.9.

   Note that a histogram is called a *frequency histogram* or a *relative frequency histogram,* depending on whether the heights of rectangles over the intervals marked on the *x*-axis are proportional to the frequencies or the relative frequencies. In both types of histograms, the width of the rectangles is equal to the class width. In fact, the only difference between the two types of histograms is that the scale used on the *y*-axis is different. This point will become apparent from the following example.

**Example 3.9  Fuel Pump Data**
The following data give the number of fuel pumps for SUVs produced per day by manufacturing companies during the month of June of a given year:

| 72 | 88 | 65 | 68 | 68 | 75 | 87 | 79 | 89 | 79 |
|----|----|----|----|----|----|----|----|----|----|
| 65 | 76 | 81 | 84 | 67 | 82 | 61 | 89 | 85 | 90 |
| 67 | 68 | 82 | 85 | 79 | 65 | 79 | 74 | 81 | 82 |

Construct a frequency histogram for the above data.

**Solution**

As mentioned earlier, we first prepare the frequency distribution table:
   1) Find the range of the data set:

$$R = 90 - 61 = 29$$

   2) Determine the number of classes to be used in the histogram:

$$m = 1 + 3.3 \log 30 = 5.87$$

**Table 3.7** Frequency distribution table for the survival time of parts.

| Class | Tally | Frequency | Relative frequency | Cumulative frequency |
|---|---|---|---|---|
| [60–65) | / | 1 | 1/30 | 1 |
| [65–70) | ///// /// | 8 | 8/30 | 9 |
| [70–75) | ///// //// | 2 | 2/30 | 11 |
| [75–80) | ///// / | 6 | 6/30 | 17 |
| [80–85) | ///// / | 6 | 6/30 | 23 |
| [85–90] | ///// // | 7 | 7/30 | 30 |
| Total | | 30 | 1 | |

After rounding up, we take the number of classes to be 6.

3) Compute the class width:

$$\text{Class width} = R/m = 29/6 = 4.83$$

Again, by *rounding up*, we have a class width of 5. Note that if we rounded down the class width, then some observations might be left out and might not belong to any class. Consequently, the total frequency would be less than $n$.

The frequency distribution table for the data in this example is as shown in Table 3.7. It is such that each number belongs to only one interval.

Having completed the frequency distribution table, we are now ready to construct a histogram. To do so, we first mark the classes on the $x$-axis and the frequencies on the $y$-axis. Remember that when we mark the classes on the $x$-axis, we must make sure there is no gap between them. Then, for each class marked on the $x$-axis, we draw a rectangle of height proportional to the frequency of the corresponding class.

We can also construct the histogram using one of the statistical packages discussed in this book. Here we show how to construct the histogram using Minitab and R.

**Minitab**

1) Enter the data in column C1.
2) From the menu bar, select **Graph** > **Histogram**. The **Histogram** dialog box appears.
3) In this dialog box, select an appropriate histogram, and click **OK**. For example, here we select **Simple**. This prompts another dialog box to appear.
4) In this dialog box, enter **C1** in the box under the **Graph variables**, and click **OK**. A histogram graph will appear in the **Session Window**.
5) To customize the number of classes, double-click on any bar of the histogram to open the **Edit Graph** dialog box. In the new dialog box, again double-click on any bar of the histogram. Another dialog box appears; in it, select **Binning**. You can then select the desired number of classes, their midpoints and lower and upper limits, etc. Click **OK**.

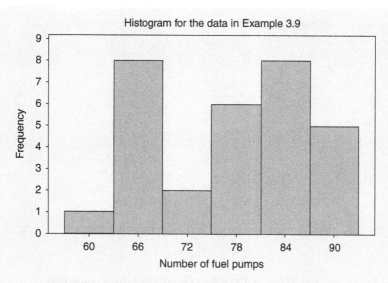

**Figure 3.7** Frequency histogram for the data in Example 3.9.

The histogram shown in Figure 3.7 appears in the **Session Window**.

**R**

Enter the data as a vector, and use the built-in "hist()" function to create the histogram by running the following code. It will construct a histogram identical to the one constructed by Minitab:

```
Data = c (72,88,65,68,68,75,87,79,89,79,65,76,81,84,67,82,61,89,85,90,67,68,82,85,79,
65,79,74,81,82)
hist(Data)
```

### 3.4.5 Line Graphs

A line graph, also known as a *time-series graph*, is commonly used to study changes that take place over time in the variable of interest. In a line graph, time is marked on the horizontal x-axis, and the variable of interest is put on the vertical *y*-axis. For illustration, we use the data in Example 3.10.

**Example 3.10  Flu Vaccine Data**

The following data give the number of flu vaccines administered by a small pharmacy in a midwestern town over a period of 12 months:

| Month | Jan. | Feb. | Mar. | April | May | June | July | Aug. | Sept. | Oct. | Nov. | Dec. |
|---|---|---|---|---|---|---|---|---|---|---|---|---|
| Flu Vaccines | 40 | 30 | 21 | 10 | 5 | 2 | 3 | 5 | 25 | 45 | 43 | 48 |

Prepare a line graph for the above data.

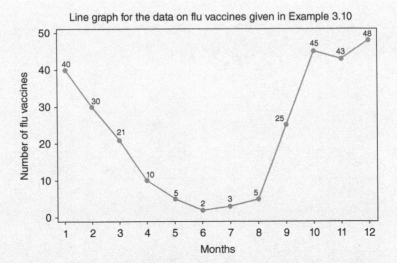

**Figure 3.8** Line graph for the data on flu vaccines given in Example 3.10.

**Solution**

To prepare the line graph, plot the previous data using the x-axis for the months and the y-axis for the flu vaccines given, and then join the plotted points with a segmented curve. The line graph for the data in this example is shown in Figure 3.8, created using Minitab (**Graph > Time Series Plot**) and following these steps:

1) Enter the months as 1, 2, 3, ... in column **C1** and the number of flu vaccines in column **C2**.
2) In the **Time Series Plot** dialog box, select **Simple** and then click **OK**.
3) A new dialog box titled **Time Series Plot: Simple** appears. In this dialog box, enter **C2** (number of flu vaccines) in the box below **Series**. Then select **Time/scale> Time > Stamp** and enter **C1** (months) in box below **Stamp columns** and click **OK**.
4) Select the option **Labels >Title**, and enter the title of the graph **Line graph....** Then select **Data Labels > Use y-value Labels** and click **OK**. Again, click **OK**. The diagram shown in Figure 3.8 will appear in the **Session Window**.

**R**

Here we use the built-in "plot()" function to construct a line graph. Using the following R code in the **Console Window**, we get a line graph that is identical to the one shown in Figure 3.8:

```
x = c (1,2,3,4,5,6,7,8,9,10,11,12)
y = c (40,30,21,10,5,2,3,5,25,45,43,48)
plot (x,y)
```

Note that in the first line, the numbers 1, 2, 3, ... represent the months Jan., Feb, Mar, ... From the line graph, we can see that the number of flu vaccines applied is seasonal: more flu vaccines are given in winter months when flu is more spread out.

## 3.4.6 Measures of Association

So far in this chapter, we have dedicated our discussion to univariate statistics because we were interested in studying only a single characteristic of a subject of interest. In the previous examples, the variable of interest was either qualitative or quantitative. Now we will focus our attention on cases involving two variables and simultaneously examine two characteristics of a subject of concern. The two variables of interest could be either qualitative or quantitative, but we will concentrate on variables that are quantitative in nature. When studying two variables simultaneously, the data obtained is known as *bivariate data*. In examining bivariate data, the first question is whether there is any association between the two variables of interest. An effective way to investigate a potential association is to prepare a graph by plotting one variable against the other: one along the horizontal scale (*x*-axis) and the second along the vertical scale (*y*-axis). Each pair of observations (*x, y*) is then plotted as a point in the *xy*-plane. This type of graph is called a *scatterplot*. A scatterplot is a very useful graphical tool because it helps us visualize the nature and strength of associations between two variables.

Example 3.11 illustrates the concept of a scatterplot and the visual assessment of the measure of association.

### Example 3.11   Medical Data

The cholesterol level and systolic blood pressure of 30 randomly selected US males in the age group 40–50 years are given in Table 3.8. Construct a scatterplot of this data and determine whether there is any association between the cholesterol levels and systolic blood pressures.

### Solution

Figure 3.9a shows the scatterplot of the data in Table 3.8. This scatterplot clearly shows that there is an upward, linear trend, and the data points are concentrated around the straight line within a narrow band. The upward trend indicates a positive association between the two variables, whereas the width of the band indicates the strength of the association, which in this case is very strong. As the association between the two variables gets stronger, the band enclosing the plotted points becomes narrower.

**Table 3.8**  Cholesterol level and systolic BP of 30 randomly selected US males ages 40–50.

| Subject | 1 | 2 | 3 | 4 | 5 | 6 | 7 | 8 | 9 | 10 |
|---|---|---|---|---|---|---|---|---|---|---|
| Cholesterol (*x*) | 195 | 180 | 220 | 160 | 200 | 220 | 200 | 183 | 139 | 155 |
| Systolic BP (*y*) | 130 | 128 | 138 | 122 | 140 | 148 | 142 | 127 | 116 | 123 |
| Subject | 11 | 12 | 13 | 14 | 15 | 16 | 17 | 18 | 19 | 20 |
| Cholesterol (*x*) | 153 | 164 | 171 | 143 | 159 | 167 | 162 | 165 | 178 | 145 |
| Systolic BP (*y*) | 119 | 130 | 128 | 120 | 121 | 124 | 118 | 121 | 124 | 115 |
| Subject | 21 | 22 | 23 | 24 | 25 | 26 | 27 | 28 | 29 | 30 |
| Cholesterol (*x*) | 245 | 198 | 156 | 175 | 171 | 167 | 142 | 187 | 158 | 142 |
| Systolic BP (*y*) | 145 | 126 | 122 | 124 | 117 | 122 | 112 | 131 | 122 | 120 |

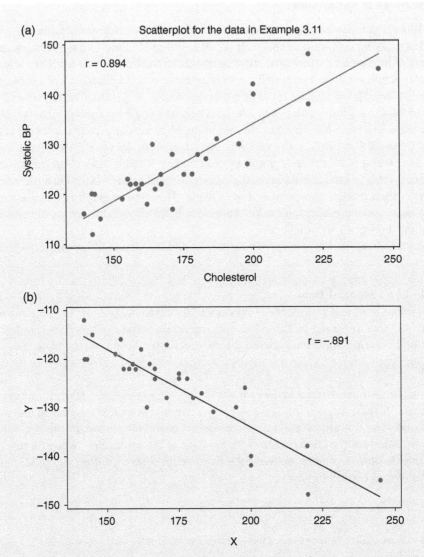

**Figure 3.9** Minitab scatterplots showing four different degrees of correlation: (a) strong positive correlation; (b) strong negative correlation; (c) positive perfect correlation; (d) negative perfect correlation. (Graphs (c) and (d) appear on the next page)

A downward trend indicates a negative association between two variables. A numerical measure of association between two numerical variables is called the *Pearson correlation coefficient* or simply the *correlation coefficient*, named after the English statistician Karl Pearson (1857–1936). The correlation coefficient between two numerical variables in sample data is usually denoted by $r$. The Greek letter $\rho$ denotes the corresponding correlation coefficient for population data. The correlation coefficient is defined as

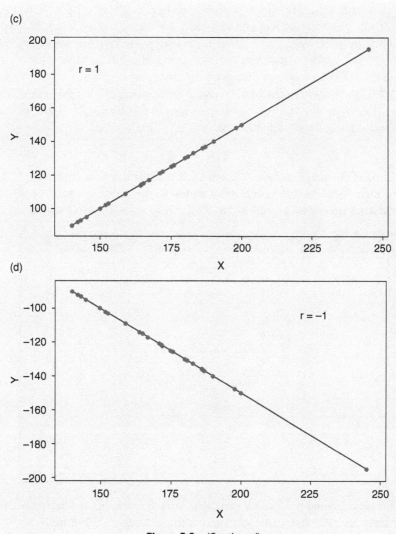

**Figure 3.9** (Continued)

$$r = \frac{\Sigma(x_i - \bar{x})(y_i - \bar{y})}{\sqrt{\Sigma(x_i - \bar{x})^2 \Sigma(y_i - y)^2}} = \frac{\Sigma x_i y_i - \dfrac{(\Sigma x_i)(\Sigma y_i)}{n}}{\sqrt{\left(\Sigma x_i^2 - \dfrac{(\Sigma x_i)^2}{n}\right)\left(\Sigma y_i^2 - \dfrac{(\Sigma y_i)^2}{n}\right)}} \qquad (3.5)$$

The correlation coefficient is a unitless measure and can take on any value in the interval [–1, +1]. As the strength of the association between the two variables grows, the absolute value of $r$ approaches 1. Thus, when there is a perfect association between the two variables, $r = 1$ or –1, depending on whether the association is positive or negative, respectively. Perfect association means that if we know the value of one variable, the value of the other variable can be determined without any error. When $r = 0$, there is no association between the two variables. As a general rule, the association is weak, moderate, or strong when the absolute value of $r$ is less than 0.3, between 0.3 and 0.7, or greater than 0.7, respectively.

Figures 3.9b–d show scatterplots for data having different correlation strengths. These graphs can be created by using any of the statistical packages in this book; here again we discuss constructing a scatterplot using Minitab and R.

**Minitab**

1) Enter the pairs of data in columns C1 and C2. Label the columns **X** and **Y**.
2) From the menu bar, select **Graph > Scatterplot**. In the dialog box that opens, select **Scatterplot With Regression** and click **OK**. This prompts the following dialog box to appear.

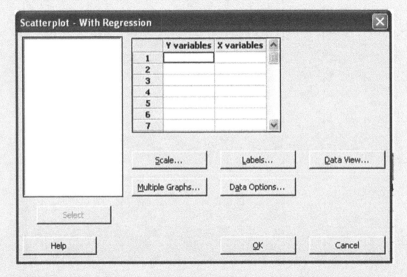

3) In this dialog box, under the X and Y variables, enter the columns of data. Choose the desired options and click **OK**. The scatterplot appears in the **Session Window**.
4) To calculate the correlation coefficient, from the menu bar, select **Stat > Basic Statistics > Correlation**.

As mentioned, the resulting scatterplot is shown in Figure 3.9a. The correlation between the cholesterol level and the systolic blood pressure calculated by using Minitab is $r = 0.894$. Using R, as shown in the next set of steps, we get exactly the same result. Thus, based on these data, we can say the there is a very strong association between cholesterol level and systolic blood pressure.

**R**

We use the built-in "plot()" function in R to generate scatterplots. The function "abline()" is used to add the trend line to the scatterplot, and the function "cor()" calculates the Pearson correlation coefficient:

```
x = c (195,180,220,160,200,220,200,183,139,155,153,164,171,143,159,167,
162,165,178,145, 245,198,156,175,171,167,142,187,158,142)
y =(130,128,138,122,140,148,142,127,116,123,119,130,128,120,121,124,
118,121,124,115, 145,126,122,124,117,122,112,131,122,120)
#To plot the data in a scatterplot
plot(x, y, main = 'Scatterplot for Cholesterol Level and Systolic Blood Pressure Data',
xlab = 'Cholesterol Level', ylab = 'Systolic Blood Pressure')
#To add a trend line
abline(lm(y ~ x))
#To calculate the Pearson correlation coefficient
cor(x, y)
```

All the results in R are identical to those obtained by using Minitab.

## 3.5 Analyzing Data Using Numerical Tools

### 3.5.1 Numerical Measures

We saw in Section 3.3 that graphical methods provide powerful tools to visualize the information contained in a data set. Perhaps you have a new appreciation for the saying "A picture is worth a thousand words"! In the following sections, we extend our knowledge from graphical methods to numerical methods (tools). Numerical methods provide information about what are commonly known as *quantitative* or *numerical measures*. The numerical methods that we are about to study are applicable to both samples as well as population data.

> **Definition 3.1** Numerical measures computed by using population data are referred to as *parameters*.

> **Definition 3.2** Numerical measures computed by using sample data are referred to as *statistics*.

Note that it is a standard practice to denote *parameters* using letters of the Greek alphabet and *statistics* with letters of the English alphabet. We divide numerical measures into two major categories: *measures of centrality* and *measures of dispersion*.

Measures of centrality give us information about the center of the data, whereas measures of dispersion provide information about the variation within the data.

### 3.5.2 Measures of Centrality

Measures of centrality are also known as *measures of central tendency*. The following measures are of primary importance:

- Mean
- Median
- Mode

The mean is also sometimes referred to as the *arithmetic mean* and is the most useful and most commonly used measure of centrality. The median is the second-most used, and mode is the least used measure of centrality.

### 3.5.2.1 Mean

The mean of a sample or a population data is calculated by taking the sum of the data measurements and then dividing the sum by the number of measurements in the sample or the population. Thus, the mean is the arithmetic average of a set of values. The mean of a sample is called the *sample mean* and is denoted by $\overline{X}$ ("X-bar"), and the population mean is denoted by the Greek letter $\mu$ ("mu"). These terms are defined numerically as follows:

$$\text{Population mean } \mu = \frac{X_1 + X_2 + \ldots + X_N}{N} = \frac{\sum X_i}{N} \tag{3.6}$$

$$\text{Sample mean } \overline{X} = \frac{X_1 + X_2 + \ldots + X_n}{n} = \frac{\sum X_i}{n} \tag{3.7}$$

where $\Sigma$ ("sigma") symbolizes the summation over all the measurements, and $N$ and $n$ denote the population size and sample size, respectively. Furthermore, note that the value of $\overline{X}$ for a particular sample is usually denoted using the lowercase letter $\bar{x}$.

**Example 3.12  Hourly Wage Data**
The following data give the hourly wages (in dollars) of some randomly selected workers in a manufacturing company:

| 8 | 6 | 9 | 10 | 8 | 7 | 11 | 9 | 8 |
|---|---|---|----|---|---|----|---|---|

Find the mean hourly wage of these workers.

**Solution**

Since the wages listed in these data are for only some of the company's workers, they represent a sample. Thus we have

$$n = 9, \Sigma x_i = 8 + 6 + 9 + 10 + 8 + 7 + 11 + 9 + 8 = 76$$

So the sample mean is

$$\bar{x} = \frac{\sum x_i}{n} = \frac{76}{9} = 8.44$$

In this example, the mean hourly wage of these employees is $8.44 per hour.

**Example 3.13  Age Data**
The following data give the ages of all the employees of a city hardware store:

| 22 | 25 | 26 | 36 | 26 | 29 | 26 | 26 |
|----|----|----|----|----|----|----|----|

Find the mean age of the employees of that hardware store.

**Solution**

Since the data give the ages of all the employees of the hardware store, we are interested in a population. Thus, we have

$$N = 8, \Sigma x_i = 22 + 25 + 26 + 36 + 26 + 29 + 26 + 26 = 216$$

So the population mean is

$$\mu = \frac{\Sigma x_i}{N} = \frac{216}{8} = 27 \text{ years}$$

In this example, the mean age of the employees of the hardware store is 27 years.

Note that even though the formulas for calculating the sample mean and population mean are very similar, it is very important to make a clear distinction between the sample mean, $\overline{X}$, and the population mean, $\mu$, for all application purposes.

Also note that sometimes a data set may include observations or measurements that are very small or very large relative to the other observations. For example, the salaries of a group of engineers in a corporation may include the salary of its CEO, who happens to be an engineer and whose salary is much larger than other engineers in that group. These very small or very large values are usually referred to as *extreme values*. If extreme values are present in the data set, the mean is not an appropriate measure of centrality, as it can be heavily influenced by the extreme values. It is important to note that any extreme value, large or small, adversely affects the mean value. In such cases, the median is a better measure of centrality since it is more robust to extreme values.

### 3.5.2.2 Median

The median is the physical center of a set of observations. We denote the median of a data set with $M_d$. To determine the median of a data set, follow these steps:

1) Arrange the measurements in the data set in ascending order, and rank them from 1 to $n$.
2) Find the rank of the median, which is equal to

$$(n + 1)/2 \text{ if } n \text{ is odd}$$

$$n/2 \text{ and } n/2 + 1 \text{ if } n \text{ is even}$$

3) Find the data value corresponding to the rank $(n + 1)/2$ of the median. This value represents the median of the data set. Note that when $n$ is even, the median is the average of two values that correspond to the two ranks $n/2$ and $(n + 1)/2$.

**Example 3.14 Rod Data**

To illustrate this method, we consider a simple example. The following data give the length of pushrods for car engines that are manufactured in a given shift:

| 50 | 44 | 54 | 48 | 52 | 55 | 49 | 46 | 56 | 50 | 53 |
|----|----|----|----|----|----|----|----|----|----|----|

Find the median length of these rods.

**Solution**

1) Write the data in ascending order, and rank them from 1 to 11 since $n = 11$. The observations in ascending order are:

|       | 44 | 46 | 48 | 49 | 50 | 50 | 52 | 53 | 54 | 55 | 56 |
|-------|----|----|----|----|----|----|----|----|----|----|----|
| Rank  | 1  | 2  | 3  | 4  | 5  | 6  | 7  | 8  | 9  | 10 | 11 |

2) Find the rank of the median:

$$\text{Rank of the median} = (n+1)/2 = (11+1)/2 = 6$$

3) Find the value corresponding to rank 6. The value corresponding to rank 6 is 50. Thus, the median length of pushrods is $M_d = 50$. This means that, at most, 50% of the pushrods are of length less than 50; and at most, 50% are of length greater than 50.

**Example 3.15  Hourly Wage Data**

The following data describe the hourly wages (in dollars) for 16 randomly selected employees of a large US manufacturing company:

| 20 | 18 | 25 | 22 | 27 | 17 | 30 | 29 | 32 | 35 | 26 | 25 | 28 | 260 | 310 | 22 |
|----|----|----|----|----|----|----|----|----|----|----|----|----|-----|-----|----|

Find the median hourly wages of these individuals.

**Solution**

1) Write all the observations in ascending order:

|      | 17 | 18 | 20 | 22 | 22 | 25 | 25 | 26 | 27 | 28 | 29 | 30 | 32 | 35 | 260 | 310 |
|------|----|----|----|----|----|----|----|----|----|----|----|----|----|----|-----|-----|
| Rank | 1  | 2  | 3  | 4  | 5  | 6  | 7  | 8  | 9  | 10 | 11 | 12 | 13 | 14 | 15  | 16  |

2) Since $n = 16$, which is even, the ranks of the median are $= 16/2$ and $16/2 + 1 = 8$ and 9.
3) Find the value corresponding to the rank 8.5. Since there are two ranks, the median is defined as the average of the values that correspond to ranks 8 and 9:

$$\text{Median } M_d = (26 + 27)/2 = 26.5$$

It is important to note the median does not have to be one of the values in the data set. When the sample size is odd, the median is the center value; and when the sample size is even, the median is the average of the two middle values when the data are arranged in the ascending order.

Finally, note that the data in the previous example contains two values – 260 and 310 –that seem to be the wages of top managers. These two large values may be considered extreme values. In this case, if we calculate the mean of these data, we find

$$\overline{X} = (17 + 18 + 20 + 22 + 22 + 25 + 25 + 26 + 27 + 28 + 29 + 30 + 32 + 35 + 260 + 310)/16 = 57.875$$

The mean value of 57.875 is much larger than the median value of 26.5, which illustrates the influence of the extreme values on the mean. In this case, the mean does not truly represent the center of these data. The median is more representative of the center of these values.

If we replace the extreme values of 260 and 310 with smaller values, say 35 and 40, then the resulting median remain unchanged, but the mean becomes $26.937. The mean is now representative of the center of these data since the extreme values were removed. Thus, it is important to remember that whenever a data set contains extreme values, the median is a better measure of centrality than the mean.

### 3.5.2.3  Mode

The mode of a data set is the value that occurs most frequently. It is the least used measure of centrality. When products are produced via mass production, such as clothes of certain sizes, rods of certain lengths, etc., the mode or the modal value is of interest. Note that in any data set, there may be no mode, one mode, or multiple modes. We denote the mode of a data set with $M_0$.

**Example 3.16**
Find the mode for the following data set:

| 7 | 18 | 15 | 16 | 20 | 37 | 49 | 20 | 7 | 12 | 41 |
|---|---|---|---|---|---|---|---|---|---|---|

**Solution**

In the given data set, each value occurs once except 7, which occurs twice. Thus, the mode for this data set is $M_0 = 7$.

**Example 3.17**
Find the mode for the following data set:

| 2 | 8 | 39 | 43 | 51 | 2 | 8 | 17 | 39 | 17 | 51 | 43 |
|---|---|---|---|---|---|---|---|---|---|---|---|

**Solution**

In this data set, each value occurs the same number of times (twice). Thus, in this data set, there is no mode.

**Example 3.18**
Find the mode for the following data set:

| 15 | 17 | 12 | 13 | 14 | 27 | 17 | 27 | 23 | 26 | 15 |
|---|---|---|---|---|---|---|---|---|---|---|

**Solution**

In this data set 15, 17, and 27 occur twice, and the rest of the values occur only once. Thus, in this example, there are three modes: $M_0 = 15, 17,$ and 27.

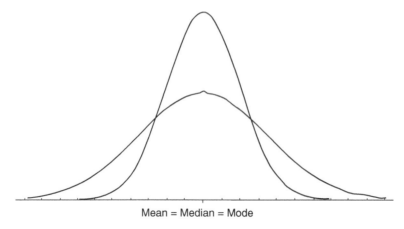

Mean = Median = Mode

**Figure 3.10** Two frequency distribution curves with equal mean, median, and mode values.

### 3.5.3 Measures of Dispersion

Sets of data are often represented by curves called *frequency distribution curves*. For example, consider the frequency distribution curves shown in Figure 3.10. Note that they are symmetric and are bell-shaped. Measures of central tendency, discussed in the previous section, give us information about the location of the physical center of distribution curves. For each of the curves in Figure 3.10, the mean, median, and mode are equal and located at the peak point of the curve. There are other forms of frequency distribution curves that are not symmetric; however, we are not going to discuss them since they are beyond the scope of this book.

Measures of central tendency do not portray the whole picture of any data set. For example, in Figure 3.10, the two frequency distributions have the same mean, median, and mode. Interestingly, however, the two distributions are significantly different from each other. The major difference between them is the variation among the values associated with each distribution. So, it is important to know about the variation among the values of a data set. Information about variation is provided by *measures of dispersion*. In this section, we will study three measures of dispersion: *range*, *variance*, and *standard deviation*.

#### 3.5.3.1 Range

The range of a data set is the easiest measure of dispersion to calculate. It is defined as follows:

$$\text{Range} = \text{Largest value} - \text{Smallest value} \tag{3.8}$$

The range is not a very efficient measure of dispersion since it takes into consideration only the largest and smallest values and does not take into account the remaining observations. For example, if a data set has 100 distinct observations, the range uses only 2 observations and ignores the remaining 98. As a rule of thumb, if a data set contains fewer than 10 observations, the range is a reasonably good measure of dispersion; but for data sets containing more than 10 observations, the range is not considered very efficient.

**Example 3.19**

The following data give the tensile strength (in psi) of a sample of material submitted for inspection:

| 8638.24 | 8550.16 | 8594.27 | 8417.24 | 8543.99 | 8468.04 | 8468.94 | 8524.41 | 8527.34 |
| --- | --- | --- | --- | --- | --- | --- | --- | --- |

Find the range for this data set.

**Solution**

The largest and smallest values in the data set are 8638.24 and 8417.24, respectively. Therefore, the range for this data set is

$$\text{Range} = 8638.24 - 8417.24 = 221.00$$

### 3.5.3.2 Variance

While range can give us some idea of the overall spread of a set of data, it does not indicate how data vary from each other – which is one of the most interesting pieces of information associated with any data. To gain a deeper understanding of variability, and to quantify how values vary from one another, we rely on more powerful indicators such as variance, which is a value that focuses on how individual observations deviate from their mean.

If the values in the data set are $x_1$, $x_2$, $x_3$, $\cdots$, $x_n$, and the mean of these data is $\bar{x}$, then $x_1 - \bar{x}$, $x_2 - \bar{x}, x_3 - \bar{x}, \cdots, x_n - \bar{x}$ are the deviations from the mean. It then seems natural to find the sum of these deviations and argue that if this sum is large, the values generally differ from each other; and if this sum is small, then the values do not differ from each other. Unfortunately, this argument does not hold, since the sum of the deviations is always zero, no matter how the values in the data set differ from each other. This is true because some of the deviations are positive, and some are negative; and when we take their sum, they cancel each other out. To remedy this, we can square these deviations and then take their sum. Squared deviations, of course, will always be positive, and the sum of the squared deviations gives a better measure of how data vary from the mean. The variance then becomes the average value of the sum of these squared deviations from the mean, $\bar{x}$. If the data set represents a population, then the deviations are taken from the population mean, $\mu$. Thus, the population variance, denoted by $\sigma^2$ ("sigma squared"), is defined as

$$\sigma^2 = \frac{1}{N} \sum_{i=1}^{N} (X_i - \mu)^2 \qquad (3.9)$$

And the sample variance, $S^2$, is defined as

$$S^2 = \frac{1}{n} \sum_{i=1}^{n} (X_i - \bar{X})^2 \qquad (3.10)$$

It is important to note, however, that the formula used in practice to calculate the sample variance $S^2$ is as given in Eq. (3.11), because it gives a better estimator of the population variance:

$$S^2 = \frac{1}{n-1}\sum_{i=1}^{n}\left(X_i - \overline{X}\right)^2 \qquad (3.11)$$

For computational purposes, Eqs. (3.12) and (3.13) give the simplified forms of Eqs. (3.9) and (3.11) for the population variance and sample variances, respectively:

$$\sigma^2 = \frac{1}{N}\left(\sum X_i^2 - \frac{\left(\sum X_i\right)^2}{N}\right) \qquad (3.12)$$

$$S^2 = \frac{1}{n-1}\left(\sum X_i^2 - \frac{\left(\sum X_i\right)^2}{n}\right) \qquad (3.13)$$

Note that one difficulty in using the variance as a measure of dispersion is that the units for measuring variance are not the same as for data values. Rather, variance is expressed as a square of the units used for the data values. For example, if the data values are dollar amounts, then the variance will be expressed in squared dollars – which is meaningless. For application purposes, we define another measure of dispersion called the *standard deviation* that is directly related to the variance and is measured in the same units as the original data values.

### 3.5.3.3 Standard Deviation
The standard deviation is obtained by taking the positive square root (with a positive sign) of the variance. The population standard deviation $\sigma$ and the sample standard deviation $S$ are defined as follows:

$$\sigma = +\sqrt{\frac{1}{N}\left[\sum X_i^2 - \frac{\left(\sum X_i\right)^2}{N}\right]} \qquad (3.14)$$

$$S = +\sqrt{\frac{1}{n-1}\left[\sum X_i^2 - \frac{\left(\sum X_i\right)^2}{n}\right]} \qquad (3.15)$$

**Example 3.20   Machining Data**
The following data give the length (in millimeters) of material chips removed during a machining operation:

| 9 | 7 | 10 | 6 | 8 | 11 | 7 | 9 | 8 | 10 |
|---|---|----|---|---|----|---|---|---|----|

Calculate the variance and the standard deviation for these data.

**Solution**

There are three simple steps involved in calculating the variance of any data:

1) Calculate $\Sigma\, X_i$, the sum of all the data values. Thus, we have

$$\Sigma\, X_i = 9 + 7 + 10 + 6 + 8 + 11 + 7 + 9 + 8 + 10 = 85$$

2) Calculate $\sum X_i^2$, the sum of the squares of all the observations:

$$\sum X_i^2 = 9^2 + 7^2 + 10^2 + 6^2 + 8^2 + 11^2 + 7^2 + 9^2 + 8^2 + 10^2 = 745$$

3) Since the sample size $n = 10$, by inserting the values $\Sigma\, X_i$ and $\sum X_i^2$ calculated in steps 1 and 2 into Eq. (3.13), we get

$$s^2 = \frac{1}{10-1}\left(745 - \frac{(85)^2}{10}\right) = \frac{1}{9}(745 - 722.5) = 2.5$$

Thus, the standard deviation is obtained by taking the square root of the variance:

$$s = \sqrt{2.5} \cong 1.58$$

Following are some important notes about the standard deviation:

- It is important to remember the value of $S^2$, and therefore of $S$, is typically greater than zero. Only when all the data values are equal will $S^2$ and $S$ be zero.
- Sometimes data values are very large, and computing the variance becomes cumbersome. In such cases, we can shift the data values by adding a constant $c$ to (or subtracting it from) each data value, and then calculate the variance/standard deviation of the shifted data. This is appropriate since the variance and standard deviation do not change when we add/subtract a constant to/from each value. For example, let $X_1, X_2, X_3, \cdots, X_n$ be a data set, and let $c \neq 0$ be any constant. Let $Y_i = X_i + c,\ i = 1, 2, 3, \cdots, n$. Then we have $\overline{Y} = \overline{X} + c$. This means the deviations of $X_i$'s from $\overline{X}$ are the same as the deviations of $Y_i$'s from $\overline{Y}$. Thus the variance and the standard deviation of $X$'s remain the same as the variance and standard deviation of $Y$'s ($S_y^2 = S_x^2$ and $S_y = S_x$).
- Multiplicative changes, or changes in scale, affect the calculation of the variance and standard deviation. For example, if $Y_i = cX_i\ (c \neq 0)$, then $\overline{Y} = c\overline{X}$, and it can also be shown that $S_y^2 = c^2 S_x^2$ and $S_y = cS_x$.

**Example 3.21**

Find the variance and the standard deviation of the following data set:

| 53 | 60 | 58 | 64 | 57 | 56 | 54 | 55 | 51 | 61 | 63 |
|----|----|----|----|----|----|----|----|----|----|----|

**Solution**

Since the numbers in the given data set are quite large, we first code our data by subtracting 50 from each data point. Then we compute the variance and the standard deviation of the coded data set: that is, (3, 10, 8, 14, 7, 6, 4, 5, 1, 11, 13). As remarked earlier, the variance and standard deviation of the coded data set are the same as those of the original data. Thus, we have

$$\Sigma X_i = 3 + 10 + 8 + 14 + 7 + 6 + 4 + 5 + 1 + 11 + 13 = 82$$

$$\sum X_i^2 = 3^2 + 10^2 + 8^2 + 14^2 + 7^2 + 6^2 + 4^2 + 5^2 + 1^2 + 11^2 + 13^2 = 786$$

so that

$$s^2 = \frac{1}{11-1}\left[786 - \frac{(82)^2}{11}\right] = \frac{1}{10}[786 - 611.27] \cong 17.473$$

$$s \cong \sqrt{17.473} \cong 4.18$$

Thus, the variance and standard deviation of the original data set are 17.473 and 4.18, respectively.

We now examine an important result known as the *empirical rule* that shows how the standard deviation provides very valuable information about a data set.

### 3.5.3.4 Empirical Rule
If data have an approximately bell-shaped distribution, the following rule, known as empirical rule, can be used to compute the percentage of data that will fall within $k$ standard deviations from the mean ($k = 1, 2, 3$):

1) About 68% of the data will fall within one standard deviation of the mean: that is, between $\mu - 1\sigma$ and $\mu + 1\sigma$.
2) About 95% of the data will fall within two standard deviations of the mean: that is, between $\mu - 2\sigma$ and $\mu + 2\sigma$.
3) About 99.7% of the data will fall within three standard deviations of the mean: that is, between $\mu - 3\sigma$ and $\mu + 3\sigma$.

Figure 3.11 illustrates the empirical rule.

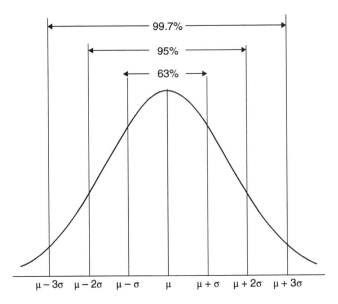

**Figure 3.11** Illustration of the empirical rule.

Note that the empirical rule is applicable for population data as well as sample data. In other words, it remains valid if we replace $\mu$ with $\overline{X}$ and $\sigma$ with $S$.

**Example 3.22  Soft Drink Data**

A machine is used to fill 16 oz soft-drink bottles. Since the amount of beverage varies slightly from bottle to bottle, it is believed that the actual amount of beverage in the bottle forms a bell-shaped distribution with a mean of 15.8 oz and standard deviation of 0.15 oz. Use the empirical rule to find what percentage of bottles contain between 15.5 and 16.1 oz of beverage.

**Solution**

From the information provided in this problem, we have

$$\mu = 15.8 \text{ oz}, \sigma = .15 \text{ oz}$$

We are interested in finding the percentage of bottles that contain between 15.5 and 16.1 oz of beverage. Comparing Figure 3.12 with Figure 3.11, we note that 15.5 and 16.1 are two standard deviations away from the mean. Thus, approximately 95% of the bottles contain between 15.5 and 16.1 oz.

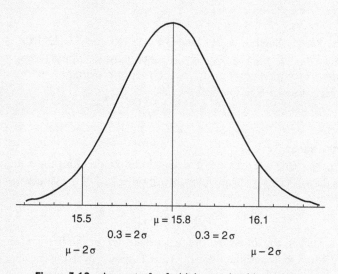

**Figure 3.12**  Amount of soft drink contained in a bottle.

**Example 3.23  Financial Data**

At the end of every fiscal year, a manufacturer writes off or adjusts its financial records to reflect the number of defective units occurring over all of its production during the year. Suppose the dollar values associated with the various units of bad production forms a bell-shaped distribution with mean $\overline{X} = \$35{,}700$ and standard deviation $S = \$2{,}500$ as shown in Figure 3.13. Find the percentage of units of bad production with a dollar value between \$28,200 and \$43,200.

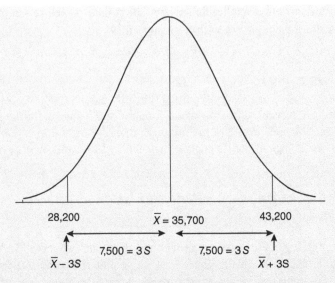

**Figure 3.13** Dollar value of units of bad production.

**Solution**

From the information provided to us, we have $\bar{X}$= \$35,700 and $S$ = \$2,500.

Since the limits \$28,200 and \$43,200 are three standard deviations away from the mean, comparing Figure 3.13 with Figure 3.11, we see that approximately 99.7% of the defective units have a value between \$28,200 and \$43,200.

### 3.5.3.5 Interquartile Range

The *interquartile range (IQR)* is a measure of dispersion that identifies the middle 50% of a population or sample of data. This range is obtained by trimming 25% of the values from the bottom and 25% from the top:

$$IQR = Q_3 - Q_1 \qquad (3.16)$$

where $Q_1$ and $Q_3$ are called the *lower* and the *upper quartiles* (or the first and the third quartiles), respectively. The method to determine these quartiles is shown by using the data in Example 3.24.

**Example 3.24  Nurse Salary Data**

Find the interquartile range for the following salary data of a group of nurses in a city hospital:

| Salary | 58 | 63 | 65 | 68 | 64 | 66 | 68 | 72 | 73 | 79 | 82 | 83 | 86 | 88 | 89 |
|--------|----|----|----|----|----|----|----|----|----|----|----|----|----|----|----|

**Solution**

In this example, the sample size is $n = 15$. To find the IQR, we first need to find the values of quartiles $Q_1$ and $Q_3$, which requires that we find their corresponding ranks when the data is arranged in ascending order. The ranks associated with quartiles $Q_1$ and $Q_3$ are found as shown here:

$$\text{Rank of } Q_1 = 25((n+1)/100) = 25((15+1)/100) = 4$$
$$\text{Rank of } Q_3 = 75((n+1)/100) = 75((15+1)/100) = 12$$

Thus, by determining the values corresponding to ranks 4 and 12, we have $Q_1 = 68$ and $Q_3 = 83$. This means the middle 50% of the nurses have a salary of between \$65,000 and \$83,000. The interquartile range in this example is

$$\text{IQR} = \$83,000 - \$68,000 = \$15,000$$

### 3.5.4 Box-and-Whisker Plot

Earlier in this chapter, we mentioned extreme values. Now we will explicitly define the criteria used to determine which values are extreme values, also known as *outliers*. A *box-and-whisker plot*, or simply a *box plot*, helps us to answer this question.

Figure 3.14 illustrates a box plot for any data set. To construct such a plot, follow these three simple steps:

1) Find the quartiles $Q_1$, $Q_2$, and $Q_3$, where quartile $Q_2$ is equal to the median of a given data set.
2) Draw a box with its outer lines standing at the first and the third quartiles. Then draw a vertical line at the second quartile to divide the box into two sections, which may or may not be of the same size.
3) From the center of the outer lines, draw straight lines extending outward up to three times the IQR, and mark them as shown in Figure 3.14. Note that the distance between points A and B, B and C, D and E, and E and F is equal to 1.5 times the distance between points A and D, which is equal to the IQR. Points S and L are, respectively, the smallest and largest data points that fall within the inner fences. The lines from A to S and D to L are called *whiskers*.

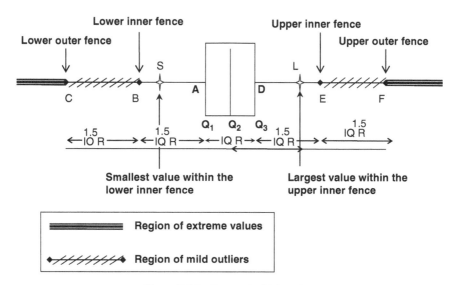

**Figure 3.14** Box-and-whisker plot.

Any data points that fall beyond the lower outer fence and upper outer fence are *extreme outliers*. These points are usually excluded from analysis.

Any data points that fall between the inner and outer fences are called *mild outliers*. These points are excluded from analysis only if we are convinced that these points are in error.

**Example 3.25  Noise Data**
The following data give noise levels measured in decibels produced by different machines in a very large manufacturing plant (a normal human conversation produces a noise level of about 75 decibels):

| 85 | 80 | 88 | 95 | 115 | 110 | 105 | 104 | 89 | 97 | 96 | 140 | 75 | 79 | 99 |
|---|---|---|---|---|---|---|---|---|---|---|---|---|---|---|

Construct a box plot, and determine whether the data set contains any outliers.

**Solution**

1) Arrange the data in ascending order, and rank them from 1 to 15 ($n = 15$):

| Data value | 75 | 79 | 80 | 85 | 88 | 89 | 95 | 96 | 97 | 99 | 104 | 105 | 110 | 115 | 140 |
|---|---|---|---|---|---|---|---|---|---|---|---|---|---|---|---|
| Rank | 1 | 2 | 3 | 4 | 5 | 6 | 7 | 8 | 9 | 10 | 11 | 12 | 13 | 14 | 15 |

2) Find the ranks of quartiles $Q_1$, $Q_2$, and $Q_3$:

$$\text{Rank of } Q_1 = 25((15+1)/100) = 4$$
$$\text{Rank of } Q_2 = 50((15+1)/100) = 8$$
$$\text{Rank of } Q_3 = 75((15+1)/100) = 12$$

Therefore,

$$Q_1 = 85, Q_2 = 96, Q_3 = 105$$
$$IQR = Q_3 - Q_1 = 105 - 85 = 20$$

And

$$(1.5) \times IQR = (1.5) \times 20 = 30$$

Figure 3.15 shows the box plot for the data. It indicates that the value 140 is a mild outlier. In this case, action should be taken to reduce the noise level in the machine that produces a noise level of 140 decibels.

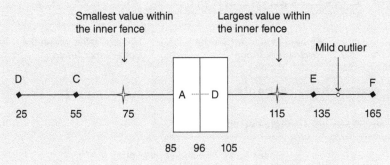

**Figure 3.15**  Box plot for the data in Example 3.25.

## 3.6 Some Important Probability Distributions

In this section, we consider four important probability distributions that are useful in the field of statistical quality control. These are the binomial distribution, hypergeometric distribution, Poisson distribution, and normal distribution. The first three are known as *discrete* probability distributions, while the last one is called a *continuous* probability distribution.

### 3.6.1 The Binomial Distribution

Among discrete probability distributions, the *binomial distribution* is very useful and important. It is associated with a binomial experiment, which has the following properties:

- The experiment consists of $n$ independent trials.
- Each trial has two possible outcomes: *success* and *failure*.
- The *probability p of success* in each trial is constant throughout the experiment. Consequently, the *probability q = 1 − p of failure* is also constant throughout the experiment.

---

**Definition 3.3**   A random variable $X$ that counts the number of successes in a binomial experiment is said to be distributed as a *binomial distribution*, and its probability function is given by

$$p(x) = P(X = x) = \binom{n}{x} p^x q^{n-x}, x = 0, 1, 2, ..., n; p + q = 1 \qquad (3.17)$$

where $n$ is the number of trials, $x$ the number of successes, and $p$ the probability of success.

Note that $\binom{n}{x}$ is referred to as "$n$ choose $x$." It represents the number of combinations when $x$ items are selected from a total of $n$ items, and is equal to

$$\binom{n}{x} = \frac{n!}{x! \ (n-x)!}$$

where $n!$ is referred to as "$n$ factorial" and is defined as $n! = n \times (n - 1) \times (n - 2) \times \cdots \times 2 \times 1$.

For example,

$$\binom{6}{4} = \frac{6!}{4! \ (6-4)!} = \frac{6 \times 5 \times 4 \times 3 \times 2 \times 1}{(4 \times 3 \times 2 \times 1)(2 \times 1)} = 15$$

Note that any probability function must satisfy the following two properties:

i) $p(x) = P(X = x) \geq 0$

ii) $\sum_x P(X = x) = 1$ \qquad (3.18)

It can easily be verified that the probability function $p(x)$ in Eq. (3.17) does satisfy these properties.

**Example 3.26**

The probability is 0.80 that a randomly selected technician will finish their project successfully. Let $X$ be the number of technicians among a randomly selected group of five technicians who will finish their project successfully. Find the probability distribution of the random variable $X$.

**Solution**

It is very clear that the random variable $X$ in this example is distributed as a binomial distribution with $n = 5$, $p = 0.8$, and $X = 0, 1, 2, 3, 4, 5$. Thus, by using the binomial probability function given in Eq. (3.17), we have

$$f(0) = \binom{5}{0}(0.80)^0(.20)^5 = 0.000, \qquad f(1) = \binom{5}{1}(0.80)^1(.20)^4 = 0.006$$

$$f(2) = \binom{5}{2}(0.80)^2(.20)^3 = 0.051, \qquad f(3) = \binom{5}{3}(0.80)^3(.20)^2 = 0.205$$

$$f(4) = \binom{5}{4}(0.80)^4(.20)^1 = 0.410, \qquad f(5) = \binom{5}{5}(0.80)^5(.20)^0 = 0.328$$

### 3.6.1.1 Binomial Probability Tables

The tables of binomial probabilities for $n = 1$ to 20 and for some selected values of $p$ are given in Table A.5 in the Appendix. Table 3.9 shows a small portion of that table. By comparing the values calculated using the probability function in Eq. (3.17) and the values given in the binomial table, we can easily see how to read the binomial distribution table. From Table 3.9, for example, we have $p(3) = P(x = 3) = 0.205$, which is equal to what we found using Eq. (3.17).

The *mean* and *standard deviation* of a binomial distribution are defined next.

**Table 3.9** Portion of Table A.5 in the Appendix for $n = 5$.

| | | *p* | | |
|---|---|---|---|---|
| *x* | .05 | - - | .80 | - - - |
| 0 | .774 | | .000 | |
| 1 | .203 | | .006 | |
| 2 | .022 | | .051 | |
| 3 | .001 | | .205 | |
| 4 | .000 | | .410 | |
| 5 | .000 | | .328 | |

---

**Definition 3.4**  The mean and standard deviation of a binomial distribution are given by

$$\text{Mean } \mu = E(X) = np \tag{3.19}$$

$$\text{Standard deviation } \sigma = \sqrt{V(X)} = \sqrt{npq} \tag{3.20}$$

---

**Example 3.27**

Find the mean and standard deviation of the random variable $X$ given in Example 3.26, where $X$ represents the number of technicians who finish their project successfully.

**Solution**

In Example 3.26, we have $n = 5$, $p = 0.8$, and $q = 1 - p = 1 - 0.8 = 0.2$. Thus, using Eqs. (3.19) and (3.20), we have

$$\mu = np = 5 \times (0.8) = 4$$

$$\sigma = \sqrt{npq} = \sqrt{5(0.8)(0.2)} = \sqrt{0.8} = 0.8944$$

### 3.6.2  The Hypergeometric Distribution

In almost all statistical applications, we deal with both samples and populations. A *population* is a collection of all conceivable individuals, elements, numbers, or entities that possess a characteristic of interest; and a portion of a population selected for study is called a *sample*. In Chapter 4, we will discuss in detail how the selection of a sample from a population can be made in two ways: *with replacement* and *without replacement*. In sampling with replacement, we select an item from a given population. However, before we select the next item, we replace the selected item in the population. This process is repeated until we select the desired number of items. On the other hand, in sampling without replacement, we do not replace the items before selecting the next item.

If sampling is done with replacement, and if we consider selecting an item from the population as a trial, then we can easily see that these trials are independent. However, if sampling is done without replacement, then these trials are not independent. That is, the outcome of any trial will depend on what happened in the previous trial(s).

Consider now a special kind of population consisting of two categories, such as a population of males and females, defectives and non-defectives, salaried and non-salaried workers, healthy and non-healthy people, successes and failures, and so on. Such populations are generally known as *dichotomized populations*. If sampling with replacement is done from a dichotomized population, then we can answer probability questions related to, for example, how many females are selected, how many salaried individuals are selected, how many healthy people are selected, etc., by using a binomial probability distribution. However, if sampling is done without replacement, then these trials are not independent, and therefore a binomial probability distribution is not applicable. To illustrate, suppose a box contains 100 parts, 5 of which are defective. We randomly select two parts, one at a time, from the box. Clearly, the probability that the first part will be defective is 5/100. However, the probability of the second part being defective depends on whether the first part was defective. That is, if the first part was defective, then the probability of the second part being defective is 4/99; but if the first part was not defective, then this probability is 5/99. Obviously, the two probabilities are different; and therefore, the two trials are not independent. In this case, we use another probability distribution, known as a *hypergeometric distribution*.

---

**Definition 3.5** A random variable $X$ is said to have a *hypergeometric* distribution if

$$P(X = x) = \frac{\binom{r}{x}\binom{N-r}{n-x}}{\binom{N}{n}}, x = a, a+1, ..., \min(r, n) \qquad (3.21)$$

where

$a = \text{Max}(0, n - N + r)$
$N = $ total number of parts, including defectives and non-defectives
$r = $ number of parts in the category of interest, say successes, which may be the number of defective parts in the population under consideration
$N - r = $ number of failures in the population
$n = $ number of trials or sample size
$x = $ number of desired successes in $n$ trials
$n - x = $ number of failures in $n$ trials

---

**Example 3.28**
A quality control engineer selects randomly 2 parts from a box containing 5 defective and 15 non-defective parts. She discards the box if one or both of the selected parts are defective. What is the probability that:

a) She will select one defective part
b) She will select two defective parts
c) She will reject the box

**Solution**
a) In this problem, we have $N = 20$, $r = 5$, $N - r = 15$, $n = 2$, and $x = 1$. Thus,

$$P(X = 1) = \frac{\binom{5}{1}\binom{15}{1}}{\binom{20}{2}} = \frac{5 \times 15}{190} = \frac{75}{190} = .3947$$

b)

$$P(X = 2) = \frac{\binom{5}{2}\binom{15}{0}}{\binom{20}{2}} = \frac{10 \times 1}{190} = .0526$$

c) The probability that she rejects the box is

$$P(X = 1) + P(X = 2) = .3947 + .0526 = .4473$$

**Example 3.29**

A manufacturer ships parts in lots of 100 parts each. The quality control department of the receiving company agrees to a sampling plan in which they will select a random sample of five parts without replacement. The lot will be accepted if the sample does not contain any defective part. What is the probability that a lot will be accepted if

a) The lot contains 10 defective parts
b) The lot contains 4 defective parts

**Solution**

a) In this problem, we have $N = 100$, $r = 10$, $N - r = 90$, $n = 5$, and $x = 0$. Therefore, the probability that the lot is accepted, or the probability that the sample does not contain any defective part, is

$$P(X = 0) = \frac{\binom{10}{0}\binom{100-10}{5-0}}{\binom{100}{5}} = \frac{\binom{10}{0}\binom{90}{5}}{\binom{100}{5}}$$

$$= \frac{\frac{10!}{0!10!} \times \frac{90!}{5!85!}}{\frac{100!}{5!95!}} = \frac{90 \cdot 89 \cdot 88 \cdot 87 \cdot 86}{100 \cdot 99 \cdot 98 \cdot 97 \cdot 96} = .5838$$

b) In this case, we have $N = 100$, $r = 4$, $N - r = 96$, $n = 5$, and $x = 0$. Therefore, the probability that the lot will be accepted is

$$P(X = 0) = \frac{\binom{4}{0}\binom{100-4}{5-0}}{\binom{100}{5}}$$

$$= \frac{\left(\frac{4!}{0!4!}\right) \cdot \left(\frac{96!}{5!91!}\right)}{\frac{100!}{5!95!}} = \frac{96 \cdot 95 \cdot 94 \cdot 93 \cdot 92}{100 \cdot 99 \cdot 98 \cdot 97 \cdot 96} = .8119$$

### 3.6.2.1 Mean and Standard Deviation of a Hypergeometric Distribution

The *mean* and *standard deviation* of a hypergeometric distribution are given by

$$\text{Mean } \mu = np \tag{3.22}$$

$$\text{Standard deviation } \sigma = \sqrt{\frac{N-n}{N-1} \cdot npq} \tag{3.23}$$

where

$N$ = total number of objects in the population
$r$ = total number of objects in the category of interest
$n$ = sample size

$$p = \frac{r}{N}$$

$$q = 1 - p = \frac{N-r}{N}$$

### Example 3.30 Computer Shipments Data

A shipment of 250 computers includes 8 computers with defective CPUs. A sample without replacement of size 20 is selected. Let $X$ be a random variable that denotes the number of computers with defective CPUs. Find the mean and the variance of $X$.

### Solution

The shipment is a dichotomized population, and the sampling is done without replacement. Thus the probability distribution of the random variable $X$ is hypergeometric. Here the category of interest is computers with defective CPUs. Using the formulas in Eqs. (3.22) and (3.23), we have

$$\mu = np = 20\left(\frac{8}{250}\right) = 0.64$$

$$\sigma = \sqrt{\frac{N-n}{N-1} \times npq} = \sqrt{\frac{250-20}{250-1} \times 20\left(\frac{8}{250}\right)\left(\frac{250-8}{250}\right)}$$

$$= \sqrt{\left(\frac{230}{249}\right) \times 20\left(\frac{8}{250}\right)\left(\frac{242}{250}\right)} = .7565$$

### Minitab

Refer to Example 3.28. Suppose, in that example, that the quality control engineer selects 10 parts. Find the probability that she will select $x$ number of defective parts, where $x = 0, 1, 2, 3, 4,$ or 5:

1) Enter the values 0, 1, 2, ..., 5 in column **C1**.
2) From the menu bar, select **Calc > Probability Distributions > Hypergeometric**.
3) In the dialog box that opens, click the circle next to **Probability**.

4) Enter **20** in the box next to **Population size**, **5** in the box next to **Event count in population** (this is the number of defective parts in the box), and **10** in the box next to **Sample size.**
5) Click the circle next to **Input column**, and type **C1** in the box next to it. Click **OK**. The desired probabilities will appear in the **Session Window** as follows:

Hypergeometric with N = 20, M = 5, and n = 10.

| x | P(X = x) |
| --- | --- |
| 0 | 0.016254 |
| 1 | 0.135449 |
| 2 | 0.348297 |
| 3 | 0.348297 |
| 4 | 0.135449 |
| 5 | 0.016254 |

**R**

In this case, we use the built-in "dhyper(x, r, N-r, n)" function in the "stats" library to calculate the hypergeometric probabilities. Here, N, r, n and x are as explained in Eq. (3.14).

The probabilities that x = 0, 1, 2, ..., 5 can be obtained by running the following R code in the R **Console Window**:

```
prob = dhyper(c(0:5), 5, 15, 10)
round(prob, 6)
# R output
[1] 0.016254 0.135449 0.348297 0.348297 0.135449 0.016254
```

### 3.6.3 The Poisson Distribution

The Poisson distribution is used whenever we are interested in finding the probability of a rare event, specifically the probability of observing a number of events occurring over a specified period of time, over a specified length measurement (such as the length of an electric wire), over a specified area, or in a specified volume, when the likelihood of these events occurring is very small. For example, we may be interested in finding the probability of a certain number of accidents occurring in a manufacturing plant over a specified period of time, the number of patients admitted to a hospital on a given day, the number of cars passing through a toll booth during rush hour, the number of customers entering a bank during lunch hour, the number of telephone calls received by a receptionist over a specified period of time, the number of defects found on a certain length of electric wire, the number of scratches over a specified area of a smooth surface, the number of holes in a paper roll, or the number of radioactive particles in a specified volume of air.

Before using the Poisson distribution, however, we must ensure that the data in question possess a common characteristic. That is, the number of events occurring over a specified period of time, specified length, specified area, etc. must satisfy the conditions of a process called the *Poisson process,* which is defined as follows.

**Definition 3.6** Let $X(t)$ denote the number of times a particular event occurs randomly in a time period $t$. Then these events are said to form a *Poisson process* having rate $\lambda (\lambda > 0)$, (that is, for $t = 1$, $X(1) = \lambda$), if

1) $X(0) = 0$.
2) The number of events that occur in any two non-overlapping intervals are independent.
3) The average number of events occurring in any interval is proportional to the size of the interval and does not depend on when or where they occur.
4) The probability of precisely one occurrence in a very small interval $(t, t + \delta t)$ of time is equal to $\lambda(\delta t)$, and the probability of two or more occurrences in such a small interval is zero.

**Definition 3.7** Let a random variable $X$ be equal to the number of events occurring according to the Poisson process. Then $X$ is said to have a *Poisson distribution* if its probability function is given by

$$P(X = x) = \frac{e^{-\lambda}\lambda^x}{x!}, x = 0, 1, 2, \ldots \tag{3.24}$$

where $\lambda > 0$ (the Greek letter lambda) is the only parameter of the distribution and $e \cong 2.71828$. It can easily be shown that Eq. (3.24) satisfies the two properties of being a probability function, that is

$$P(X = x) \geq 0$$
$$\sum_x P(X = x) = 1$$

Note: The binomial distribution when $p$ is very small and $n$ is very large, such that $\lambda = np$ as $n \to \infty$, can be approximated by a Poisson distribution. Note that as a rule of thumb, the approximation is good when $\lambda = np < 10$.

**Example 3.31**
It is known from experience that 4% of the parts manufactured at a plant are defective. Use a Poisson approximation of the binomial distribution to find the probability that in a lot of 200 parts manufactured at that plant, 7 parts will be defective.

**Solution**

Since $n = 200, p = .04$, we have $np = 200(.04) = 8 (< 10)$, and Poisson approximation should give a satisfactory result. Thus, for example, using Eq. (3.24) we get

$$P(X = 7) = f(7) = \frac{e^{-8}(8)^7}{7!} = \frac{(.00093355)(2097152)}{5040} = 0.1396$$

### 3.6.3.1 Mean and Standard Deviation of a Poisson Distribution

$$\text{Mean: } \mu = E(X) = \lambda \qquad (3.25)$$

$$\text{Variance: } \sigma^2 = V(X) = \lambda \qquad (3.26)$$

$$\text{Standard deviation: } \sigma = \sqrt{V(X)} = \sqrt{\lambda} \qquad (3.27)$$

### 3.6.3.2 Poisson Probability Tables

The tables of Poisson probabilities for various values of $\lambda$ are given in Table A.6 in the Appendix. We illustrate the use of these tables with the following example.

**Example 3.32**

The average number of accidents occurring in a manufacturing plant over a period of one year is equal to two. Find the probability that during any given year, five accidents will occur.

**Solution**

To use the Poisson table, first find the values of $x$ and $\lambda$ for which the probability is being sought. Then the desired probability is the value at the intersection of the row corresponding to $x$ and column corresponding to $\lambda$. Thus, for example, in Table 3.10, the probability $P(x = 5)$ when $\lambda = 2$ is 0.036. Note that this probability can also be found by using the probability distribution formula given in (3.24).

We can also determine binomial and Poisson probabilities by using one of the statistical packages discussed in this book. Here we illustrate the procedure for determining these probabilities using Minitab and R.

**Table 3.10** Portion of Table A.6 from the Appendix.

| | | $\lambda$ | | |
|---|---|---|---|---|
| $x$ | 1.1 | - - - | 1.9 | 2.0 |
| 0 | .333 | | .150 | .135 |
| 1 | .366 | | .284 | .271 |
| 2 | .201 | | .270 | .271 |
| 3 | .074 | | .171 | .180 |
| 4 | .020 | | .081 | .090 |
| 5 | .005...... | | .031 | **.036** |
| 6 | .001 | | .010 | .012 |
| 7 | .000 | | .003 | .004 |
| 8 | .000 | | .000 | .001 |
| 9 | .000 | | .000 | .000 |

**Minitab (Binomial Distribution)**

We illustrate this procedure using Example 3.26 and by following these steps:

1) Enter the values 0, 1, 2, ..., 5 in column C1.
2) From the menu bar, select **Calc > Probability Distributions > Binomial**.
3) In the dialog box, click the circle next to **Probability**.
4) Enter **5** (the number of trials) in the box next to **Number of trials** and **0.8** (the probability of success) in the box next to **Event Probability**.
5) Click the circle next to **Input column**, and type **C1** in the box next to it. Click **OK**. The desired probabilities will appear in the **Session Window** as follows:

**Binomial probabilities with $n = 5$ and $p = 0.8$.**

| x | $P(X = x)$ |
|---|------------|
| 0 | 0.00032 |
| 1 | 0.00640 |
| 2 | 0.05120 |
| 3 | 0.20480 |
| 4 | 0.40960 |
| 5 | 0.32768 |

**R**

R has a built-in "dbinom(x, size, prob)" function that is used to calculate binomial probabilities. Here $x$ is the number of successes, "size" is the total number of trials, and "prob" is the probability of success in a single trial. Probabilities that $x = 0, 1, 2, 3, ..., 5$ are obtained by running the following R code in the R **Console Window**:

```
prob = dbinom(c(0:5), 5, .8)
round(prob, 6)
#R output
[1] 0.00032 0.00640 0.05120 0.20480 0.40960 0.32768
```

**Minitab (Poisson Distribution)**

We illustrate this procedure using $\lambda = 4$ and by following these steps:

1) Enter the values 0, 1, 2, ..., 5 in column C1.
2) From the menu bar, select **Calc > Probability Distributions > Poisson**.
3) In the dialog box, click the circle next to **Probability**.
4) Enter **4** (the mean value, which is equal to $\lambda$) in the box next to **Mean**.
5) Click the circle next to **Input column**, and type **C1** in the box next to it.
6) Click **OK**. The desired probabilities will appear in the **Session Window** as follows:

| **Poisson Probabilities with mean = 4.** | |
|---|---|
| *x* | *P(X = x)* |
| 0 | 0.018316 |
| 1 | 0.073263 |
| 2 | 0.146525 |
| 3 | 0.195367 |
| 4 | 0.195367 |
| 5 | 0.156293 |

**R**

In this case, we use the built-in "dpois(x, lambda)" function to calculate the Poisson probabilities. Here *x* denotes a value of the random variable *X*, and $\lambda$ is the mean of the Poisson distribution. Probabilities that $x = 0, 1, 2, 3, ..., 5$ are obtained by running the following R code in the R **Console Window**:

```
prob = dpois(c(0:5), lambda=4)
round(prob, 6)
# R output
[1]  0.018316  0.073263  0.146525  0.195367  0.195367  0.156293
```

### 3.6.4  The Normal Distribution

The *normal distribution* forms the basis of modern statistical theory. The normal distribution is the most widely used probability distribution in applied statistics. In fact, normal distributions appear in many situations in everyday life. For example, the tearing strength of paper, survival time of a part, time taken by a programmer to neutralize a new computer virus, volume of a chemical compound in a one-pound container, compressive strength of a concrete block, length of rods, time taken by a commuter to travel from home to work, and heights and weights of people can all be modeled by a normal distribution. Another useful application is the *central limit theorem,* which posits that the sample mean, $\overline{X}$, of a random sample is approximately normally distributed. In applied statistics, the normal distribution and central limit theorem are of paramount importance, and we will see some of these applications in subsequent chapters.

A normal probability distribution is defined as follows:

**Definition 3.8**   A random variable $X$ is said to be distributed as a *normal probability distribution* if its probability density function is given by

$$f(x) = \frac{1}{\sqrt{2\pi}\sigma} e^{-(x-\mu)^2/2\sigma^2} \quad -\infty < x < +\infty \qquad (3.28)$$

where $-\infty < \mu < +\infty$ and $\sigma > 0$ are the two parameters of the distribution, and they represent, respectively, the mean and the standard deviation of the distribution. Also, note that $\pi \cong 3.14159$ and $e \cong 2.71828$.

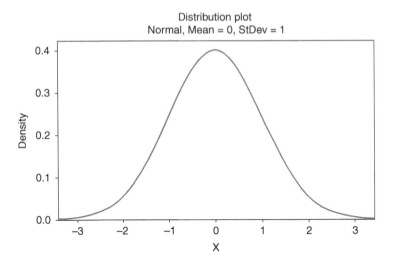

**Figure 3.16**  The normal density function curve with μ = 0 and σ = 1.

Some key characteristics of the normal density function are as follows:

- The density curve is bell-shaped and completely symmetric about its mean, $\mu$.
- The specific shape of the curve – its height and spread – is determined by its standard deviation $\sigma$.
- The tails of the density function curve extend from $-\infty$ to $+\infty$.
- The total area under the curve is 1.0. However, 99.74% of the area falls within three standard deviations of the mean, $\mu$.
- The area under the normal curve to the right of $\mu$ is 0.5, and the area to the left of $\mu$ is also 0.5. Figure 3.16 shows the normal density function curve of a random variable $X$ with mean $\mu = 0$ and standard deviation $\sigma = 1$.

Since 99.74% of the area under a normal curve with mean $\mu$ and standard deviation $\sigma$ falls between $\mu - 3\sigma$ and $\mu + 3\sigma$, the total distance of $6\sigma$ between $\mu - 3\sigma$ and $\mu + 3\sigma$ is usually considered the range of the normal distribution. Figures 3.17 and 3.18 show how changing the mean $\mu$ and the standard deviation $\sigma$ affect the location and the shape of a normal curve.

From Figure 3.18, we can observe an important phenomenon of the normal distribution: as the standard deviation $\sigma$ becomes smaller and smaller, the total probability (area under the curve) becomes more and more concentrated about $\mu$. We will see later that this property of the normal distribution is very useful in making inferences about populations.

To calculate normal probabilities, we need to introduce a new random variable called a standardized random variable. A *standard normal random variable*, denoted by $Z$, is defined as follows:

$$Z = \frac{X - \mu}{\sigma} \tag{3.29}$$

The new random variable $Z$ is also distributed normally, but with mean 0 and standard deviation equal to 1. The distribution of the random variable $Z$ is generally known as the *standard normal distribution*.

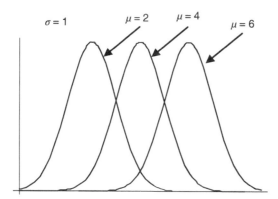

**Figure 3.17**   Curves representing the normal density function with different means but the same standard deviation.

**Definition 3.9**   The normal distribution with mean 0 and standard deviation 1 is known as the *standard normal distribution* and is usually written as $N(0,1)$.

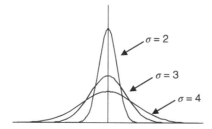

**Figure 3.18**   Curves representing the normal density function with different standard deviations but the same mean.

The values of the standard normal random variable $Z$, denoted by the lowercase letter $z$, are called *z-scores*. In Figure 3.19, the points marked on the $x$-axis are the $z$-scores. The probability of the random variable $Z$ falling in an interval $(a, b)$ is shown by the shaded area under the standard normal curve in Figure 3.20. This probability is determined by using a standard normal distribution table (see Table A.7 in the Appendix).

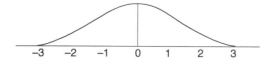

**Figure 3.19**   The standard normal density function curve.

**Figure 3.20**   Probability ($a \leq Z \leq b$) under the standard normal curve.

**Table 3.11** A portion of standard normal distribution table, Table A.7, from the Appendix.

| Z | .00 | .01 | .02 | .03 | .04 | .05 | .06 | .07 | .08 | .09 |
|---|---|---|---|---|---|---|---|---|---|---|
| 0.0 | .0000 | .0040 | .0080 | .0120 | .0160 | .0199 | .0239 | .0279 | .0319 | .0359 |
| 0.1 | .0398 | .0438 | .0478 | .0517 | .0557 | .0596 | .0636 | .0675 | .0714 | .0753 |
| . | . | | | | | | | | | |
| . | . | | | | | | | | | |
| . | . | | | | | | | | | |
| 1.0 | .3413 | .3438 | .3461 | .3485 | .3508 | .3531 | .3554 | .3577 | .3599 | .3621 |
| 1.1 | .3643 | .3665 | .3686 | .3708 | .3729 | .3749 | .3770 | .3790 | .3810 | .3830 |
| . | . | | | | | | | | | |
| . | . | | | | | | | | | |
| . | . | | | | | | | | | |
| 1.9 | .4713 | .4719 | .4726 | .4732 | .4738 | .4744 | .4750 | .4756 | .4761 | .4767 |
| 2.0 | .4772 | .4778 | .4783 | .4788 | .3793 | .4798 | .4803 | .3808 | .4812 | .4817 |

The standard normal distribution Table A.7 in the Appendix lists the probabilities of the random variable $Z$ for its values between $z = 0.00$ and $z = 3.09$. A small portion of this table is reproduced in Table 3.11. The entries in the body of the table are the probabilities $P(0 \leq Z \leq z)$, where $z$ is a point in the interval $(0, 3.09)$. These probabilities are also shown by the shaded area under the normal curve given at the top of Table A.7 in the Appendix. To read this table, we mark the row and the column corresponding to the value of $z$ to the first and the second decimal point, respectively. Then the entry at the intersection of that row and column is the probability $P(0 \leq Z \leq z)$. For example, the probability $P(0 \leq Z \leq 2.09)$ is found by marking the row corresponding to $z = 2.0$ and column corresponding to $z = .09$ (note that $z = 2.09 = 2.0 + .09$) and locating the entry at the intersection of the marked row and column, which in this case is .4817. The probabilities for negative values of $z$ are found, due to the symmetric property of the normal distribution, by finding the probabilities of the corresponding positive values of $z$. For example,

$$P(-1.54 \leq Z \leq -0.50) = P(0.50 \leq Z \leq 1.54)$$

**Example 3.33**
Use the standard normal distribution table, Table A.7 in the Appendix, to find the following probabilities:

a) $P(1.0 \leq Z \leq 2.0)$
b) $P(-1.50 \leq Z \leq 0)$
c) $P(-2.2 \leq Z \leq -1.0)$

**Solution**
a) From Figure 3.21, it is clear that

$$P(1.0 \leq Z \leq 2.0) = P(0 \leq Z \leq 2.0) - P(0 \leq Z \leq 1.0)$$
$$= .4772 - .3413 = 0.1359$$

b) Since the normal distribution is symmetric about the mean, which in this case is 0, the probability of $Z$ falling between $-1.5$ and 0 is the same as the probability of $Z$ falling between 0 and 1.5. Figure 3.22 also supports this assertion. Using Table A.7 in the Appendix, we have the following:

$$P(-1.50 \leq Z \leq 0) = P(0 \leq Z \leq 1.50) = 0.4332$$

c) By using the same argument as in (b) and using Table A.7 (also see Figure 3.23), we get

$$P(-2.2 \leq Z \leq -1.0) = P(1.0 \leq Z \leq 2.2) = P(0 \leq Z \leq 2.2) - P(0 \leq Z \leq 1.0)$$
$$= .4861 - .3413 = 0.1448$$

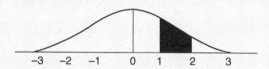

**Figure 3.21** Shaded area equal to $P(1 \leq Z \leq 2)$

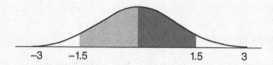

**Figure 3.22** Two shaded areas showing $P(-1.50 \leq Z \leq 0) = P(0 \leq Z \leq 1.50)$.

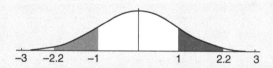

**Figure 3.23** Two shaded areas showing $P(-2.2 \leq Z \leq -1.0) = P(1.0 \leq Z \leq 2.2)$.

**Example 3.34**

Suppose a quality characteristic of a product is normally distributed with mean $\mu = 18$ and standard deviation $\sigma = 1.5$. The specification limits furnished by the customer are $(15, 21)$. Determine what percentage of the product meets the specifications set by the customer.

**Solution**

Let the random variable $X$ denote the quality characteristic of interest. Then $X$ is normally distributed with mean $\mu = 18$ and standard deviation $\sigma = 1.5$. So, we are interested in finding the percentage of product with the characteristic of interest within the limits $(15, 21)$, which is given by

$$100 \cdot P(15 \leq X \leq 21) = 100P\left(\frac{15-18}{1.5} \leq \frac{X-18}{1.5} \leq \frac{21-18}{1.5}\right)$$

$$= 100 \, P(-2.0 \leq Z \leq 2.0)$$

$$= 100 \, [P(-2.0 \leq Z \leq 0) + P(0 \leq Z \leq 2.0)]$$

$$= 100 \times 2P(0 \leq Z \leq 2.0)$$

$$= 100 \times 2(.47725) = 95.45\%$$

Thus, in this case, the percentage of product that will meet the specifications set by the customer is 95.54%. Note that the probability can also be determined by using one of the statistical packages discussed in this book. Here we illustrate the procedure for determining these probabilities using Minitab and R.

**Minitab**

1) Enter the values **15** and **21** in column C1.
2) From the menu bar, select **Calc > Probability Distribution > Normal**.
3) In the dialog box that appears, click the circle next to **Cumulative probability**.
4) Enter **18** (the value of the mean) in the box next to **Mean** and **1.5** (the value of the standard deviation) in the box next to **Standard deviation**.
5) Click the circle next to **Input column**, and type **C1** in the box next to it. Then click **OK**.

The following results will appear in the **Session Window**. Thus $P(15.0 \leq X \leq 21.0) = P$ $(X \leq 21.0) - P(X \leq 15.0) = 0.977250 - 0.022750 = 0.95$.

**Normal with mean = 18 and standard deviation = 1.5.**

| x | $P(X \leq x)$ |
|---|---|
| 15 | 0.022750 |
| 21 | 0.977250 |

**R**

R has a built-in cumulative normal distribution function "pnorm(x, mean, sd)", where "x" is the value of the random variable X and "mean" and "sd" are the mean and the standard deviation of the normal distribution, respectively:

```
pnorm(21, 18, 1.5)
[1] 0.9772499
pnorm(15, 18, 1.5)
[1] 0.02275013
```

In many statistical applications, we select a random sample from a given population and assume that the given population follows a normal distribution. In such cases, it becomes important that we verify that the assumption of normality is valid. This verification can be easily done by using one of the statistical packages discussed in this book.

## Example 3.35

Use one of the software packages discussed in this text to verify if the following data come from a normal population:

| | | | | | | | | | |
|---|---|---|---|---|---|---|---|---|---|
| 33 | 40 | 36 | 35 | 45 | 34 | 23 | 27 | 24 | 54 |
| 35 | 24 | 30 | 35 | 39 | 24 | 33 | 41 | 47 | 53 |
| 34 | 33 | 40 | 44 | 42 | 41 | 43 | 52 | 46 | 36 |
| 53 | 33 | 27 | 48 | 45 | 54 | 33 | 51 | 31 | 43 |
| 27 | 27 | 44 | 28 | 25 | 48 | 49 | 45 | 22 | 46 |

## Solution

In this example, we use Minitab and R to verify whether the previous data come from a normal population.

**Minitab**

To achieve our goal, we follow these steps:

1) Enter the data in column C1.
2) From the menu bar, select **Stat > Basic Statistic > Normality Test** to open the **Normality Test** dialog box.
3) Enter **C1** in the box next to **Variable**. Under **percentile lines**, check the circle next to **None**. And under **Normality Test**, check one of the circles next to **Anderson-Darling, Ryan-Joiner**, or **Kolmogorov-Smirnov**. In this example, check the circle next to **Anderson-Darling** and then click **OK**. The normal probability graph appears in the **Session Window**. If you like, put the title of the problem in the box next to **Title.**

The small box in the upper-right corner in Figure 3.24 provides various statistics. The last statistic gives the *p*-value. In general, if the *p*-value is >0.05, we have sufficient evidence to assume

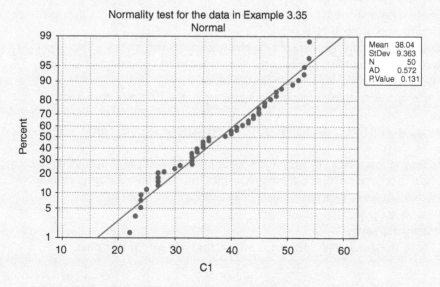

**Figure 3.24** Minitab normal probability plot for the data in Example 3.35.

that the data come from a normal population. In this example, we can say that the data come from a normal population. Also, note that the normality probability graph in Figure 3.24 shows that the data points fall almost on a straight line, which further confirms that the data come from a normal population.

**R**

To perform the normality test using R, we use the built-in "qqnorm()" function, shown as follows in the R **Console Window**. R produces a normal probability plot identical to the one produced by Minitab.

```
Data = c (33, 40, 36, 35, 45, 34, 23, 27, 24, 54, 35, 24, 30, 35, 39, 24, 33, 41, 47, 53,
34, 33, 40, 44, 42, 41, 43, 52, 46, 36, 53, 33, 27, 48, 45, 54, 33, 51, 31, 43,
27, 27, 44, 28, 25, 48, 49, 45, 22, 46)
qqnorm(Data)
```

## Review Practice Problems

1  Determine whether each of the following scenarios would be recorded as qualitative or quantitative data:
   a) Time needed for a technician to finish a project
   b) Number of days a patient stays in a hospital after bypass surgery
   c) Average number of cars passing through a toll booth per hour
   d) Types of beverages served by a restaurant
   e) Size of a rod used in a project
   f) Condition of a home for sale (excellent, good, fair, bad)
   g) Heights of basketball players
   h) Dose of medication prescribed by a physician for patients
   i) Recorded temperatures at a tourist destination during the month of January
   j) Ages of persons waiting in a physician's office
   k) The speed of a vehicle crossing George Washington Bridge in New York

2  Referring to Problem 1, classify the data in each case as nominal, ordinal, interval, or ratio.

3  A consumer protection agency conducts opinion polls to determine the quality (excellent, good, fair, bad) of products of interest imported from China. Suppose the agency conducted a poll in which 500 randomly selected individuals were interviewed face to face.
   a) What is the population of interest?
   b) What is the sample?
   c) Classify the variable of interest as nominal, ordinal, interval, or ratio.

**4**  The following data represent the number of employees of a large company who called in sick during the month of November, when the flu season starts:

| 13 | 10 | 31 | 14 | 23 | 27 | 26 | 24 | 28 | 23 |
|----|----|----|----|----|----|----|----|----|----|
| 12 | 11 | 26 | 27 | 17 | 32 | 26 | 29 | 21 | 26 |
| 19 | 11 | 10 | 32 | 29 | 17 | 20 | 17 | 27 | 29 |

a) Prepare a dot plot of the data.

b) Summarize what you learn about the data from the dot plot.

**5**  The following data represent the number of patients admitted to a city hospital over 40 days.

| 22 | 15 | 36 | 33 | 32 | 19 | 28 | 31 | 33 | 20 |
|----|----|----|----|----|----|----|----|----|----|
| 31 | 21 | 23 | 39 | 38 | 20 | 36 | 17 | 31 | 23 |
| 30 | 30 | 35 | 29 | 39 | 22 | 20 | 38 | 17 | 27 |
| 17 | 31 | 16 | 31 | 16 | 21 | 18 | 31 | 30 | 27 |

a) Prepare a frequency distribution table for these data.

b) On how many days were more than 20 patients admitted?

**6**  The following data give the number of accidents per week in a manufacturing plant during a period of 30 weeks:

| 3 | 6 | 6 | 2 | 3 | 4 | 3 | 5 | 1 | 1 |
|---|---|---|---|---|---|---|---|---|---|
| 6 | 6 | 1 | 4 | 3 | 1 | 2 | 1 | 6 | 4 |
| 1 | 2 | 4 | 1 | 1 | 6 | 6 | 2 | 6 | 1 |

a) Construct a single-valued frequency distribution table for these data. A frequency distribution table is called *single-valued* when each category is represented by a single number.

b) Construct a bar chart for the data.

c) During how many weeks was the number of accidents no more than two?

**7**  Prepare a pie chart for the data in Problem 6.

**8**  The first cases of coronavirus (COVID-19) were found in China during January 2020. This virus later spread to many countries. The following data show the number of cases discovered in 18 different countries:

| 3 | 6 | 6 | 2 | 3 | 4 | 3 | 5 | 1 | 6 | 6 | 1 | 4 | 3 | 1 | 4 | 5 | 6 |
|---|---|---|---|---|---|---|---|---|---|---|---|---|---|---|---|---|---|

a) Construct a dot plot for the data.

b) Summarize what you learn about the data from the dot plot.

**9**  Refer to Problem 8.

a) Construct a single-valued frequency distribution table for the data.

b) In how many countries was the number of coronavirus cases less than 2?

c) In how many countries was the number of coronavirus cases at least 3?

**10** A manufacturer of a part is interested in finding the lifespan of the part. A random sample of 30 parts gave the following lifespans (in months):

| 20 | 25 | 22 | 17 | 45 | 39 | 34 | 24 | 28 | 16 | 20 | 35 | 23 | 16 | 39 |
|----|----|----|----|----|----|----|----|----|----|----|----|----|----|----|
| 22 | 27 | 31 | 22 | 20 | 21 | 35 | 18 | 23 | 31 | 43 | 16 | 25 | 44 | 26 |

Use a statistical package to construct frequency and relative frequency histograms for the data.

**11** The following are the annual salaries of 30 new engineering graduates in thousands of dollars:

| 57 | 50 | 49 | 40 | 40 | 44 |
|----|----|----|----|----|----|
| 53 | 49 | 56 | 59 | 55 | 65 |
| 57 | 59 | 40 | 44 | 55 | 49 |
| 59 | 54 | 53 | 48 | 49 | 56 |
| 57 | 47 | 48 | 55 | 45 | 56 |

a) Construct a frequency histogram for the data
b) Construct box plot for the data, and determine whether there are any mild or extreme outliers.

**12** The following data show the number of defective items in 40 shipments received at a manufacturing facility.

| 6 | 4 | 4 | 4 | 2 | 6 | 4 | 4 | 3 | 5 |
|---|---|---|---|---|---|---|---|---|---|
| 4 | 3 | 3 | 3 | 2 | 3 | 2 | 3 | 3 | 3 |
| 4 | 3 | 3 | 2 | 4 | 5 | 2 | 5 | 4 | 4 |
| 2 | 3 | 6 | 5 | 5 | 4 | 4 | 2 | 2 | 2 |

a) Construct a bar chart for the frequency distribution.
b) Construct a dot plot for the number of defective items received by the manufacturing facility.
c) What percentage of the shipments contained at least three defective items?

**13** The following are the R&D budgets in millions of dollars of five different oil companies in Houston, Texas:

| | |
|-----------|------|
| Company A | 11.5 |
| Company B | 16.4 |
| Company C | 13.2 |
| Company D | 18.5 |
| Company E | 17.5 |

a) Construct a bar graph of the data.
b) Construct a pie chart of the data.

**14** The following data provide the lengths in centimeters of 60 randomly selected small rods from a production line. Construct a histogram for the data using class intervals [40–44), [34–48), [48–52), [52–56), [56–60), [60–64), [64–68), and [68–72]. Determine the percentage of rods with length greater than or equal to 48 cm.

| 55 | 41 | 59 | 54 | 62 | 63 | 61 | 70 | 61 | 64 |
|----|----|----|----|----|----|----|----|----|----|
| 44 | 63 | 62 | 54 | 49 | 65 | 62 | 63 | 42 | 41 |
| 54 | 61 | 52 | 42 | 48 | 56 | 53 | 46 | 66 | 65 |
| 65 | 67 | 43 | 45 | 43 | 60 | 46 | 66 | 49 | 70 |
| 49 | 65 | 50 | 66 | 53 | 69 | 41 | 69 | 61 | 57 |
| 51 | 60 | 60 | 47 | 61 | 41 | 43 | 47 | 59 | 46 |

**15** The following data give the amount of gasoline sold (in gallons) on a weekday by 60 randomly selected gas stations in Houston. Construct a frequency histogram of these data using eight class intervals.

| 513 | 524 | 503 | 572 | 544 | 556 | 534 | 546 | 555 | 529 |
|-----|-----|-----|-----|-----|-----|-----|-----|-----|-----|
| 535 | 574 | 515 | 510 | 536 | 565 | 566 | 570 | 519 | 572 |
| 512 | 539 | 533 | 563 | 572 | 523 | 561 | 505 | 534 | 545 |
| 568 | 550 | 539 | 525 | 576 | 541 | 532 | 501 | 556 | 520 |
| 522 | 562 | 567 | 517 | 568 | 512 | 540 | 527 | 505 | 552 |
| 513 | 554 | 534 | 510 | 560 | 577 | 501 | 579 | 576 | 534 |

**16** The following data give the number of radioactive particles emitted per minute by a radioactive substance. Construct a relative frequency histogram for the data using seven classes.

| 71 | 77 | 60 | 64 | 80 | 48 | 51 | 82 | 48 | 57 |
|----|----|----|----|----|----|----|----|----|----|
| 50 | 52 | 46 | 94 | 74 | 49 | 46 | 78 | 57 | 82 |
| 93 | 53 | 70 | 77 | 84 | 54 | 60 | 89 | 66 | 52 |
| 74 | 89 | 63 | 89 | 86 | 89 | 57 | 74 | 52 | 65 |
| 77 | 65 | 47 | 82 | 47 | 65 | 92 | 53 | 68 | 57 |

**17** Refer to Problem 16. Sort the previous data in ascending order, and then determine what percent of the time the radioactive substance emits:
a) More than 70 radioactive particles per minute
b) Less than 60 radioactive particles per minute
c) Between 60 and 80 (inclusive) radioactive particles per minute

**18** Use the data in Problem 16 to do the following:
a) Construct a frequency distribution table with 10 classes, i.e. [45–50), [50–55), [55–60), [60–65), [65–70), [70–75), [75–80), [80–85), [85–90), [90–95].
b) Calculate the relative frequency of each classes
c) Calculate the cumulative frequency of each class.

**19** Construct a scatterplot for the following data obtained in a study of the age and total cholesterol level of 10 randomly selected people.

| Person | Age | Cholesterol |
|---|---|---|
| 1 | 39 | 148 |
| 2 | 44 | 165 |
| 3 | 48 | 170 |
| 4 | 52 | 162 |
| 5 | 50 | 155 |
| 6 | 57 | 177 |
| 7 | 57 | 178 |
| 8 | 65 | 180 |
| 9 | 59 | 168 |
| 10 | 61 | 172 |

**20** Determine the correlation coefficient for the data in Problem 19.

**21** Construct a scatterplot for the following data obtained for 12 randomly selected math majors. The data consists of the numbers of hours they study each day and their respective CGPAs. Find the correlation coefficient between the hours the student's study and their CGPAs.

| Student | Hours of study | CGPA |
|---|---|---|
| 1 | 8 | 3.42 |
| 2 | 7 | 3.40 |
| 3 | 9 | 3.36 |
| 4 | 8 | 3.55 |
| 5 | 6 | 3.10 |
| 6 | 9 | 3.78 |
| 7 | 7 | 3.64 |
| 8 | 8 | 3.67 |
| 9 | 6 | 3.44 |
| 10 | 5 | 2.80 |
| 11 | 9 | 3.83 |
| 12 | 9 | 2.70 |

**22** The number of vehicles in millions manufactured by the big three automakers over the past seven years is shown here. Construct and interpret a time series graph or a line graph for the data.

| Year | Number of vehicles |
|---|---|
| 2004 | 46.93 |
| 2005 | 44.76 |
| 2006 | 42.30 |
| 2007 | 40.33 |
| 2008 | 38.53 |
| 2009 | 37.20 |
| 2010 | 37.73 |

**23** The following data give the US dollar and Chinese yuan ratio for 2005–2010. Construct and interpret a time series or line graph for these data.

| Year | Dollar/Yuan |
|------|-------------|
| 2005 | 8.5 |
| 2006 | 8.0 |
| 2007 | 7.7 |
| 2008 | 6.9 |
| 2009 | 6.9 |
| 2010 | 6.7 |

**24** During winter months in northern Maine, homeowners heat their homes using wooden logs. This usually causes more home fires than occur in summer months. The following data give the number of patients who were treated for second- or third-degree fire burns in a city hospital during the months of November and December in a certain year.

| | | | | | | | | | |
|---|---|---|---|---|---|---|---|---|---|
| 4 | 5 | 5 | 5 | 4 | 2 | 4 | 3 | 2 | 4 |
| 6 | 3 | 3 | 6 | 6 | 3 | 6 | 4 | 5 | 6 |
| 6 | 3 | 4 | 5 | 5 | 4 | 5 | 6 | 4 | 6 |
| 5 | 3 | 5 | 3 | 4 | 6 | 6 | 5 | 4 | 3 |
| 6 | 3 | 4 | 3 | 5 | 5 | 5 | 6 | 5 | 5 |
| 6 | 4 | 4 | 4 | 4 | 6 | 5 | 5 | 6 | 3 |

a) Construct a pie chart for the data.
b) Construct a bar chart for the data.
c) Comment on what you learn from the graphs in (a) and (b).

**25** Consider the following quantitative data, which represent the number of defective parts received by a customer in 40 shipments.

| | | | | | | | | | |
|---|---|---|---|---|---|---|---|---|---|
| 6 | 8 | 7 | 4 | 8 | 9 | 3 | 3 | 7 | 6 |
| 6 | 4 | 4 | 7 | 4 | 4 | 6 | 3 | 6 | 6 |
| 3 | 5 | 5 | 6 | 3 | 4 | 4 | 5 | 4 | 7 |
| 7 | 6 | 5 | 4 | 3 | 4 | 5 | 6 | 7 | 7 |

a) Construct a pie chart for the data.
b) Construct a bar chart for the data.
c) Comment on what you learn from the graphs in (a) and (b).

**26** Use one of the statistical packages discussed in this book to construct a bar chart and a pie chart for the following categorical data.

| | | | | | | | |
|---|---|---|---|---|---|---|---|
| 2 | 3 | 5 | 4 | 2 | 3 | 5 | 4 |
| 6 | 6 | 5 | 4 | 2 | 3 | 4 | 5 |
| 4 | 2 | 5 | 6 | 4 | 2 | 3 | 6 |
| 5 | 4 | 2 | 3 | 5 | 6 | 4 | 5 |

**27** The following data represent the amounts of contaminants, in grams, retrieved from 30 sources of water.

| | | | | | | | | | |
|---|---|---|---|---|---|---|---|---|---|
| 117.9 | 117.7 | 121.9 | 116.8 | 118.9 | 121.2 | 119.0 | 117.5 | 120.1 | 122.6 |
| 120.1 | 124.1 | 120.1 | 118.4 | 117.2 | 121.7 | 122.2 | 122.0 | 121.2 | 120.4 |
| 119.8 | 121.6 | 118.1 | 119.3 | 121.1 | 119.6 | 117.9 | 119.4 | 120.8 | 122.1 |

a) Determine the mean and median of the data.
b) Determine the standard deviation of the data.

**28** The following data show the tread depth, in mm, of 20 tires selected randomly from a large shipment of recently received tires.

| | | | | | | | | | |
|---|---|---|---|---|---|---|---|---|---|
| 6.28 | 7.06 | 6.50 | 6.76 | 6.82 | 6.92 | 6.86 | 7.15 | 6.57 | 6.48 |
| 6.64 | 6.94 | 6.49 | 7.14 | 7.16 | 7.10 | 7.08 | 6.48 | 6.40 | 6.54 |

a) Find the mean and median tread depth.
b) Find the variance and standard deviation of the tread depth.
c) If the desired tread depth on these tires is 7 mm, what can you say about the quality of the tires?

**29** The following data give the number of parts manufactured in 36 (eight-hour) shifts.

| | | | | | | | | |
|---|---|---|---|---|---|---|---|---|
| 55 | 58 | 46 | 58 | 49 | 46 | 41 | 60 | 59 |
| 41 | 59 | 42 | 40 | 44 | 42 | 58 | 46 | 46 |
| 46 | 40 | 51 | 59 | 48 | 46 | 42 | 43 | 56 |
| 48 | 41 | 54 | 56 | 57 | 48 | 43 | 49 | 43 |

a) Find the mean, mode, and median of parts produced.
b) Prepare a box plot for the data, and comment on the shape of the distribution.

**30** The following data give the lengths, in cm, of 24 randomly selected rods.

| | | | | | | | |
|---|---|---|---|---|---|---|---|
| 21 | 18 | 21 | 18 | 20 | 18 | 18 | 19 |
| 19 | 20 | 20 | 20 | 19 | 18 | 21 | 18 |
| 19 | 22 | 19 | 18 | 22 | 18 | 22 | 26 |

a) Construct a box plot of the data.
b) Does the data set contain any outliers?

**31** The following data provide the number of Six Sigma Green Belt engineers working in 36 randomly selected US manufacturing companies.

| | | | | | | | | |
|---|---|---|---|---|---|---|---|---|
| 73 | 64 | 80 | 67 | 73 | 78 | 66 | 78 | 59 |
| 79 | 74 | 75 | 73 | 66 | 63 | 62 | 61 | 58 |
| 65 | 76 | 60 | 79 | 62 | 63 | 71 | 75 | 56 |
| 78 | 73 | 75 | 63 | 66 | 71 | 74 | 64 | 43 |

a) Find the quartiles $Q_1$, $Q_2$, and $Q_3$ of the data.
b) Determine the number of data points that fall inclusively between the first and third quartiles.
c) Construct a box plot for the data and determine whether the data set contains any outliers.

**32** Consider the following two data sets.

**Data set I:**

| 28 | 29 | 24 | 25 | 26 | 23 | 24 | 29 | 29 | 24 |
|----|----|----|----|----|----|----|----|----|----|
| 28 | 23 | 27 | 26 | 21 | 20 | 25 | 24 | 30 | 28 |
| 29 | 28 | 22 | 26 | 30 | 21 | 26 | 27 | 25 | 23 |

**Data set II:**

| 58 | 46 | 48 | 60 | 43 | 57 | 47 | 42 | 57 | 43 |
|----|----|----|----|----|----|----|----|----|----|
| 56 | 59 | 52 | 53 | 41 | 58 | 43 | 50 | 49 | 49 |
| 60 | 57 | 54 | 51 | 46 | 60 | 44 | 55 | 43 | 52 |

a) For each data set, find its mean and standard deviation.
b) Find the coefficient of variation (CV, defined as the ratio of the standard deviation to the mean; it measures the variability of a data set) for each data set.
c) Use the result of (b) to determine which data set has greater relative variability.

**33** Find the mean and standard deviation for the data set in Problem 31, and then determine the number of data points that fall in the intervals $(\bar{x} - s, \bar{x} + s)$, $(\bar{x} - 2s, \bar{x} + 2s)$, and $(\bar{x} - 3s, \bar{x} + 3s)$. Assuming that the distribution of this data set is bell-shaped, use the empirical rule to find the number of data points that you would expect to fall in these intervals. Compare the two results, and comment on the possible shape of the distribution of the data set.

**34** The salaries of engineers working in a manufacturing company have a mean of $55,600 with a standard deviation of $4,500. Assuming that the distribution of salaries is bell-shaped, estimate the intervals of salaries within which about 68%, 95%, and 99.7% of all engineers' salaries are expected to fall.

**35** Which measure of central tendency, mean or median, would be a better measure for the following data? Justify your choice.

| 54 | 48 | 48 | 55 | 46 | 40 | 50 | 60 | 281 |
|----|----|----|----|----|----|----|----|-----|
| 54 | 53 | 45 | 58 | 45 | 43 | 42 | 42 | 50 |
| 42 | 41 | 49 | 52 | 50 | 57 | 56 | 55 | 140 |

**36** The following data show the gestational age in weeks of 40 children born in a city hospital.

| 39 | 39 | 36 | 35 | 40 | 36 | 40 | 40 | 39 | 37 |
|----|----|----|----|----|----|----|----|----|----|
| 38 | 35 | 38 | 37 | 40 | 35 | 39 | 40 | 36 | 35 |
| 36 | 36 | 35 | 38 | 40 | 37 | 37 | 38 | 36 | 36 |
| 38 | 40 | 36 | 40 | 38 | 37 | 38 | 37 | 39 | 40 |

a) Compute the mean, median, standard deviation, and coefficient of variation of the data.
b) Construct a box plot for the data.
c) What percentage of the gestational ages are within one standard deviation of the mean? Within two standard deviations? Within three standard deviations?

**37** The following data give the number of patients treated in an urgent care center each day during a period of two months. Use a statistical package to construct a box plot for these data and determine whether the data set contains any outliers.

| | | | | | | | | | |
|---|---|---|---|---|---|---|---|---|---|
| 50 | 35 | 50 | 35 | 31 | 35 | 33 | 30 | 33 | 47 |
| 36 | 34 | 35 | 40 | 30 | 50 | 39 | 50 | 49 | 42 |
| 42 | 30 | 50 | 50 | 38 | 30 | 37 | 36 | 40 | 43 |
| 45 | 39 | 43 | 49 | 36 | 32 | 42 | 42 | 30 | 43 |
| 37 | 49 | 43 | 33 | 32 | 37 | 49 | 31 | 40 | 33 |
| 43 | 31 | 36 | 48 | 41 | 33 | 49 | 43 | 48 | 95 |

**38** Consider the following data.

| | | | | | | | |
|---|---|---|---|---|---|---|---|
| 87 | 76 | 79 | 68 | 71 | 79 | 79 | 73 |
| 86 | 67 | 82 | 78 | 74 | 87 | 74 | 86 |

a) Determine the quartiles ($Q_1, Q_2, Q_3$) for the data.
b) Find the interquartile range.
c) What percentage of data fall between $Q_1$ and $Q_3$?
d) What percentage of data should you expect to fall between $Q_1$ and $Q_3$?
e) Compare the results of (c) and (d), and comment.

**39** In a study of productivity among Six Sigma Green Belt and non–Six Sigma Green Belt technicians, two random samples from each group are taken from eight different industries and their productivities determined. The following data give the results obtained from the two studies.

| SSGB | 114 | 106 | 113 | 114 | 114 | 105 | 110 | 109 |
|---|---|---|---|---|---|---|---|---|
| Non SSGB | 105 | 94 | 95 | 93 | 101 | 100 | 98 | 100 |

a) Find the mean and standard deviation for each data set.
b) Find the coefficient of variation for each data set, and determine which data set has more variability.

**40** Consider the following data set. Use technology to find the mean, median, and standard deviation of the data.

| | | | | | | | | | |
|---|---|---|---|---|---|---|---|---|---|
| 56 | 48 | 37 | 54 | 48 | 39 | 63 | 37 | 72 | 61 |
| 60 | 42 | 44 | 75 | 47 | 42 | 67 | 34 | 69 | 65 |
| 39 | 58 | 57 | 63 | 36 | 41 | 46 | 53 | 47 | 76 |
| 73 | 61 | 51 | 49 | 38 | 43 | 53 | 66 | 72 | 70 |

**41** Refer to the data in Problem 40. Using technology, test whether these data come from a normal population.

**42** Refer to the data in Problem 37. Using one of the statistical packages discussed in this book, test whether these data come from a normal population.

# 4

# Sampling Methods

## 4.1 Introduction

The science of sampling is as old as civilization. When trying a new cuisine, for example, we take only a small bite to decide whether the taste of the food is to our liking. However, modern advances in sampling techniques have only taken place starting in the twentieth century. Now sampling is a matter of routine, and the effects of the outcomes can be felt in our day-to-day lives. Most decisions about government policies, marketing, trade, and manufacturing are based on the outcomes of sampling conducted in different fields.

There are various types of sampling. The particular type of sampling used for a given situation depends on factors such as the composition of the population, objectives of the sampling, or simply time and budgetary constraints. Since sampling is an integral part of statistical quality control, we dedicate this chapter to the study of various types of sampling and estimation problems.

## 4.2 Basic Concepts of Sampling

The primary objective of sampling is to use the information contained in a sample taken from some population to make inferences about a certain population parameter, such as the *population mean, population total, population proportion,* or *population variance.* To make such inferences in the form of *estimates* of parameters that otherwise are unknown, we collect data from the population of interest. The aggregate of these data constitutes a *sample.* Each data point in the sample provides us with some information about the population parameter. However, since collecting each data point costs time and money, it is important to keep a balance. Too small a sample may not provide enough information to obtain good estimates, and too large a sample may result in a waste of resources. Thus, it is very important that in any sampling procedure, an appropriate sampling scheme, normally known as the *sample design*, is put in place. Ultimately, we will use an appropriate sample to determine an estimate of a parameter value. Before doing this, however, we first need to determine a parameter's corresponding *estimator.*

> **Definition 4.1**    An *estimator* is a function of *observable random variables* and possibly *known constants.* By assigning certain values to the observable random variable, we obtain what is referred to as an *estimate.*

*Statistical Quality Control: Using Minitab, R, JMP, and Python*, First Edition. Bhisham C. Gupta.
© 2021 John Wiley & Sons, Inc. Published 2021 by John Wiley & Sons, Inc.
Companion website: www.wiley.com/go/college/gupta/SQC

The sample size required for the estimate is usually determined by two factors: (i) the degree of precision desired for the estimate and (ii) any budgetary restrictions. If $\theta$ is the population parameter of interest, $\hat{\theta}$ is an *estimator* of $\theta$, and $E$ is the margin of error (a *bound* on the error of estimation, where the error of estimation is defined as the difference between a point estimate and the parameter it estimates) of estimation, then the *sample size* is usually determined by specifying the value of E and the probability with which we will achieve that value of E.

In this chapter, we discuss four different *sample designs*: simple random sampling, stratified random sampling, systematic random sampling, and cluster random sampling from a finite population. Before studying these sample designs, we first introduce some common terminology used in sampling theory.

---

**Definition 4.2**   A *population* is a collection of all conceivable individuals, elements, numbers, or entities that possess some characteristic of interest.

---

For example, if we are interested in the ability of employees of a company with a specific job title to perform specific job functions, the population may be defined as all employees with the specific job title working at all sites of the company. However, if we are interested in the ability of employees with a specific job title or classification to perform specific job functions at a particular location, the population may be defined as all employees with the specific job title working only at that particular site or location. Populations, therefore, are shaped by the point or level of interest.

Populations can be finite or infinite. A population where all the elements are *easily identifiable* is considered *finite,* and a population where all the elements are *not easily identifiable* is considered *infinite*. For example, a batch or lot of production is normally considered a finite population, whereas all the products that may be produced on a certain manufacturing line would normally be considered infinite.

It is important to note that in statistical applications, the term *infinite* is used in the relative sense. For instance, if we are interested in studying the products produced or service delivery iterations occurring over a given period of time, the population may be considered finite or infinite, depending on our frame of reference.

In most statistical applications, studying every element of a population is not only time-consuming and expensive, but also potentially impossible. For example, if we are interested in studying the average lifespan of a particular kind of electric bulb manufactured by a company, then we cannot study the entire population without using every bulb and thus destroying the population. Simply put, in almost all studies, we end up studying only a small portion of the population, or a *sample*.

---

**Definition 4.3**   The population from which a sample is selected is called a *sampled population*.

---

**Definition 4.4**   The *target population* is the population about which we want to make inferences based on the information contained in a sample.

---

Normally, these two populations, the *sampled population* and the *target population*, coincide with each other since every effort is made to ensure the sampled population is the same as the target population. However, situations do arise when the sampled population does not cover the entire target population. In such cases, conclusions made about the sampled population are not usually applicable to the target population.

It is important, before taking a sample, to divide the *target population* into non-overlapping units called *sampling units*. Sampling units in a given population may not always be the same and are determined by the sample design chosen. For example, in sampling voters in a metropolitan area, the sampling units might be an individual voter, the voters in a family, or all voters living in a city block. Similarly, in sampling parts from a manufacturing plant, sampling units might be each individual part or a box containing several parts.

---

**Definition 4.5** A list of all sampling units from a population is called the *sampling frame*.

---

Before selecting a sample, we must decide about the method of measurement or how information from the sample is to be collected and recorded. Commonly used methods include personal interviews, telephone interviews, physical measurements, direct observations, and mailed questionnaires. No matter what method of measurement is used, it is very important that the person taking the sample know what measurements are to be taken – in other words, we must know the objective of sampling so that all the pertinent information relative to the objective is collected.

People collecting samples are usually called *field workers*. All field workers should be well acquainted with what measurements to take and how to take them. Good training of all field workers is an important and essential part of any sampling procedure. The accuracy of the measurements, which affects the final results, depends on how well the field workers are trained.

It is quite common to first select a small sample and examine it carefully, a process called a *pretest*. The pretest allows us to make any necessary improvements in the questionnaire or method of measurement, and to eliminate any difficulties experienced in taking the measurements or using a questionnaire. The pretest also allows for the reviewing of the quality of the measurements or, in case of questionnaires, the quality and quantity of the returns.

Once the data are collected, the next step is to describe how to manage the data and get the desired information, or how to organize, summarize, and analyze these data. Some of these goals can be achieved by using methods of analyzing data using graphical and numerical methods, as previously discussed in Chapters 3.

## 4.2.1 Introducing Various Sampling Designs

Next, we give a brief description of various sample designs: *simple random sampling, stratified random sampling, systematic random sampling*, and *cluster random sampling*. Later in this chapter, we will present some results pertaining to these sample designs.

The most commonly used sample design is *simple random sampling*. The simple random sampling design consists of selecting $n$ (sample size) sampling units in such a way that each sampling unit has the same chance of being selected. If the population is finite, say of size $N$, then the simple random sampling design may be defined as selecting $n$ sampling units in such a way that each sample of size $n$, out of $\binom{N}{n}$ possible samples, has the same chance of being selected.

---

**Example 4.1**
Suppose a Six Sigma Green Belt engineer wants to take a sample of some machine parts manufactured during a shift at a given plant. Since the parts from which the engineer wants to take the sample are manufactured during the same shift at the same plant, it is safe to assume that all parts are similar. Hence, in this case, a *simple random sampling design* seems to be most appropriate.

The second sampling design we now look at is the *stratified random sampling design*, which may give improved results while costing the same amount of money as simple random sampling. Stratified random sampling is most appropriate when a population can be divided into various non-overlapping groups called *strata* such that the sampling units in each stratum are similar but may vary from stratum to stratum. Each stratum is treated as a population by itself, and a simple random sample is taken from each of these subpopulations or strata.

In manufacturing, this type of situation may arise quite often. For instance, in Example 4.1, if the sample is taken from a population of parts that were manufactured either in different plants or during different shifts, then stratified random sampling would be more appropriate than simple random sampling. In addition, there is an advantage of administrative convenience. For example, if the machine parts in the example are manufactured in plants located in different parts of the country, then stratified random sampling would be beneficial. Each plant may have a sampling department that can conduct the sampling within that plant. We note that to obtain better results in this case, the sampling departments in all the plants should communicate with each other before starting sampling to ensure that the same sampling norms are followed.

The third kind of sampling design is *systematic random sampling*. The systematic random sampling procedure is quite simple and is particularly useful in manufacturing processes when the sampling is done *on line*. Under this scheme, the first item is selected randomly, and then every $m^{th}$ item manufactured is selected after that until we have a sample of the desired size. Systematic sampling is not only easy but also, under certain conditions, more precise than simple random sampling.

The last sampling design we look at is *cluster random sampling*. In cluster sampling, each sampling unit is a group or cluster of smaller units. In the manufacturing environment, this sampling scheme is particularly useful since it is not easy to prepare a list of each part that would constitute a sampling frame. On the other hand, it may be much easier to prepare a list of boxes, where each box contains many parts. Hence a cluster random sample is merely a simple random sample of these boxes. Another advantage of cluster sampling is that by selecting a simple random sample of only a few clusters, we can obtain a large sample of smaller units at minimal cost.

As mentioned earlier, the primary objective of sampling is to make inferences about population parameters using information contained in a sample taken from that population. In the remainder of this chapter, we present inferential results pertaining to each sample design just discussed. However, we first give some basic definitions that are useful in studying estimation problems.

---

**Definition 4.6**   The value assigned to a population parameter based on the information contained in a sample is called a *point estimate,* or simply an *estimate* of the population parameter. The sample statistic used to calculate such an estimate is called an *estimator.*

---

**Definition 4.7**   An interval formed by a pair of values obtained by using the information contained in a sample such that the interval contains the true value of an *unknown population parameter* with certain probability is called a *confidence interval*. The probability with which it contains the true value of the unknown parameter is called the *confidence coefficient.*

---

**Definition 4.8**   The difference between an estimate of a parameter and its true value is called an *error of estimation*. The maximum absolute value of the error of estimation with a given probability is called the *margin of error*, which is usually denoted by $E$.

## 4.3 Simple Random Sampling

Simple random sampling is the most basic form of sampling design. A simple random sample can be drawn from an infinite or a finite population. There are two techniques to take a simple random sample from a finite population: sampling with replacement and sampling without replacement. Simple random sampling with replacement from a finite population and a simple random sampling from an infinite population have the same statistical properties.

> **Definition 4.9**  When sampling without replacement from a finite population of size $N$, a random sample is called a *simple random sample* if each sample of size $n$ of the $\binom{N}{n}$ possible samples has the same chance of being selected.

> **Definition 4.10**  When sampling from an infinite population or sampling with replacement from a finite population, a random sample is called a *simple random sample*. Note that in simple random sampling each sampling unit has the same chance of being included in the sample.

A simple random sample can be taken by using tables of random numbers: a series of digits 0, 1, 2, …, 9 arranged in columns and rows, as shown in Table 4.1.

Note that these random numbers can easily be obtained by using one of the statistical packages discussed in this book. For example, using Minitab, random numbers are easily obtained by following these steps:

1) From the menu bar, select **Calc** > **Random Data** > **Integer** to open the **Integer Distribution** dialog box.
2) In this dialog box, enter the number of rows in the box next to **Number of rows of data to generate**. For example, to generate 10 rows of data, enter the number **10**.
3) Enter the names of columns, say, C1 to C10, in the box below **Store in column (s)**.
   This will generate 10 columns of data and store them in columns C1–C10 of the worksheet.
4) In the box next to **Minimum value**, enter a value like **000000**.
5) In the box next to **Maximum value**, enter a value like **999999**.
6) Click **OK**.

A table of random numbers will appear in the **Session Window**. Note that this table will be similar to the one in Table 4.1, but not identical. A table of random numbers is also given in Table A.1 in the Appendix.

The first step in selecting a simple random sample from a population of 10,000 units, say, using a table of random numbers, is to label all the sampling units from 0000 to 9999. Once all the sampling units are labeled, then we select the same number of digits (columns) from the table of random numbers that match the number of digits in $(N - 1)$. Third, we read down $n$ ($\leq N$) numbers from the selected columns, starting from any arbitrary point and ignoring any repeated numbers (if the sampling is being done with replacement, then we do not ignore the repeated numbers). The last step for taking a simple random sample of size $n$ is to select the $n$ sampling units with labels matching the selected numbers.

Sometimes, in a manufacturing environment, it may not be possible to label all the parts, in which case selecting a random sample by using random numbers may not be feasible. For example, suppose we want to select a simple random sample of 50 ball bearings from a large lot of ball bearings. In such a case, it is difficult and economically unfeasible to label every single ball bearing. The

**Table 4.1** Table of random numbers.

| | | | | | | | | | |
|---|---|---|---|---|---|---|---|---|---|
| 016893 | 173302 | 709329 | 373112 | 411553 | 870482 | 298432 | 785928 | 15238 | 302439 |
| 164110 | 255869 | 638942 | 665488 | 814927 | 077599 | 257517 | 501645 | 639820 | 693343 |
| 357495 | 414180 | 762729 | 212781 | 389738 | 715639 | 504993 | 844249 | 154735 | 171102 |
| 114628 | 226683 | 790935 | 591252 | 450523 | 455716 | 382408 | 963741 | 875864 | 122066 |
| 618892 | 660732 | 694668 | 062300 | 331206 | 911707 | 967560 | 083523 | 576468 | 875153 |
| 861834 | 266477 | 677869 | 865628 | 240623 | 052629 | 029081 | 198717 | 363996 | 477864 |
| 207228 | 719322 | 932208 | 558299 | 540073 | 740870 | 175883 | 628462 | 943245 | 784317 |
| 736803 | 170693 | 651684 | 733645 | 127119 | 933714 | 702257 | 407230 | 699936 | 249002 |
| 309483 | 946316 | 036759 | 504195 | 527099 | 284019 | 792768 | 308640 | 007903 | 444429 |
| 255239 | 359157 | 407303 | 490615 | 417491 | 983928 | 776484 | 606597 | 040410 | 524521 |

method described is more to develop statistical theory. In practice, to select simple random samples, we sometimes use a convenient method so that no particular sampling unit is given any preferential treatment. For instance, to select a simple random sample of 50 ball bearings from a large lot, we may simply mix up all the ball bearings and then select 50 of them from various spots. This sampling technique is sometimes called *convenience sampling*.

### 4.3.1 Estimating the Population Mean and Population Total

Let $y_1, y_2, ..., y_N$ be the sampling units of a population under consideration, and let $y_1, y_2, ..., y_n$ be a simple random sample from this population. We denote some of the population parameters and sample statistics as follows:

$\mu$: population mean
$\sigma^2$: population variance
$T$: population total
$N$: population size
$\bar{y}$: sample mean
$s^2$: sample variance
$n$: sample size

For a simple random sample, the sample mean $\bar{y}$ and sample variance $s^2$ are defined as follows:

$$\bar{y} = \frac{y_1 + y_2 + ... + y_n}{n} = \frac{\sum_{i}^{n} y_i}{n} \tag{4.1}$$

$$s^2 = \frac{1}{n-1}\left(\sum_{i}^{n}(y_i - \bar{y})^2\right) = \frac{1}{n-1}\left(\sum_{1}^{n} y_i^2 - \frac{(\sum y_i)^2}{n}\right) \tag{4.2}$$

We use the sample mean $\bar{y}$ as an estimator of the population mean $\mu$, or

$$\hat{\mu} = \bar{y} \tag{4.3}$$

Note that in simple random sampling, the sample mean $\bar{y}$ is an unbiased estimator of the population mean $\mu$. That is,

$$E\left(\bar{y}\right) = \mu \tag{4.4}$$

To assess the accuracy of the estimator $\bar{y}$ of $\mu$, it is important to find the variance of $\bar{y}$, which will allow us to find the margin of error.

The variance of $\bar{y}$ for a simple random sampling without replacement from a finite population of size $N$ is given by

$$V\left(\bar{y}\right) = \frac{\sigma^2}{n}\left(\frac{N-n}{N-1}\right) \tag{4.5}$$

If $N$ is *large* or $n < 0.05N$, then we have

$$V\left(\bar{y}\right) \cong \frac{\sigma^2}{n}\left(1 - \frac{n}{N}\right) \tag{4.5a}$$

In practice, we usually do not know the population variance, $\sigma^2$; therefore, it becomes necessary to use an estimator of $V(\bar{y})$, which is given by

$$\hat{V}(\bar{y}) = \frac{s^2}{n}\left(1 - \frac{n}{N}\right) \tag{4.6}$$

For a simple random sample taken either from an infinite population or from a finite population when sampling is done with replacement, the variance $V(\bar{y})$ is given by

$$V\left(\bar{y}\right) = \frac{\sigma^2}{n} \tag{4.7}$$

and its estimator is given by

$$\hat{V}(\bar{y}) = \frac{s^2}{n} \tag{4.8}$$

If either the population is normal or the sample size is large ($n \geq 30$), then the margin of error of estimation of the population mean with probability $(1 - \alpha)$ is given by

$$E = \pm z_{\alpha/2}\sqrt{V(\bar{y})} \tag{4.9}$$

If the population variance $\sigma^2$ is unknown, then we replace $V(\bar{y})$ in Eq. (4.9) with its estimator, $\hat{V}(\bar{y})$, or simply replace $\sigma^2$ with $s^2$.

**Example 4.2  Steel Casting Data**

Suppose that we are interested in studying the mean yield point of steel castings manufactured by a large manufacturer. The following data give the yield point in units of 1000 pounds per square inch (psi) of a simple random sample of size 10:

| 83.37 | 86.21 | 83.44 | 86.29 | 82.80 | 85.29 | 84.86 | 84.20 | 86.04 | 89.60 |
|-------|-------|-------|-------|-------|-------|-------|-------|-------|-------|

a) Find an estimate of the population mean, $\mu$.
b) Find the value of $E$, the margin of error of estimation with 95% probability.

**Solution**

The sample mean for the given data in this example is

$$\bar{y} = \frac{1}{10}(83.37 + 86.21 + 83.44 + 86.29 + 82.80 + 85.29 + 84.86$$
$$+ 84.20 + 86.04 + 89.60)$$
$$= \frac{1}{10}(852.1) = 85.21$$

The sample variance for these data is

$$s^2 = \frac{1}{(10-1)}\left(83.37^2 + 86.21^2 + \ldots + 89.60^2 - \frac{(83.37 + 86.21 + \ldots + 89.60)^2}{10}\right)$$
$$= \frac{1}{9}(35.595) = 3.955 \Rightarrow s = 1.989$$

Thus, an estimate for the population mean $\mu$ is given by

$$\hat{\mu} = 85.21$$

(b) Since the population under investigation consists of all steel castings manufactured by the manufacturer, it can be considered infinite. Thus, the margin of error for estimating the population mean with (say) 95% probability using Eqs. (4.8) and (4.9) is given by

$$E = \pm z_{0.025}\frac{s}{\sqrt{n}} = \pm 1.96\left(\frac{1.989}{\sqrt{10}}\right) = \pm 1.233$$

Note that an unbiased estimator of the finite population total $T$ is given by

$$\hat{T} = N\bar{y} \tag{4.10}$$

The variance of estimator $\hat{T}$ is given by

$$N^2 V(\bar{y}) \tag{4.11}$$

The standard error of $\hat{T}$ is given by

$$N\sqrt{V(\bar{y})} \tag{4.12}$$

If the population variance $\sigma^2$ is unknown, then we replace $V(\bar{y})$ with its estimator $\hat{V}(\bar{y})$ in Eqs. (4.11) and (4.12). Assuming the sample size is large ($n \geq 30$), it can be shown that the margin of error of estimation of the population total with probability $(1 - \alpha)$ is

$$
\pm N \times (\textit{margin of error of estimation for population mean})
$$

$$
= \pm N \times z_{\alpha/2} \sqrt{\hat{V}(\bar{y})}
$$

(4.13)

Note: As we saw in Chapter 3, this sample mean and sample variance can easily be found by using one of the statistical packages discussed in this book.

### Example 4.3 Manufacturing Time

A manufacturing company has developed a new device for the Army. The company received a defense contract to supply 25,000 pieces of this device to the army. To meet contractual obligations, the HR department wants to estimate the number of workers that the company needs to hire. This can be accomplished by estimating the number of manpower hours needed to manufacture 25,000 devices. The following data show the number of hours spent by 15 randomly selected workers to manufacture 15 pieces of such a device. (We assume that all the parts manufactured by each worker meet the specifications set by the Army.)

| 8.90 | 8.25 | 6.50 | 7.25 | 8.00 | 7.80 | 9.90 | 8.30 | 9.30 | 6.95 | 8.25 | 7.30 | 8.55 | 7.55 | 7.70 |
|------|------|------|------|------|------|------|------|------|------|------|------|------|------|------|

Estimate the total manpower hours needed to manufacture 25,000 devices, and determine the margin of error of estimation with 95% probability.

### Solution

In this example, we first estimate the mean manpower hours needed to manufacture one device and determine the margin of error with 95% probability. The sample mean for the given data in this example is

$$
\bar{y} = \frac{1}{15}(8.90 + 8.25 + 6.50 + \dots + 8.55 + 7.55 + 7.70)
$$

$$
= \frac{1}{15}(120.50) = 8.033
$$

The sample variance for these data is

$$
s = \sqrt{\frac{1}{(15-1)}\left(8.90^2 + 8.25^2 + \dots + 7.70^2 - \frac{(8.90 + 8.25 + \dots + 7.70)^2}{15}\right)^{1/2}}
$$

$$
= 0.900
$$

Thus, an estimate for the population mean $\mu$ is

$$
\hat{\mu} = 8.033
$$

Using Eq. (4.9), we can see that the margin of error for estimating the population mean with 95% probability is

$$E = \pm z_{0.025}\frac{s}{\sqrt{n}} = \pm 1.96\left(\frac{0.900}{\sqrt{15}}\right) = \pm 0.455$$

Now, suppose that $T$ is the total manpower hours needed to manufacture 25,000 devices. Then, using Eqs. (4.10) and (4.13), the estimate for $T$ is

$$\hat{T} = N \times \bar{y} = 25,000 \times 8.033 = 200,825$$

and the margin of error of estimation for the total manpower hours is

$$\pm 25,000 \times 0.455 = \pm 11,375$$

### 4.3.2 Confidence Interval for the Population Mean and Population Total

Once we have estimates for the population mean, population total, and corresponding margins of error of estimation with probability $(1 - \alpha)$, then finding a confidence interval with confidence coefficient $(1 - \alpha)$ for the population mean and population total is very simple. For example, to find a confidence interval with confidence coefficient $(1 - \alpha)$ for the population mean, we use

$$\hat{\mu} \pm \text{margin of error for estimating } \mu$$

or

$$\hat{\mu} \pm z_{\alpha/2}\sqrt{\hat{V}(\bar{y})} \tag{4.14}$$

And a confidence interval with confidence coefficient $(1 - \alpha)$ for the population total is given by

$$\hat{T} \pm \text{margin of error for estimating } T$$

or

$$\hat{T} \pm N \times z_{\alpha/2}\sqrt{\hat{V}(\bar{y})} \tag{4.15}$$

**Example 4.4**

Reconsider the data in Example 4.3, and find a 95% confidence interval for the population mean $\mu$ and the population total $T$. (Assume that the population is normally distributed)

**Solution**

Since the sample size is small, we use Eqs. (4.14) and (4.15) but replace $z_{\alpha/2}$ with $t_{n-1,\alpha/2}$, where $(n-1)$ is the degrees of freedom (see Table A.9 in the Appendix). Thus we have a 95% confidence for population mean $\mu$ as

$$\hat{\mu} \pm \text{margin of error for estimating } \mu = \hat{\mu} \pm t_{14,0.025}\sqrt{\hat{V}(\bar{y})}$$

$$\text{or } 8.033 \pm 0.498 = (7.535, 8.531)$$

and a 95% confidence interval for the population total $T$ is

$$\hat{T} \pm \text{margin of error for estimating } T = 200\,825 \pm 12\,450 = (188\,375, 213\,275)$$

### 4.3.3 Determining Sample Size

The *sample size* needed to estimate the *population mean* with a margin of error $E$ with probability $(1 - \alpha)$ is given by

$$n = \frac{z_{\alpha/2}^2 N \sigma^2}{(N-1)E^2 + \sigma^2 z_{\alpha/2}^2} \tag{4.16}$$

where $N$ is the population size. In practice, the population variance $\sigma^2$ is usually unknown. Thus, to find the sample size $n$, we need to find an estimate $s^2$ for the population variance $\sigma^2$. This can be achieved by using the data from a pilot study or a previous survey. So, by replacing the population variance $\sigma^2$ with its estimator $s^2$, the *sample size* needed to estimate the population mean with a margin of error $E$ with probability $(1 - \alpha)$ is given by

$$n = \frac{z_{\alpha/2}^2 N s^2}{(N-1)E^2 + s^2 z_{\alpha/2}^2} \tag{4.17}$$

If the population size $N$ is large, then the factor $(N-1)$ in Eq. (4.17) is usually replaced by $N$, which follows from Eq. (4.6).

Note: The *sample size* needed to estimate the *population total* with a margin of error $E'$ with probability $(1 - \alpha)$ can be found by using Eq. (4.16) or (4.17), and replacing $E$ with $(E'/N)$.

**Example 4.5**

Suppose that in Example 4.3, we would like the margin of error for estimating the population mean to be 0.25 with probability 0.99. Determine the appropriate sample size to achieve this goal.

**Solution**

From the information provided to us, we have

$$E = 0.25, z_{\alpha/2} = 2.575$$

And from Example 4.3, we have

$$N = 25,000 \text{ and } s = 0.900$$

Plugging these values into Eq. (4.17) and replacing the factor $(N-1)$ in the denominator with $N$, we get

$$n = \frac{(2.575)^2 \times (25.000) \times (0.900)^2}{25,000(0.25)^2 + (0.900)^2 (2.575)^2} \cong 86$$

Note that the sample size for estimating the population total with a margin of error 6250 $(= 25{,}000 \times 0.25)$ will again be equal to 86. To attain a sample of size 86, we take another simple random sample of size 71, which is usually known as a *supplemental sample*. Then, to have a sample of size 86, we combine the supplemental sample with the one we had already taken, which was of size 15.

## 4.4 Stratified Random Sampling

It is not uncommon in a manufacturing process for a characteristic of interest to vary from country to country, region to region, plant to plant, or machine to machine. For example, a Six Sigma Green Belt engineer could be interested in finding the longevity of parts that are being manufactured in different regions of the United States. Regional variation could be due to differences in raw materials, worker training, or variations in machines. In such cases, when the population is heterogeneous, we get more precise estimates by using a sample design known as *stratified random sampling*. In stratified random sampling, the population is divided into different non-overlapping groups called *strata*. These strata constitute the entire population: the population within each stratum is as homogeneous as possible, and the population between strata varies. A stratified random sample is taken by selecting a simple random sample from each stratum.

Some of the advantages of stratified random sampling over simple random sampling are the following.

1) It provides more precise estimates for population parameters than simple random sampling with the same sample size.
2) It is more convenient to administer, which may result in a lower cost for sampling.
3) It provides a simple random sample for each subgroup or stratum, and these are homogeneous. Therefore, the samples can be very useful for studying each individual subgroup separately without incurring any extra cost.

Before we discuss how to use the information obtained from a stratified random sample to estimate population parameters, we discuss briefly the process of generating a stratified random sample. To generate a stratified random sample:

1) Divide the sampling population of $N$ units into non-overlapping subpopulations or strata of $N_1$, $N_2$, ..., $N_K$ units, respectively. These strata constitute the entire population of $N$ units. That is, $N = N_1 + N_2 + ... + N_K$.
2) Select independently from each stratum a simple random sample of size $n_1$, $n_2$, ..., $n_K$ so that $n = n_1 + n_2 + ... + n_K$ is the total sample size.
3) To make full use of stratification, the strata sizes $N_1$, $N_2$, ..., $N_K$ must be known.

### 4.4.1 Estimating the Population Mean and Population Total

Let $y_{11}, y_{12}, ..., y_{1N_1}; y_{21}, y_{22}, ..., y_{2N_2}; ...; y_{K1}, y_{K2}, ..., y_{kN_K}$ be the sampling units in the 1, 2, ..., $K^{th}$ stratum, respectively, and let $y_{11}, y_{12}, ..., y_{1n_1}; y_{21}, y_{22}, ..., y_{2n_2}; ...; y_{K1}, y_{K2}, ..., y_{kn_K}$ be the respective simple random samples of sizes $n_1$, $n_2$, ..., $n_K$ from them. Let $\mu$ and $T$ be the population mean and population total, and let $\mu_i$, $T_i$, $\sigma_i^2$ be the respective population mean, total, and variance of the $i^{th}$ stratum. Then we have

$$T = \sum_{i=1}^{K} T_i = \sum_{i=1}^{K} N_i \mu_i \qquad (4.18)$$

$$\mu = T/N = \frac{1}{N} \sum_{i=1}^{K} N_i \mu_i \qquad (4.19)$$

Let $\bar{y}_i$ be the sample mean of the simple random sample from the $i^{th}$ stratum. Then, from the previous section, we know that $\bar{y}_i$ is an unbiased estimator of the mean $\mu_i$ of the $i^{th}$ stratum. An unbiased estimator of the population mean $\mu$, denoted as $\bar{y}_{st}$, is given by

$$\hat{\mu} = \bar{y}_{st} = \frac{1}{N} \sum_{i=1}^{K} N_i \bar{y}_i \qquad (4.20)$$

where

$$\bar{y}_i = \frac{1}{n_i} \sum_{j=1}^{n_i} y_{ij} \qquad (4.21)$$

An estimator of the population total $T$ is given by

$$\hat{T} = \sum_{i=1}^{K} \hat{T}_i = \sum_{i=1}^{K} N_i \hat{\mu}_i = \sum_{i=1}^{K} N_i \bar{y}_i = N \times \bar{y}_{st} \qquad (4.22)$$

Recall that from Eq. (4.5), we have

$$V(\bar{y}_i) = \frac{\sigma_i^2}{n_i} \left( \frac{N_i - n_i}{N_i - 1} \right) \qquad (4.23)$$

Thus, from Eqs. (4.20) and (4.23), it follows that the variance of the estimator $\bar{y}_{st}$ of the population mean, $\mu$, is given by

$$V(\bar{y}_{st}) = \frac{1}{N^2} \sum_{i=1}^{K} N_i^2 V(\bar{y}_i) = \frac{1}{N^2} \sum_{i=1}^{K} N_i^2 \frac{\sigma_i^2}{n_i} \left( \frac{N_i - n_i}{N_i - 1} \right) \qquad (4.24)$$

For large $N_i$, we have

$$V(\bar{y}_{st}) \cong \frac{1}{N^2} \sum_{i=1}^{K} N_i^2 \frac{\sigma_i^2}{n_i} \left( 1 - \frac{n_i}{N_i} \right) \qquad (4.24a)$$

An estimator of the variance in Eq. (4.24a) is given by

$$\hat{V}(\bar{y}_{st}) = \frac{1}{N^2} \sum_{i=1}^{K} N_i^2 \frac{s_i^2}{n_i} \left(1 - \frac{n_i}{N_i}\right) \tag{4.25}$$

From Eqs. (4.22) and (4.25), it follows that the variance of the estimator of the population total $T$ is given by

$$V(\hat{T}) = N^2 V(\bar{y}_{st}) = \sum_{i=1}^{K} N_i^2 \frac{\sigma_i^2}{n_i} \left(\frac{N_i - n_i}{N_i - 1}\right) \tag{4.26}$$

An estimator of the variance in Eq. (4.26) is given by

$$\hat{V}(\hat{T}) = N^2 \hat{V}(\bar{y}_{st})$$
$$= \sum_{i=1}^{K} N_i^2 \frac{s_i^2}{n_i} \left(1 - \frac{n_i}{N_i}\right) \tag{4.27}$$

If either the population is normal or the sample size is large ($n \geq 30$), then the margin of error of estimation of the population mean with probability $(1 - \alpha)$ is given by

$$E = \pm z_{\alpha/2} \sqrt{V(\bar{y}_{st})} = \pm z_{\alpha/2} \frac{1}{N} \sqrt{\sum_{i=1}^{K} N_i^2 \frac{\sigma_i^2}{n_i} \left(\frac{N_i - n_i}{N_i - 1}\right)} \tag{4.28}$$

If the sample sizes relative to the stratum sizes are small, then we have

$$E = \pm z_{\alpha/2} \sqrt{V(\bar{y}_{st})} = \pm z_{\alpha/2} \frac{1}{N} \sqrt{\sum_{i=1}^{K} N_i^2 \frac{\sigma_i^2}{n_i}} \tag{4.28a}$$

When the strata variances are unknown, which is usually the case, the stratum variance $\sigma_i^2$ in Eq. (4.28) is replaced by its estimator, $s_i^2$. Then the margin of error of estimation of the population mean with probability $(1 - \alpha)$ is given by

$$E = \pm z_{\alpha/2} \frac{1}{N} \sqrt{\sum_{i=1}^{K} N_i^2 \frac{s_i^2}{n_i} \left(1 - \frac{n_i}{N_i}\right)} \tag{4.29}$$

If the sample sizes relative to the stratum sizes are small, then we have

$$E = \pm z_{\alpha/2} \frac{1}{N} \sqrt{\sum_{i=1}^{K} N_i^2 \frac{s_i^2}{n_i}} \tag{4.29a}$$

The margin of error of estimation of the population total with probability $(1 - \alpha)$ is given by

$$E' = \pm z_{\alpha/2} \sqrt{\sum_{i=1}^{K} N_i^2 \frac{\sigma_i^2}{n_i} \left( \frac{N_i - n_i}{N_i - 1} \right)} \qquad (4.30)$$

If the sample sizes relative to the stratum sizes are small, then we have

$$E' = \pm z_{\alpha/2} \sqrt{\sum_{i=1}^{K} N_i^2 \frac{\sigma_i^2}{n_i}} \qquad (4.30a)$$

Again, when the strata are large and variances are unknown, the stratum variance $\sigma_i^2$ in Eq. (4.30) is replaced by its estimator, $s_i^2$. Then the margin of error for the estimation of the population total with probability $(1 - \alpha)$ is given by

$$E' = \pm z_{\alpha/2} \sqrt{\sum_{i=1}^{K} N_i^2 \frac{s_i^2}{n_i} \left( 1 - \frac{n_i}{N_i} \right)} \qquad (4.31)$$

If the sample sizes relative to the stratum sizes are small, then we have

$$E' = \pm z_{\alpha/2} \sqrt{\sum_{i=1}^{K} N_i^2 \frac{s_i^2}{n_i}} \qquad (4.31a)$$

### 4.4.2 Confidence Interval for the Population Mean and Population Total

Assuming that the population is normal, we have from Eqs. (4.29) and (4.31) the margins of error of estimation for the population mean and population total with probability $(1 - \alpha)$. Thus, a confidence interval with confidence coefficient $(1 - \alpha)$ for the population mean and population total is given, respectively, by

$$\hat{\mu} \pm z_{\alpha/2} \frac{1}{N} \sqrt{\sum_{i=1}^{K} N_i^2 \frac{s_i^2}{n_i} \left( 1 - \frac{n_i}{N_i} \right)} \qquad (4.32)$$

and

$$\hat{T} \pm z_{\alpha/2} \sqrt{\sum_{i=1}^{K} N_i^2 \frac{s_i^2}{n_i} \left( 1 - \frac{n_i}{N_i} \right)} \qquad (4.33)$$

Notes: An estimate $\bar{y}$ of the population mean, obtained from a simple random sample, coincides with estimate $\bar{y}_{st}$ obtained from a stratified random sample if in every stratum, $n_i/N_i$ is equal to $n/N$.

If the sample size in every stratum relative to the stratum size is small ($n_i/N_i < 5\%$), then the factor $n_i/N_i$ in Eqs. (4.25) through (4.33) is usually ignored.

### Example 4.6   Labor Cost Data

Suppose a manufacturing company is producing parts in its facilities located in three different countries, including the United States. The labor costs, raw material costs, and other overhead expenses vary tremendously from country to country. To meet the target value, the company is interested in estimating the average cost of a part it will produce during a certain period. The number of parts expected to be produced during that period in its three facilities are $N_1 = 8,900$, $N_2 = 10,600$, and $N_3 = 15,500$. Thus the total number of parts that are expected to be produced during that period in all its facilities is $N = 8,900 + 10,600 + 15,500 = 35,000$. To achieve its goal, the company calculated the cost of 12 randomly selected parts from facility 1, 12 parts from facility 2, and 16 parts from facility 3. These efforts produced the following data (in dollars).

Sample from stratum 1:

$$6.48, 6.69, 7.11, 6.15, 7.09, 7.27, 7.58, 6.49, 6.32, 6.47, 6.63, 6.90$$

Sample from stratum 2:

$$10.06, 10.25, 11.03, 11.18, 10.29, 9.33, 10.42, 9.34, 11.06, 9.78, 10.54, 11.45$$

Sample from stratum 3:

$$24.72, 24.77, 25.64, 25.65, 26.09, 24.70, 25.05, 23.21, 24.00, 26.00, 26.21, 26.00,$$
$$24.21, 26.11, 24.63, 26.38$$

Use these data to determine the following:

a) Estimation of the population mean and population total
b) Margin of error of estimation of the population mean and population total with probability 95%
c) Confidence interval for the population mean and population total with confidence coefficient 95%

### Solution

In our example, each facility constitutes a stratum. So first, we determine the sample mean and sample standard deviations for each stratum:

$$\bar{y}_1 = \frac{1}{12}(6.48 + 6.69 + 7.11 + \ldots + 6.63 + 6.90) = 6.765$$

$$\bar{y}_2 = \frac{1}{12}(10.06 + 10.25 + 11.03 + \ldots + 10.54 + 11.45) = 10.394$$

$$\bar{y}_3 = \frac{1}{16}(24.72 + 24.77 + 25.64 + \ldots + 24.63 + 26.38) = 25.211$$

$$s_1^2 = \frac{1}{12-1}\left((6.48^2 + 6.69^2 + \dots + 6.90^2) - \frac{(6.48 + 6.69 + \dots + 6.90)^2}{12}\right)$$
$$= 0.182$$

$$s_1 = \sqrt{0.1823} = 0.427$$

$$s_2^2 = \frac{1}{12-1}\left((10.06^2 + 10.25^2 + \dots + 11.45^2) - \frac{(10.06 + 10.25 + \dots + 11.45)^2}{12}\right)$$
$$= 0.488$$

$$s_2 = \sqrt{0.4277} = 0.698$$

$$s_3^2 = \frac{1}{16-1}\left((24.72^2 + 24.77^2 + \dots + 26.38^2) - \frac{(24.72 + 24.77 + \dots + 26.38)^2}{16}\right)$$
$$= 0.874$$

$$s_3 = \sqrt{0.874} = 0.935$$

Now, using Eqs. (4.20) and (4.22), the estimates of the population mean and population total are, respectively,

$$\hat{\mu} = \bar{y}_{st} = \frac{1}{35,000}(8,900 \times 6.765 + 10,600 \times 10.394 + 15,500 \times 25.211)$$
$$= \$16.00$$

$$\hat{T} = N \times \bar{y}_{st} = 35000 \times 16 = \$560,000$$

Using Eqs. (4.29a) and (4.31a), and noting that the sample sizes relative to the strata sizes are small, the margins of error of estimation for the population mean and population total with probability 95% are, respectively,

$$\pm 1.96 \times \frac{1}{35,000}\sqrt{8,900^2\frac{0.182}{12} + 10,600^2\frac{0.488}{12} + 15,500^2\frac{0.874}{16}} = \pm \$0.243$$

and

$$\pm 1.96\sqrt{8,900^2\frac{0.182}{12} + 10,600^2\frac{0.488}{12} + 15,500^2\frac{0.874}{16}} = \pm \$8,505$$

Using Eqs. (4.32) and (4.33), the confidence intervals for population mean and population total with confidence coefficient 95% are, respectively,

$$16.00 \pm 0.243 = (\$15.757, \$16.243)$$

and

$$560,000 \pm 8,505 = (\$551495.00, \$568505.00)$$

### 4.4.3 Determining Sample Size

The *sample size* needed to estimate the *population mean* with a margin of error $E$ with probability $(1 - \alpha)$ is given by

$$n = \frac{z_{\alpha/2}^2 \sum_{i=1}^{K} N_i^2 \sigma_i^2 / w_i}{N^2 E^2 + z_{\alpha/2}^2 \sum_{i=1}^{K} N_i \sigma_i^2} \qquad (4.34)$$

where $w_i$ is the fraction of the total sample size allocated to the $i^{th}$ stratum, or $w_i = n_i/n$. In the case of proportional allocation, $w_i = N_i/N$. Also note that if the stratum population variance $\sigma_i^2$ is not known, it is replaced by the corresponding sample variance $s_i^2$, which can be found from a pilot study or historical data from a similar study.

**Example 4.7**

Continuing Example 4.6, determine an appropriate sample size if a Six Sigma Green Belt engineer at the company is interested in attaining the margin of error of estimation of \$0.10 with probability 95%. We assume that the sample allocation is proportional.

**Solution**

From Example 4.6, we have

$$w_1 = 8900/35000 = 0.254, w_2 = 10600/35000 = 0.303,$$
$$w_3 = 15500/35000 = 0.443$$

Since strata variances are unknown, plugging the values $E = 0.1$, sample variance $s_i^2, N, N_i$, and $w_i; i = 1, 2, 3$ into Eq. (4.34), we get

$$n \simeq 222$$

Thus, in Example 4.6, to achieve the margin of error of 10 cents with probability 95%, the company must take samples of sizes 57, 67, and 98 from strata 1, 2, and 3, respectively (these sizes are obtained by using the formula $n_i = n \times w_i$). This means we will need supplemental samples of sizes 45, 55, and 82 from strata 1, 2, and 3, respectively.

**Notes:**

1) The *sample size* needed to estimate the *population total* with a margin of error $E'$ with probability $(1 - \alpha)$ can be found using Eq. (4.33) by simply replacing $E$ with $(E'/N)$.
2) An optimal allocation of sample size to each stratum depends on three factors: the total number of sampling units in each stratum, variability of observations within each stratum, and cost of taking an observation from each stratum. The details on this topic are beyond the scope of this book; for more details, refer to Cochran (1977), Govindarajulu (1999), Lohr (1999), or Scheaffer et al. (2006).

## 4.5   Systematic Random Sampling

Systematic random sampling design may be the easiest method of selecting a random sample. Suppose that the elements in the sampled population are numbered 1 to $N$. To select a systematic random sample of size $n$, we select an element at random from the first $k$ ($\leq N/n$) elements, and then every $k$th element is selected until we get a sample of size $n$. For example, if $k = 20$, and if the first element selected is the 15th element, then the other elements, which will be included in the sample, are 35, 55, 75, 95, and so on. This procedure identifies a complete *systematic random sample* and is usually known as an *every $k$th systematic sample.* Note that if the population size is unknown, we cannot accurately choose $k$; we randomly select some approximate value of $k$ such that our sample size is equal to the desired value of $n$.

A major advantage of a systematic sample over a simple random sample is that a systematic sample design is easier to implement, particularly if the sampled population has some kind of natural ordering. This is particularly true when a sample is taken directly from a production or assembly line. Another advantage of systematic sampling is that workers collecting the samples do not need any special training, and sampled units are spread evenly over the population of interest.

If the population is in random order, then a systematic random sample provides results comparable to those of a simple random sample. Moreover, a systematic sample usually covers the entire sampled population uniformly, which may provide more information about the population than provided by a simple random sample. However, if the sampling units of a population are in some kind of periodic or cyclical order, the systematic sample may not be a representative sample. For instance, if every fifth part produced by a machine is defective, and if $k = 5$, then the systematic random sample will contain either all defective or all non-defective parts, depending on whether the first part selected was defective or not. In this case, a systematic sample will not be a representative sample. As another example, suppose we want to estimate a worker's productivity, and we decide to take samples of their productivity every Friday afternoon. If the worker's productivity is lower on Fridays, as is often the case, we might be underestimating their true productivity.

If there is a linear trend in the data, then a *systematic random sample* mean provides a more precise estimate of the population mean than that of a *simple random sample,* but less precise than a *stratified random sample.* However, as the underlying conditions change, the precision of different sample designs may also change. In other words, in terms of precision, we cannot conclude that any one particular sample design is better than another. For further details about precision, refer to Cochran (1977), Govindarajulu (1999), Lohr (1999), or Scheaffer et al. (2006).

### 4.5.1   Estimating the Population Mean and Population Total

If our sample is an *every $k$th systematic random sample,* then clearly there are $k$ possible systematic samples from a population of size $N$ ($k \times n = N$), which we write as

$$y_{11}, y_{12}, ..., y_{1n}; y_{21}, y_{22}, ..., y_{2n}; y_{31}, y_{32}, ..., y_{3n}; y_{k1}, y_{k2}, ..., y_{kn}$$

where $y_{ij}$ denotes the $j$th ($j = 1, 2, ..., n$) element of the $i$th ($i = 1, 2, ..., k$) systematic sample. We denote the mean of the $i$th sample as $\bar{y}_i$ and the mean of a randomly selected systematic sample as $\bar{y}_{sy}$, where the subscript $sy$ means a systematic sample was used. We use $\bar{y}_{sy}$ equal to $\bar{y}_i$, the mean of a systematic sample, that is randomly selected from the list of all systematics samples, where

$$\bar{y}_i = \frac{\sum\limits_{j=1}^{n} y_{ij}}{n}$$

Thus, we have

$$\hat{\mu} = \bar{y}_{sy} = \bar{y}_i \tag{4.35}$$

which is an unbiased estimator of the population mean $\mu$.

Note that the mean of a systematic sample is more precise than the mean of a simple random sample if and only if the variance within the systematic samples is greater than the population variance as a whole. That is,

$$S^2_{wsy} > S^2 \tag{4.36}$$

where

$$S^2_{wsy} = \frac{1}{k(n-1)} \sum_{i=1}^{k} \sum_{j=1}^{n} \left( y_{ij} - \bar{y}_i \right) \tag{4.37}$$

$$S^2 = \frac{1}{(N-1)} \sum_{i=1}^{k} \sum_{j=1}^{n} \left( y_{ij} - \bar{Y} \right) \tag{4.38}$$

and

$$\bar{Y} = \frac{1}{N} \sum_{i=1}^{k} \sum_{j=1}^{n} y_{ij} \tag{4.39}$$

The estimated variance of $\bar{y}_{sy}$ is given by

$$\hat{V}\left(\bar{y}_{sy}\right) = \frac{s^2_{sy}}{n} \left( 1 - \frac{n}{N} \right) \tag{4.40}$$

where

$$s^2_{sy} = \frac{1}{n-1} \sum_{i=1}^{n} \left( y_i - \bar{y}_{sy} \right)^2 \tag{4.41}$$

An estimator of the population total $T$ is given by

$$\hat{T} = N \times \bar{y}_{sy} \tag{4.42}$$

and an estimated variance of $\hat{T}$ is given by

$$\hat{V}(\hat{T}) = N^2 \hat{V}(\bar{y}_{sy}) = N^2 \left(\frac{s_{sy}^2}{n}\right)\left(1 - \frac{n}{N}\right) = N(N-n)\frac{s_{sy}^2}{n} \tag{4.43}$$

Sometimes a manufacturing scenario may warrant stratifying the population and then taking a systematic sample within each stratum. For example, if there are several plants and we want to collect samples on line from each of these plants, then we are first stratifying the population and then taking a systematic sample within each stratum.

The margins of error with probability $(1 - \alpha)$ for estimating the population mean and total are given by, respectively,

$$E = \pm z_{\alpha/2}\sqrt{\hat{V}(\bar{y}_{sy})} = \pm z_{\alpha/2}\sqrt{\frac{s_{sy}^2}{n}\left(1 - \frac{n}{N}\right)} \tag{4.44}$$

and

$$E' = \pm z_{\alpha/2}\sqrt{\hat{V}(\hat{T})} = \pm z_{\alpha/2}\sqrt{N(N-n)\frac{s_{sy}^2}{n}} \tag{4.45}$$

### 4.5.2 Confidence Interval for the Population Mean and Population Total

Using Eqs. (4.44) and (4.45), the confidence intervals for the population mean and total with confidence $(1 - \alpha)$ are given by, respectively,

$$\hat{\mu} \pm z_{\alpha/2}\sqrt{\frac{s_{sy}^2}{n}\left(1 - \frac{n}{N}\right)} \tag{4.46}$$

$$\hat{T} \pm z_{\alpha/2}\sqrt{N(N-n)\frac{s_{sy}^2}{n}} \tag{4.47}$$

### 4.5.3 Determining the Sample Size

The *sample size* needed to estimate the *population mean* with a margin of error $E$ with probability $(1 - \alpha)$ is given by

$$n = \frac{z_{\alpha/2}^2 N s_{sy}^2}{(N-1)E^2 + s_{sy}^2 z_{\alpha/2}^2} \tag{4.48}$$

The *sample size* needed to estimate the *population total* with a margin of error $E'$ with probability $(1 - \alpha)$ can be found from Eq. (4.48) by replacing $E$ with $(E'/N)$.

**Example 4.8   Timber Volume Data**
A pulp and paper mill plans to buy 1000 acres of timberland for wood chips. However, before closing the deal, the company is interested in determining the mean timber volume per lot of one-fourth of an acre. To achieve its goal, the company conducted an every 100th systematic random sample and obtained the data given in the following table. Estimate the mean timber volume, $\mu$, per lot and the total timber volume $T$ in 1000 acres. Determine a 95% margin of error of estimation for estimating the population mean $\mu$ and the population total $T$, and then find 95% confidence intervals for the mean, $\mu$, and the total, $T$.

| Lot sampled | 57 | 157 | 257 | 357 | 457 | 557 | 657 | 757 |
|---|---|---|---|---|---|---|---|---|
| Volume cu. feet | 850 | 935 | 780 | 1150 | 940 | 898 | 956 | 865 |
| Lot sampled | 857 | 957 | 1057 | 1157 | 1257 | 1357 | 1457 | 1557 |
| Volume cu. feet | 1180 | 1240 | 1150 | 980 | 960 | 1470 | 950 | 860 |
| Lot sampled | 1657 | 1757 | 1857 | 1957 | 2057 | 2157 | 2257 | 2357 |
| Volume cu. feet | 1080 | 1186 | 1090 | 869 | 870 | 870 | 960 | 1058 |
| Lot sampled | 2457 | 2557 | 2657 | 2757 | 2857 | 2957 | 3057 | 3157 |
| Volume cu. feet | 1080 | 980 | 960 | 880 | 1030 | 1150 | 1200 | 1070 |
| Lot sampled | 3257 | 3357 | 3457 | 3557 | 3657 | 3757 | 3857 | 3957 |
| Volume cu. feet | 980 | 950 | 780 | 930 | 650 | 700 | 910 | 1300 |

**Solution**

Form the given data, we have

$$\hat{\mu} = \bar{y}_{sy} = \frac{1}{40}(850 + 935 + 780 + \dots + 1300) = \frac{1}{40}(39697) = 992.425 \text{ cubic feet}$$

$$\hat{T} = N \times \bar{y}_{sy} = 4000 \times 992.425 = 3969700 \text{ cubic feet}$$

To find the margin of error with 95% probability for the estimation of the population mean, $\mu$, and population total, $T$, we need to determine the estimated variances of $\hat{\mu}$ and $\hat{T}$. We first need to find the sample variance:

$$s_{sy}^2 = \frac{1}{40-1}\left(\begin{array}{c}(850^2 + 935^2 + 780^2 + \dots + 1300^2)\\ -\frac{(850 + 935 + 780 + \dots + 1300)^2}{40}\end{array}\right)$$

$$= \frac{1}{39}(40446211.0 - 39396295.225) = \frac{1}{39}(1049915.775) = 26920.9173$$

Thus the estimated variances of $\hat{\mu}$ and $\hat{T}$ are given by

$$\hat{V}(\bar{y}_{sy}) = \frac{s_{sy}^2}{n}\left(1-\frac{n}{N}\right) = \frac{26920.9173}{40}\left(1-\frac{40}{4000}\right) = 666.293$$

$$\hat{V}(\hat{T}) = N^2\hat{V}(\bar{y}_{sy}) = 4000^2 \times 666.293 = 10660688000.0$$

Using Eqs. (4.44) and (4.45), the margins of error of estimation for the population mean and population total with probability 95% are, respectively,

$$\pm z_{\alpha/2}\sqrt{\hat{V}(\bar{y}_{sy})} = \pm 1.96\sqrt{666.293} = 50.59 \text{ cubic feet}$$

and

$$\pm z_{\alpha/2}\sqrt{\hat{V}(\hat{T})} = \pm 1.96\sqrt{10660688000} = 202360 \text{ cubic feet}$$

Using Eqs. (4.46) and (4.47), the confidence intervals for the population mean and population total with confidence coefficient 95% are, respectively,

$$992.425 \pm 50.59 = (941.835, 1043.015)$$

and

$$3969700 \pm 202360 = (3767340, 4172060)$$

**Example 4.9**

Continuing from Example 4.8, determine an appropriate sample size if the manager of the pulp and paper company is interested in attaining the margin of error of 25 cubic feet with probability 0.95.

**Solution**

Substituting the values of $s_{sy}^2$, $N$, $E$, and $z_{\alpha/2}$ in Eq. (4.48), we have

$$n = \frac{z_{\alpha/2}^2 N s_{sy}^2}{(N-1)E^2 + s_{sy}^2 z_{\alpha/2}^2}$$

$$= \frac{1.96^2 \times 4000 \times 26920.9173}{(4000-1) \times 25^2 + 26920.9173 \times 1.96^2} \cong 159$$

Note that in Example 4.8, the sample size was 40 and the margin of error for estimating the mean timber per lot was 50.59 cubic feet with probability 95%. Thus, to attain the margin of error of 25 cubic feet, we will have to take a sample of size of at least 159.

Note: In systematic sampling, normally it is not possible to take only a supplemental sample to achieve a sample of full size, which in this example is 159. However, it can be possible if we choose the values of $n$ and $k$ as $n_1$ and $k_1$ (say) such that $N = n_1 k_1$ and $k = r k_1$, where $r$ is an integer. Furthermore, we must keep the first randomly selected unit the same. Thus, in this example, we take a sample of size 160, $k_1 = 25$, $r = 4$, and keep first randomly selected lot as lot number 57; then our new sample will consist of lot numbers 7, 32, 57, 82, 107, 132, 157, and so on. Obviously, the original sample is a part of the new sample, which means in this case that we can take only a supplemental sample of size 120 instead of taking a full sample of size 159.

## 4.6 Cluster Random Sampling

In all the sampling designs discussed so far, we have assumed that the sampling units in a sampled population are such that the sampling frame can be prepared inexpensively.

However, this may not always be true. That is, to prepare a good frame listing is very expensive, or not all sampling units are easily accessible. Another scenario could be that the sampling units are so spread out that observing each sampling unit is expensive and/or time-consuming. In such cases, we prepare a sampling frame consisting of larger sampling units, called *clusters,* such that each cluster consists of several original sampling units (subunits) of the sampled population. Then we take a simple random sample of clusters and observe all the subunits in the selected clusters. This technique of sampling is known as a *cluster random sampling design* or simply *cluster sampling design.*

Cluster sampling is not only cost-effective but also a time-saver, since collecting data from adjoining units is cheaper, easier, and quicker than if the sampling units are spread out. In manufacturing, for example, it may be easier and less expensive to randomly select some boxes, each of which contains several parts, instead of randomly selecting individual parts. Note that although conducting cluster sampling may be cost-effective, it may be less efficient in terms of precision compared to simple random sampling when the sample sizes are the same. Also, the efficiency of cluster sampling may further decrease if we let the cluster sizes increase.

Cluster samplings may be a one-stage or two-stage procedure. In one-stage cluster sampling, we examine all the sampling subunits within the selected clusters. In two-stage cluster sampling, we examine only a portion of sampling subunits, which are selected from each selected cluster using simple random sampling. Furthermore, in cluster sampling, the cluster sizes may or may not be of the same. Normally, in field sampling, it is not feasible to have clusters of equal sizes. For example, in a sample survey of a large metropolitan area, city blocks may be considered clusters. If the sampling subunits are households or people, then it will be nearly impossible to have the same number of households or people in every block. However, in industrial sampling, we can always have clusters of equal sizes; for example, boxes containing the same number of parts may be considered clusters.

### 4.6.1 Estimating the Population Mean and Population Total

Let $N$ be the number of clusters in the sampled population, with the $i$th cluster having $m_i$ sampling subunits. We then take a simple random sample of $n$ clusters. Let $y_{ij}$ be the observed value of the characteristic of interest of the $j$th subunit in the $i$th cluster, $j = 1, 2, ..., m_i; i = 1, 2, ..., n$. Then we have the following:

$$\text{Total of all observations in the } i\text{th cluster} = y_i = \sum_{j=1}^{m_i} y_{ij}$$

$$\text{Total number of subunits in the sample} = m = \sum_{i=1}^{n} m_i$$

$$\text{Average cluster size in the sample} = \overline{m} = \frac{1}{n}\sum_{i=1}^{n} m_i = \frac{m}{n}$$

$$\text{Total number of subunits in the population} = M = \sum_{i=1}^{N} m_i$$

$$\text{Average cluster size in the population} = \overline{M} = \frac{1}{N}\sum_{i=1}^{N} m_i = \frac{M}{N}$$

Then an estimator of the population mean, $\mu$ (average value of the characteristic of interest per subunit), is given by

$$\hat{\mu}_c = \bar{y} = \frac{\sum\limits_{i=1}^{n} y_i}{m} \tag{4.49}$$

and its estimated variance is given by

$$\hat{V}(\hat{\mu}_c) = \hat{V}(\bar{y}) = \left(\frac{N-n}{Nn\overline{M}^2}\right)\left(\frac{1}{n-1}\sum_{i=1}^{n}(y_i - m_i\bar{y})^2\right) \tag{4.50}$$

If $\overline{M}$ is unknown, then it is replaced by $\overline{m}$, so the estimated variance of $\hat{\mu}_c$ is

$$\hat{V}(\hat{\mu}_c) = \hat{V}(\bar{y}) = \left(\frac{N-n}{Nn\overline{m}^2}\right)\left(\frac{1}{n-1}\sum_{i=1}^{n}(y_i - m_i\bar{y})^2\right) \tag{4.51}$$

An estimator of the population total $T$ (total value of the characteristic of interest for all subunits in the population) is given by

$$\hat{T} = M \times \bar{y} = \frac{M}{m}\sum_{i=1}^{n} y_i \tag{4.52}$$

and its estimated variance is given by

$$\hat{V}(\hat{T}) = M^2\hat{V}(\bar{y}) = N^2\overline{M}^2\hat{V}(\bar{y})$$

Substituting the value of $\hat{V}(\bar{y})$ from Eq. (4.50), we have

$$\hat{V}(\hat{T}) = N\left(\frac{N}{n} - 1\right)\left(\frac{1}{n-1}\sum_{i=1}^{n}(y_i - m_i\bar{y})^2\right) \tag{4.53}$$

The margins of error with probability $(1-\alpha)$ for estimating the population mean and population total are given, respectively, by

$$E = \pm z_{\alpha/2}\sqrt{\hat{V}(\bar{y})} = \pm z_{\alpha/2}\sqrt{\left(\frac{N-n}{Nn\overline{m}^2}\right)\left(\frac{1}{n-1}\sum_{i=1}^{n}(y_i - m_i\bar{y})^2\right)} \tag{4.54}$$

and

$$E' = \pm z_{\alpha/2}\sqrt{\hat{V}(\hat{T})} = \pm z_{\alpha/2}\sqrt{N\left(\frac{N}{n} - 1\right)\left(\frac{1}{n-1}\sum_{i=1}^{n}(y_i - m_i\bar{y})^2\right)} \qquad (4.55)$$

### 4.6.2 Confidence Interval for the Population Mean and Population Total

Using Eqs. (4.54) and (4.55), the confidence intervals for the population mean and population total with confidence $(1 - \alpha)$ are given, respectively, by

$$\bar{y} \pm z_{\alpha/2}\sqrt{\left(\frac{N-n}{Nn\bar{m}^2}\right)\left(\frac{1}{n-1}\sum_{i=1}^{n}(y_i - m_i\bar{y})^2\right)} \qquad (4.56)$$

and

$$\frac{M}{m}\sum_{i=1}^{n}y_i \pm z_{\alpha/2}\sqrt{N\left(\frac{N}{n} - 1\right)\left(\frac{1}{n-1}\sum_{i=1}^{n}(y_i - m_i\bar{y})^2\right)} \qquad (4.57)$$

**Example 4.10    Repair Cost of Hydraulic Pumps**
The quality manager of a company that manufactures hydraulic pumps is investigating the cost of warranty claims per year for one specific pump model. The pump model being investigated is typically installed in six applications that include foodservice operations (i.e. pressurized drink dispensers), dairy operations, soft-drink bottling operations, brewery operations, wastewater treatment, and light commercial sump water removal. The quality manager cannot determine the exact warranty repair cost for each pump; however, through company warranty claims data, she can determine the total repair cost and the number of pumps used by each industry for these applications. In this case, the quality manager decides to use cluster sampling, using each industry as a cluster, and she selects a random sample of $n = 15$ (clusters) from the $N = 120$ industries (clusters) to whom she provides the pumps. The data on total repair cost per industry and number of pumps owned by each industry are provided in the following table. Estimate the average repair cost per pump and the total cost incurred by the 120 industries over a period of 1 year. Determine a 95% confidence interval for the population mean and population total.

Table and solution follow.

| Sample # or cluster # | Number of pumps used in the selected industry | Total repair cost during one year of warranty period, in dollars |
|---|---|---|
| 1 | 5 | 250 |
| 2 | 7 | 338 |
| 3 | 8 | 290 |
| 4 | 3 | 160 |
| 5 | 9 | 390 |
| 6 | 11 | 460 |
| 7 | 5 | 275 |
| 8 | 8 | 358 |
| 9 | 6 | 285 |
| 10 | 4 | 215 |
| 11 | 6 | 220 |
| 12 | 10 | 310 |
| 13 | 9 | 320 |
| 14 | 7 | 275 |
| 15 | 5 | 280 |

**Solution**

To estimate the population-mean $\mu$, we proceed as follows:

$$m = \text{total number of pumps in the sample} = 5 + 7 + 8 + \dots + 5 = 103$$

$$\overline{m} = \text{average cluster size in the sample} = \frac{1}{15}(103) = 6.867$$

$$y = \text{total repair cost for the selected clusters during one year} = \$4426.00$$

Thus an estimate of the population mean, $\mu$, is

$$\hat{\mu} = \overline{y} = \frac{4426.00}{103} = \$42.97$$

Since we do not have information about the sizes of all the clusters, we need to estimate the total number of subunits in the population: that is,

$$\hat{M} = N \times \hat{\overline{M}} = N \times \overline{m} = 120 \times 6.867 = 824.04 \cong 824$$

Thus an estimate of the total number of pumps owned by all the 120 industries is 824.

Therefore, an estimator of the population total, $T$ (total repair cost for the manufacturer during one year of the warranty period in dollars), is given by

$$\hat{T} = \hat{M} \times \hat{\mu} = 824 \times 42.97 = \$35407.28$$

Now, to determine a 95% confidence interval for the population mean and population total, we first need to determine $s^2$, which is given by

$$\frac{1}{n-1}\sum_{i=1}^{n}(y_i - m_i\overline{y})^2$$

That is,

$$s^2 = \frac{1}{15-1}\left((250 - 5 \times 42.97)^2 + (338 - 7 \times 42.97)^2 + \dots + (280 - 5 \times 42.97)^2\right)$$

$$= \frac{1}{14}(38185.3464) = 2727.5247$$

Now, using Eq. (4.56), a 95% confidence interval for the population mean is

$$42.97 \pm 1.96\sqrt{\frac{120-15}{120 \times 15 \times 6.867^2}(2727.5247)} = 42.97 \pm 3.6 = (39.37, 46.57)$$

Similarly, using Eq. (4.57), a 95% confidence interval for the population total is

$$824 \times 42.97 \mp 1.96\sqrt{120\left(\frac{120}{15} - 1\right)(2727.5247)} = 35407.28 \mp 2966.74$$

$$= (32440.54, 38374.02)$$

### 4.6.3 Determining the Sample Size

The sample size needed to estimate the population mean with a margin of error $E$ with probability $(1 - \alpha)$ is given by

$$n \geq \frac{z_{\alpha/2}^2 N s^2}{N\overline{M}^2 E^2 + z_{\alpha/2}^2 s^2} \tag{4.58}$$

where

$$s^2 = \frac{1}{n-1}\sum_{i=1}^{n}(y_i - m_i\bar{y})^2$$

The sample size needed to estimate the population total with a margin of error $E'$ with probability $(1 - \alpha)$ can be found from Eq. (4.58) by replacing $E$ with $(E'/M)$.

**Example 4.11**

Continuing from Example 4.10, determine the sample size if the manufacturer wants to estimate the mean repair cost per pump with a margin of error \$5.00 with 95% probability.

**Solution**

Using Eq. (4.58), the desired sample size is

$$n \geq \frac{(1.96)^2(120)(2727.5247)}{(120)(6.867)^2(5)^2 + (1.96)^2(2727.5247)} = 8.275$$

Thus the sample size should be $n = 9$. Note that to determine the sample size, we needed $s^2$ and $\bar{m}$, both of which require a sample and thus a sample size, which is what we are trying to identify. To resolve this issue, we evaluate theses quantities either by using some similar data, if already available, or by taking a smaller sample.

## Review Practice Problems

1 Select and list all possible simple random samples of size n = 3 from a population {2, 3, 4, 5, 6}. Find the population mean, $\mu$, and population variance, $\sigma^2$. Find the sample mean, $\bar{y}$, for all samples generated; find $E(\bar{y})$ and $V(\bar{y})$; and then verify the results given in Eqs. (4.4) and (4.5).

2 Determine $s^2$ for each sample generated in Problem 1. Then, using the value of $\sigma^2$ obtained in Problem 1, verify the following theoretical result:

$$E(s^2) = \frac{N}{N-1}\sigma^2$$

Note here that $N = 5$.

3 During a given shift, a technician takes a simple random sample of 16 bolts from a production line (assume that the production line produces a very large number of bolts in each shift) and takes measurements of their lengths, in centimeters, which are as follows:

| |
|---|
| 9.32  9.90  9.72  9.66  9.27  9.07  9.61  9.93  9.03  9.60  9.76  9.43  9.39  9.59  9.61  9.76 |

Find the sample mean, sample standard deviation, and a 95% margin of error.

4 Refer to Problem 3. How large a sample should be taken if we want to estimate the population mean, $\mu$, with a margin of error 0.06 with probability 95%? Assume that the population standard deviation is known to be equal to 0.28 cm.

5 A random sample of 100 residents from a town of 6000 residents is selected, and electric meter readings are taken for a given month. The mean and standard deviation for the sample are determined to be $\bar{y} = 875$ kw and $s = 150$ kw hours. Find an estimate of the population mean, $\mu$, and the total consumption in kw hours by the town.

6 Refer to Problem 5. Find 95% and 99% confidence intervals for the population mean. Assume that the population is normal.

7 Suppose we are given a population of 10,000 parts from a manufacturing company, and they are labeled 0000, 0001, ..., 9999. Now suppose we want to select a random sample of parts from the given population. Use Table A.1 from the Appendix to determine a list of 15 parts that would constitute a random sample.

8 Using the data in Problem 3, find 95% and 99% confidence intervals for the mean bolt length. Assume that the lengths of all of the bolts (population) are normally distributed.

9 Refer to Problem 5. Suppose that the consumption of a random sample of 30 household during the peak month of one summer is as given here:

| 1491 | 1252 | 1153 | 1366 | 1141 | 1322 | 1121 | 1439 | 1332 | 1175 |
|------|------|------|------|------|------|------|------|------|------|
| 1299 | 1175 | 1278 | 1340 | 1457 | 1415 | 1149 | 1228 | 1435 | 1361 |
| 1265 | 1217 | 1243 | 1409 | 1175 | 1299 | 1273 | 1299 | 1486 | 1440 |

Find 95% confidence intervals for (a) the mean and (b) the total electric consumption by the town during that month. Assume that the monthly electric consumption by all the residents in that town is normally distributed.

**10** Suppose in Example 4.6 that the company's engineer decides to increase the examination period by almost 40% so that the number of parts expected to be produced during that period in three facilities are $N_1 = 12{,}500$, $N_2 = 15{,}000$, and $N_3 = 22{,}500$.

Thus the total number of parts that are expected to be produced during that period in all the facilities is $N = 12{,}500 + 15{,}000 + 22{,}500 = 50{,}000$. To achieve the company's goal, the engineer decides to randomly calculate the cost of 16 parts from facility 1, 16 parts from facility 2, and 25 parts from facility 3, located in the United States. These efforts produce the following data (in dollars).

Sample from stratum 1:

8.58, 8.69, 9.10, 8.15, 9.19, 9.27, 9.18, 8.69, 8.36, 8.37, 8.63, 8.90, 8.80,

8.56, 9.30, 9.55

Sample from stratum 2:

13.26, 13.45, 15.03, 15.28, 14.29, 12.33, 13.62, 12.54, 14.36, 12.78, 14.54, 14.95,

14.65, 14.90, 14.70, 14.60

Sample from stratum 3:

38.70, 38.77, 39.64, 39.70, 38.80, 37.70, 39.05, 37.21, 38.00,

39.20, 38.21, 38.00, 38.21, 39.11, 38.63, 40.38, 39.00, 38.87, 38.90, 39.05, 39.20,

40.00, 40.10, 41.00, 40.50

Use these data to estimate the population mean and population total.

**11** Use the data in Problem 10 to determine the margin of error of estimation for the population mean and population total with probability 99%. Then determine 99% confidence intervals for both the population mean and population total.

**12** Use the data in Problem 10 to determine an appropriate sample size if the company is interested in attaining a $0.08 margin of error of estimation of the population mean with probability 95%.

**13** A four-year college has an enrollment of 5000 undergraduate students. Lately it has received several complaints that it has become too expensive for the community where it is located. These complaints are affecting its PR in the community. To improve its PR, the college is considering cutting some costs. However, before taking that drastic step, the college decides to determine how much money each student spends per year for tuition, clothing, books, board, and lodging. To achieve this goal, the administration decides to take a systematic random sample of one in every 200 students, so that they can find the total money (in thousands of dollars) that each student spends per year. To select the desired sample, the college prepares an alphabetical list of all 5000 students. The first student randomly selected was number 68. The results obtained from the selected sample are as given in the following table:

| Student | Money spent | Student | Money spent | Student | Money spent | Student | Money spent | Student | Money spent |
|---|---|---|---|---|---|---|---|---|---|
| 68 | 58 | 1068 | 59 | 2068 | 59 | 3068 | 58 | 4068 | 58 |
| 268 | 56 | 1268 | 58 | 2268 | 58 | 3268 | 59 | 4268 | 56 |
| 468 | 58 | 1468 | 57 | 2468 | 60 | 3468 | 60 | 4468 | 56 |
| 668 | 55 | 1668 | 58 | 2668 | 56 | 3668 | 57 | 4668 | 57 |
| 868 | 56 | 1868 | 55 | 2868 | 55 | 3868 | 60 | 4868 | 55 |

a) Use these data to estimate the mean (average) amount of money spent per student.
b) Use these data to estimate the total amount of money spent by all of the students.

**14** Using the data in Problem 13, determine the margins of error of estimation for the population mean and population total with probability 95%. Then determine 95% confidence intervals for the population mean and population total.

**15** Use the data in Problem 13 to determine an appropriate sample size if the college administration is interested in attaining a $200 margin of error of estimation for the population mean with probability 95%.

**16** An employee of a social media company is interested in estimating the per capita income in a small rural town in the state of Iowa. The town does not have any systematic listing of all residents, but it is divided into small rectangular blocks, which are marked with numbers from 1 to 225. Since no listing of all the residents other than the number of blocks is available, cluster sampling seemed to be the best scenario, using the blocks as clusters. Each block has only a small number of residents. Cluster sampling is conducted by selecting a random sample of 20 blocks (clusters) and interviewing every resident in the sampled clusters. The data collected is shown in the following table:

| Sample or cluster # | Number of residents, $n_i$ | Total income per cluster, $x_i$ | Sample or cluster # | Number of residents, $n_i$ | Total income per cluster, $x_i$ |
|---|---|---|---|---|---|
| 1 | 8 | 248 000 | 11 | 8 | 242 000 |
| 2 | 12 | 396 000 | 12 | 6 | 175 000 |
| 3 | 9 | 286 000 | 13 | 9 | 268 000 |
| 4 | 11 | 315 000 | 14 | 10 | 297 000 |
| 5 | 10 | 270 000 | 15 | 7 | 204 000 |
| 6 | 6 | 168 000 | 16 | 6 | 187 000 |
| 7 | 8 | 262 000 | 17 | 8 | 260.000 |
| 8 | 10 | 255 000 | 18 | 5 | 145 000 |
| 9 | 9 | 275 000 | 19 | 9 | 292 000 |
| 10 | 7 | 208 000 | 20 | 7 | 216 000 |

a) Use these data to estimate per capita income.
b) Use these data to estimate the total income of all the residents in the town.

**17** Using the data in Problem 16, determine 99% confidence intervals for the population mean and population total.

**18** Use the data in Problem 16 to determine an appropriate sample size if the employee is interested in attaining a margin of error of $500 with probability 95%.

**19** Use the data in Example 4.6 to determine an appropriate sample size if the company is interested in attaining a $0.20 margin of error of estimation for the mean with probability 95%.

**20** Refer to Example 4.6. Suppose that, to cut the manufacturing cost of the parts, the company makes some changes at all facilities and then repeats the study. Assuming that the stratum sizes and sample sizes from each stratum are the same as in Example 4.6, the data obtained from the new study are as follows.
Sample from stratum 1

5.68 6.91 6.23 6.18 5.09 6.11 6.41 5.93 6.63 5.50 5.58 5.79

Sample from stratum 2

8.90 9.73 8.39 9.82 9.20 9.42 8.56 8.89 8.44 8.84 8.58 8.31

Sample from stratum 3

23.34 22.91 21.84 21.54 22.71 22.32 22.96 22.57 23.81 23.56 23.39 23.39

21.36 21.34 23.41 22.50

Use these data to determine the following:
a) Estimates for the population mean and the population total
b) The margin of error of estimation of the population mean and population total with probability 95%
c) Confidence intervals for the population mean and the population total with confidence coefficient 95%

**21** Refer to Problem 20. Determine an appropriate sample size if the company is now interested in attaining a $0.15 margin of error of estimation for the mean with probability 99%.

**22** Refer to Problem 13. After cutting some annual costs, the college finds that the number of complaints has decreased significantly, so it decides to repeat the earlier study. In the new study, the first student selected from the first stratum is number 38. The data collected in the new study is shown in the following table:

| Student | Money spent | Student | Money spent | Student | Money spent | Student | Money spent | Student | Money spent |
|---|---|---|---|---|---|---|---|---|---|
| 38 | 52 | 1038 | 51 | 2038 | 52 | 3038 | 49 | 4038 | 51 |
| 238 | 51 | 1238 | 51 | 2238 | 50 | 3238 | 53 | 4238 | 48 |
| 438 | 48 | 1438 | 50 | 2438 | 53 | 3438 | 54 | 4438 | 48 |
| 638 | 48 | 1638 | 50 | 2638 | 50 | 3638 | 50 | 4638 | 50 |
| 838 | 50 | 1838 | 48 | 2838 | 47 | 3838 | 53 | 4838 | 47 |

a) Use this data to estimate the mean amount of money spent per student.
b) Use this data to estimate the total amount of money spent by all the students.

**23** Compare the results obtained in Problem 22 with those obtained in Problem 13, and comment.

**24** Use the data in Problem 22 to determine the margins of error of estimation for the population mean and population total with probability 95%. Then determine 95% confidence intervals for the population mean and population total.

**25** Use the data in Problem 22 to determine an appropriate sample size if the college administration is interested in attaining a $150 margin of error of estimation for the population mean with probability 99%.

# 5

# Phase I Quality Control Charts for Variables

## 5.1 Introduction

As we discussed in earlier chapters, the concept of quality is centuries old. However, the concept of statistical quality control (SQC) is less than a century old. SQC is merely a set of interrelated tools such as statistical process control (SPC), design of experiments (DOE), acceptance sampling, and other tools that are used to monitor and improve quality. SPC is one of the tools included in SQC and consists of several tools useful for process monitoring. The quality control chart, which is the center of our discussion in this chapter, is a tool within SPC. Walter A. Shewhart from Bell Telephone Laboratories was the first person to apply statistical methods to quality improvement when he presented the first *quality control chart* in 1924. In 1931, Shewhart wrote a book on SQC, *Economic Control of Quality of Manufactured Product*, which was published by D. Van Nostrand Co., Inc. The publication of this book set the notion of SQC in full swing. In the initial stages, the acceptance of the notion of SQC in the United States was almost negligible. However, during World War II, armed services adopted statistically designed sampling inspection schemes, and the concept of SQC started to penetrate American industry.

The growth of modern SQC in America has been very slow. At first, American manufacturers believed that quality and productivity couldn't go hand in hand. They argued that producing goods of high quality would cost more because high quality would require buying higher-grade raw materials, hiring more qualified personnel, and providing more training to workers. They also believed that producing higher-quality goods would slow productivity since better-qualified and better-trained workers would not compromise quality to meet their daily quotas. However, such arguments overlook the cost associated with defective items: the time and money required to rework or discard defective items that are not fixable. Because of this miscalculation, starting in the 1960s, American companies started losing market share. Furthermore, the industrial revolution in Japan proved that American manufacturers were wrong. A famous American statistician, Dr. W. Edward Deming, wrote the following in 1986: "Improvement of quality transfers waste of man-hours and machine-time into manufacturing good products and better services." Furthermore, Deming posited that these efforts lead to a chain reaction: "lower costs, better competitive position, and happier people on the job, jobs and more jobs." We discussed such a chain reaction in Section 1.2.3.

*Statistical Quality Control: Using Minitab, R, JMP, and Python*, First Edition. Bhisham C. Gupta.
© 2021 John Wiley & Sons, Inc. Published 2021 by John Wiley & Sons, Inc.
Companion website: www.wiley.com/go/college/gupta/SQC

## 5.2 Basic Definition of Quality and Its Benefits

Although we discussed the concept of quality in Chapter 1, here we put more emphasis on achieving quality. According to Deming, a product is of good quality if it meets the needs of a customer, and the customer is happy to have bought that product. The customer may be internal or external, an individual or a corporation. If a product meets the needs of the customer, the customer is bound to buy that product again. On the other hand, if a product does not meet the needs of a customer, then they will not buy that product, even if they are an internal customer. Consequently, that product will be deemed of bad quality, and the product is bound to eventually fail and go out of the market.

Other components of a product's quality are its reliability, required maintenance, and availability of service. In evaluating the quality of a product, its attractiveness and rate of depreciation also play an important role.

As described by Deming in one of his telecast conferences, the benefits of better quality are numerous. First and foremost, it enhances the overall image and reputation of the company by meeting the needs of customers and thus making them happy. A happy customer is bound to tell the story of a good experience with the product to friends, relatives, and neighbors. Therefore, the company gets first-hand publicity without spending a dime, which results in increased sales and profits. Higher profits then lead to higher stock prices and ultimately a higher net worth for the company. Similarly, better quality provides workers with satisfaction and pride in their workmanship, and a satisfied worker goes home happy, which makes their family happy. A happy family boosts the worker's morale, which means more dedication and greater company loyalty. Another benefit of better quality is decreased cost. This is due to the reduced need for rework, meaning less scrap, fewer raw materials used, and fewer man-hours and machine-hours wasted. All this ultimately means increased productivity, a better competitive position, and ultimately higher sales and higher market share.

On the other hand, losses due to poor quality are enormous. Poor quality not only affects sales and competitive position, but also carries with it high hidden costs that usually are not calculated or known with precision. These costs include unusable product, product sold at discounted prices, and so on. In most companies, the accounting department provides only minimal information to quantify actual losses incurred due to poor quality. This lack of awareness could lead to company managers failing to take appropriate actions to improve quality. Lastly, poor product quality brings down the morale of the workers, the consequences of which cannot be quantified but are enormous.

## 5.3 Statistical Process Control

We defined *process* in Definition 1.1 as a series of actions or operations performed in producing manufactured or non-manufactured products. A *process* may also be defined as a combination of workforce, equipment, raw material, methods, and environment that work together to produce output. The quality of the final product, however, depends upon how well the process is designed and how well it is executed. But note that no matter how perfectly a process is designed and how well it is executed, no two items produced by a process are identical. The difference between two items is called *variation*. Such variations occur due to two causes: (i) *common* causes (or *random* causes) and (ii) *special* causes (or *assignable* causes).

As mentioned earlier, the first attempt to understand and remove variation in a scientific way was made by Walter A. Shewhart. Shewhart recognized that *special* or *assignable causes* are due to *identifiable* sources, which can be systematically identified and eliminated. *Common* or *random causes*, on the other hand, are not due to identifiable sources, but rather are inherited by the process. Therefore, common causes cannot be eliminated without costly measures such as redesigning the process, replacing old machines with new machines, or renovating the whole or part of the system. Note that in our future discussions, a process is considered in statistical control when only common causes are present. Deming (1975) states, "But a state of statistical control is not a natural state for a manufacturing process. It is instead an achievement, arrived at by elimination, one by one, by determined efforts, of special causes of excessive variation."

Since process variation cannot be entirely eliminated, controlling process variation imperative. If process variation is controlled, then the process becomes predictable; otherwise, the process is unpredictable. To achieve this goal, Shewhart introduced a type of chart called *quality control charts*. Shewhart control charts are generally used in phase I implementation of SPC. In phase I, a process usually is strongly influenced by special or assignable causes and, consequently, is experiencing excessive variation or significant shifts.

Quality control charts can be divided into two major categories: (i) control charts for variables and (ii) control charts for attributes. In this chapter, we study control charts for variables.

---

**Definition 5.1** A quality characteristic that can be expressed numerically or measured on a numerical scale is called a *variable*.

---

Some examples of a variable are the length of a tie rod, tread depth of a tire, compressive strength of a concrete block, tensile strength of a wire, shearing strength of a paper, concentration of a chemical, diameter of a ball bearing, amount of liquid in 16 oz can, and so on.

In any process, there are some measurable or observable quality characteristics that define the quality of the process. As mentioned earlier in this chapter, we consider only processes that possess measurable characteristics. If all such characteristics are behaving in the desired manner, then the process is considered stable, and it will produce products of good quality – the produced products will meet the *specifications* set by the customer. The specifications set by the customer are usually expressed in terms of two limits: the *lower specification limit* (LSL) and the *upper specification limit* (USL). For example, during the repair process for a car engine's timing chain, the mechanic needs to use a 30 mm bolt to fix the crankshaft. However, the mechanic knows that a bolt of length 28–32 mm will also work correctly. The LSL and USL values are 28 mm and 32 mm, respectively.

It is important to note that the scenario of a stable process occurs only if common causes or random causes are present in the process. If any of the characteristics in a process are not behaving as desired, then the process is considered unstable: it cannot produce good-quality products, and the produced products will not meet the specifications set by the customer. In such cases, in addition to common causes, *special causes* or *assignable causes* are also present in the process.

A characteristic of a process is usually determined by two parameters: its mean and standard deviation. Thus to verify whether a characteristic is behaving in a desired manner, we need to verify that these two parameters are under control; we can do so by using quality control charts. In addition to quality control charts, several other tools are valuable in achieving process stability. The set of all these tools, including control charts, constitutes an integral part of statistical process control. SPC is very useful in any process related to the manufacturing, service, and retail industries. This set of tools consists of seven parts, often known as *the magnificent seven*:

- Histogram or stem-and-leaf diagram
- Scatter diagram
- Check sheet
- Pareto chart
- Cause-and-effect diagram (also known as a fishbone or Ishikawa diagram)
- Defect-concentration diagram
- Control charts

These SPC tools form a simple but very powerful structure for quality improvement. Once workers become thoroughly familiar with these tools, management must get involved to implement SPC for an ongoing quality-improvement process. Management must create an environment where these tools become part of day-to-day production or service processes. The implementation of SPC without management's involvement, cooperation, and support is bound to fail. Thus in addition to discussing these tools, we will also explore some of the questions that arise while implementing SPC techniques.

Every job, whether in a manufacturing company or a service company, involves a process. As described in Chapter 1, each process consists of a certain number of steps. Thus no matter how well the process is planned, designed, and executed, there is always some potential for *variability*. In some cases, this variability may be very low, while in other cases, it may be very high. If the variability is very low, then it is usually caused by common causes that inherited and not easily controllable. If the variability is too high, then in addition to common causes, there are probably *special causes* or *assignable causes* in the process as well.

The first two tools in the previous list are discussed in Chapter 3. In this chapter and the next two chapters, we discuss the remainder of these tools.

### 5.3.1 Check Sheets

To improve the quality of a product, management must try to reduce the variation of all the quality characteristics; that is, the process must be brought to a stable condition. In any SPC procedure used to stabilize a process, it becomes essential to know precisely what types of defects affect the quality of the final product. The check sheet is an important tool to achieve this goal. We discuss this tool with the help of a real-life example.

**Example 5.1   Paper Mill Data**
In a paper mill, a high percentage of paper rolls are discarded due to various types of defects. To identify these defects and their frequency, a study is launched. This study is done over four weeks. The data is collected daily and summarized in the form of Table 5.1, called the *check sheet*.

The summary data not only give the total number of different types of defects but also provide a significant source of trends and patterns of defects. These trends and patterns can help find possible causes for any particular type of defect(s). Note that the column totals in the table show the number of defects (rolls of paper rejected) occurring daily, whereas the row totals show the number of defects by defect type occurring over four weeks. It is important to remark here that this type of data becomes more meaningful if a logbook of all changes (changes in raw material, calibration of machines, training of workers or new workers hired, etc.) is well maintained.

**Table 5.1** Check sheet summarizing study data over a period of four weeks.

| Defect type | 1 | 2 | 3 | 4 | 5 | 6 | 7 | 8 | 9 | 10 | 11 | 12 | 13 | 14 | 15 | 16 | 17 | 18 | 19 | 20 | 21 | 22 | 23 | 24 | 25 | 26 | 27 | 28 | Total |
|---|---|---|---|---|---|---|---|---|---|---|---|---|---|---|---|---|---|---|---|---|---|---|---|---|---|---|---|---|---|
| **Corrugation** | 1 |  | 2 |  | 1 |  | 3 |  | 1 | 1 | 2 | 2 |  | 1 |  | 2 |  | 1 |  | 1 | 3 | 2 |  | 1 | 2 | 1 | 1 | 1 | 28 |
| **Streaks** |  |  | 1 |  |  |  | 1 |  |  | 1 |  |  |  |  | 1 |  | 1 |  | 2 |  |  |  | 1 |  |  | 1 | 1 |  | 10 |
| **Pinholes** | 1 |  |  | 1 | 1 | 1 |  |  |  |  | 1 |  |  | 1 |  | 1 |  |  |  |  | 1 |  |  | 1 |  |  |  |  | 9 |
| **Dirt** |  | 1 |  |  |  |  |  | 1 |  | 1 |  |  | 1 |  |  |  |  |  | 1 |  |  |  |  |  |  | 2 |  |  | 7 |
| **Blistering** |  |  | 2 |  |  |  | 2 |  |  |  |  | 1 |  |  |  | 2 |  |  |  |  | 1 |  |  | 2 |  |  |  |  | 11 |
| **Others** | 1 |  |  |  | 1 |  |  |  |  |  |  |  |  |  | 1 |  |  | 1 |  |  |  |  | 1 |  |  |  |  |  | 5 |
| **Total** | 3 | 1 | 5 | 1 | 3 | 1 | 6 | 1 | 1 | 3 | 3 | 3 | 1 | 2 | 2 | 5 | 1 | 2 | 3 | 1 | 5 | 2 | 2 | 4 | 2 | 4 | 2 | 1 | 70 |

### 5.3.2 Pareto Chart

The *Pareto chart* is a very useful tool used whenever we want to quickly and visually learn more about attribute data. A Pareto chart, named after its inventor, Vilfredo Pareto, an Italian economist who died in 1923, is simply a bar chart of attribute data (say, defects) in descending order of frequency. For example, consider the data on defective rolls in Example 5.1. These data is plotted in Figure 5.1 with frequency totals (row totals) of occurrence of each defect starting from highest to lowest frequency versus the defect types.

The chart allows the user to easily identify which defects occur more frequently or less frequently. Thus the user can prioritize the use of resources to eliminate those defects that most affect the quality of the product. For instance, the Pareto diagram in Figure 5.1 indicates that 40% of paper-roll rejections are due to a corrugation problem; corrugation and blistering together are responsible for 55.7% of the rejected paper; and corrugation, blistering, and streaks are responsible for 70%. Thus to reduce overall rejections, the company should first attempt to eliminate or reduce defects due to corrugation, then blistering, then streaks, and so on. By eliminating these three types of defects, the company could change the percentage of rejected paper dramatically and reduce losses.

It is important to note that if we can eliminate more than one defect simultaneously, then we should consider eliminating them even though some of them are occurring less frequently. Furthermore, after one or more defects are either eliminated or reduced, we should collect the data again and reconstruct the Pareto chart to determine whether the priority has changed. Note that in the example, the others category may include defects such as porosity, grainy edges, wrinkles, or brightness, none of which occur very frequently. Thus if we have very limited resources, we should not use our resources on this category until other defects are eliminated.

Sometimes, defects are not equally important. This is particularly true when some defects are life-threatening, while others are merely a nuisance or inconvenience. It is quite common to allocate weights to each defect and then plot the weighted frequencies versus defects to construct a Pareto chart. For example, suppose a product has five types of defects, denoted by A, B, C, D, and E: A is life-threatening, B is not life-threatening but very serious, C is serious, D is somewhat serious, and E is a nuisance. Suppose we assign a weight of 100 to A, 75 to B, 50 to C, 20 to D, and 5 to E. The data collected over a period of study are as shown in Table 5.2.

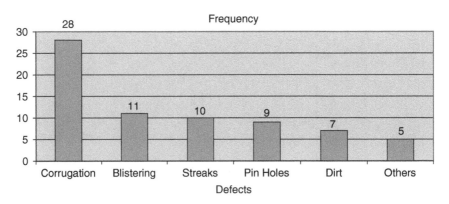

**Figure 5.1** Pareto chart for the data in Example 5.1.

**Table 5.2** Frequencies and weighted frequencies when different types of defects are not equally important.

| Defect Type | Frequency | Weighted Frequencies |
|---|---|---|
| A | 5 | 500 |
| B | 6 | 450 |
| C | 15 | 750 |
| D | 12 | 240 |
| E | 25 | 125 |

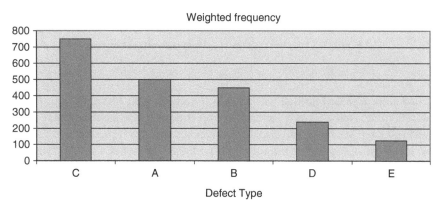

**Figure 5.2** Pareto chart when weighted frequencies are used.

In Figure 5.2, note that the Pareto Chart using weighted frequencies presents a completely different picture than if we had used the actual frequencies. That is, by using weighted frequencies, the order of priority of removing the defects is C, A, B, D, E, whereas without using the weighted frequencies, this order would have been E, C, D, B, A.

### 5.3.3 Cause-and-Effect (Fishbone or Ishikawa) Diagrams

In an SPC-implementing scheme, identifying and isolating the causes of any particular problem(s) is vital. A very effective tool for this is the *cause-and-effect* diagram. Because of its shape, this diagram is also known as a *fishbone* diagram or, sometimes, an *Ishikawa* diagram (a name referencing its inventor). Japanese manufacturers have used this diagram widely to improve the quality of their products. To prepare a cause-and-effect diagram, it is quite common to use a technique commonly known as *brainstorming*: a form of creative collaborative thinking. This technique, therefore, is used in a team setting. The team usually includes personnel from the production, inspection, purchasing, and design departments, along with management, and may additionally include any other members associated with the product. Some of the rules used in a brainstorming session are the following:

1) Each team member makes a list of ideas.
2) The team members sit around a table and take turns reading one idea at a time.
3) As the ideas are read, a facilitator displays them on a board so that all team members can see them.

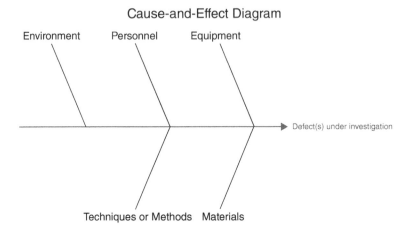

**Figure 5.3** An initial form of a cause-and-effect diagram.

4) Steps 2 and 3 are repeated until all of the ideas have been read and displayed.
5) Cross-questioning is allowed only for clarification concerning a team member's idea.
6) When all ideas have been read and displayed, the facilitator asks each team member if they have any new ideas. This procedure continues until no team member can think of new ideas.

Once all the ideas are presented by using the brainstorming technique, the next step is to analyze them. The cause-and-effect diagram is a graphical technique to do so. The initial structure of a cause-and-effect diagram is as shown in Figure 5.3.

The five spines in Figure 5.3 indicate the five significant factors or categories that could be the cause(s) of defect(s). In most workplaces, whether manufacturing or non-manufacturing, the causes of problems usually fall into one or more of these categories.

Using a brainstorming session, the team brings up all possible causes in each category. For example, in the environment category, a cause could be management's unwillingness to release funds for

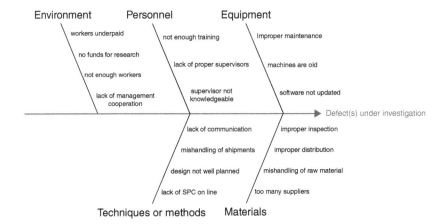

**Figure 5.4** A complete cause-and-effect diagram.

research, change suppliers, or cooperate internally. In the personnel category, a cause could be lack of proper training for workers, supervisors who do not help solve problems, lack of communication between workers and supervisors, workers who are afraid to ask their supervisors questions, supervisors who lack knowledge of the problem, or workers who are hesitant to report problems.

Once all possible causes in each significant category are listed in the cause-and-effect diagram, the next step is to isolate and eliminate one or more common causes. In the previous case, a complete cause-and-effect diagram may appear as shown in Figure 5.4.

### 5.3.4 Defect-Concentration Diagrams

A *defect-concentration diagram* is a visual representation of the product under study that depicts all defects. This diagram helps workers to determine whether there are any patterns or particular locations where defects occur and what kinds of defects are occurring (minor or major). The patterns or locations may help workers to find the specific cause(s) for such defects. The diagram must show the product from different angles. For example, if the product is shaped like a rectangular prism and defects are found on the surface, then the diagram should show all six faces, clearly indicating the location of the defects. In Figure 5.5, the two diagonally opposite edges are damaged, which clearly could have happened during transport or while moving the item from the production area to the storage area.

A defect concentration diagram proved to be of great use when my daughter made a claim with a transportation company. In 2001, I shipped a car from Boston, Massachusetts to my daughter in San Jose, California. After receiving the car, she found that the front bumper's paint was damaged. She filed a claim with the transportation company for the damage, but the company turned it down, simply stating that the damage was not caused by the company. Fortunately, a couple of days later, she found similar damage symmetrically opposite under the back bumper. She again called the company and explained that this damage had clearly been caused by the belts that were used to hold the car during transport. This time the company could not turn down her claim since she could prove scientifically, using a defect-concentration diagram, that the transportation company caused the damage.

In addition to the magnificent seven tools, we will study another tool called a *run chart*. Run charts are useful in studying specific patterns and trends that may occur during any production process.

### 5.3.5 Run Charts

In any SPC procedure, it is essential to detect trends that may be present in the data set. Run charts help to identify such trends by plotting data over a specified period. For example, if the proportion of nonconforming parts produced from shift to shift is perceived to be a problem, we may plot the

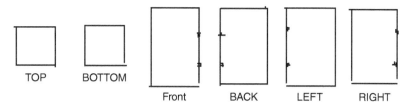

**Figure 5.5** A damaged item shaped like a rectangular prism.

**Table 5.3**  Percentage of nonconforming units in different shifts over a period of 30 shifts or 10 days.

| Shift number | 1 | 2 | 3 | 4 | 5 | 6 | 7 | 8 | 9 | 10 |
|---|---|---|---|---|---|---|---|---|---|---|
| % Nonconforming | 5 | 9 | 12 | 7 | 12 | 4 | 11 | 7 | 15 | 3 |
| Shift number | 11 | 12 | 13 | 14 | 15 | 16 | 17 | 18 | 19 | 20 |
| % Nonconforming | 5 | 8 | 2 | 5 | 15 | 4 | 6 | 15 | 5 | 5 |
| Shift number | 21 | 22 | 23 | 24 | 45 | 46 | 47 | 28 | 29 | 30 |
| % Nonconforming | 8 | 6 | 10 | 15 | 7 | 10 | 13 | 4 | 8 | 14 |

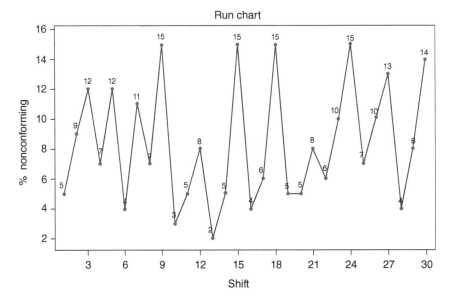

**Figure 5.6**  Run chart.

number of nonconforming parts against the shifts for a certain period to determine whether there are any trends. Trends usually help us to identify the causes of nonconformance. This type of chart is particularly useful when data are collected from a production process over a specified period.

A run chart for the data in Table 5.3 is shown in Figure 5.6. Here we present the percentage of nonconforming in different shifts over 10 days starting with the morning shift on day 1.

From this run chart, we can easily see that the percentage of nonconforming units is least for morning shifts and highest for night shifts. There are some problems during evening shifts, too, but they are not as severe as those at night. Since such trends or patterns are usually caused by special or assignable causes, the run chart will prompt management to explore how the various shifts differ. Does the quality of the raw material differ from shift to shift? Is there inadequate training of workers in the later shifts? Are evening and late shift workers more susceptible to fatigue? Do environmental problems increase in severity as the day wears on?

Deming (1982, p. 312) points out that sometimes the frequency distribution of a set of data does not give an accurate picture of the data, whereas a run chart can bring out the real problems in the data. The frequency distribution gives us the overall picture of the data but does not show us any trends or patterns that may be present in the process on a short-term basis.

*Control charts* are perhaps the most essential part of SPC. We first study some basic concepts of control charts in the following section. Then we will dedicate the rest of this chapter and the next two chapters exclusively to the study of various kinds of control charts.

## 5.4    Control Charts for Variables

A control chart is a simple but powerful statistical device used to keep a process predictable. As noted by Duncan (1986), a control chart is a device for (i) describing in concrete terms what a state of statistical control is, (ii) attaining control, and (iii) judging whether control has been attained. Control charts let us avoid unnecessary process adjustments since they can distinguish between background noise variation and real, abnormal variations. Process adjustments made due to background noise variation can be very expensive, as they can unnecessarily interrupt the production process, introduce more variation, or cause other similar problems.

Before control charts are implemented in any process, it is important to understand the process and to have a concrete plan for future actions required to improve the process.

### 5.4.1    Process Evaluation

Process evaluation includes the study of not only the final product but also all the *intermediate steps or outputs* that describe the actual operating state of the process. For example, in the paper production process, wood chips and pulp may be considered intermediate outputs. If the data on process evaluation is collected, analyzed, and interpreted correctly, it can show where and when corrective action is necessary to make the entire process work more efficiently. Process evaluation also includes the study of the cost of implementing control charts. This includes the cost of sampling, producing defectives, interrupting the production process, and taking corrective actions to fix any out-of-control warning.

### 5.4.2    Action on the Process

Action on the process is essential for any process since it prevents the production of an out-of-specification product. An *action on the process* may include changes in raw material, operator training, change of some equipment, change of design, or other measures. Effects of such actions on a process should be monitored closely, and further action should be taken if necessary.

### 5.4.3    Action on the Output

If the process evaluation does not indicate that any action on the process is necessary, the last action on the process is to ship the final product to its destination. Note that some people believe action on the output consists of sampling plans and discarding out-of-specification product that has already been produced. Obviously, such an action on the output is futile and expensive. We are interested in correcting the output before it is produced. This goal can be achieved through the use of control charts.

### 5.4.4    Variation

No process can produce two products that are exactly alike or possess precisely the same characteristics. Any process is bound to contain some sources of variation. A difference between two products may be very large, moderate, very small, or virtually undetectable, but there is always some difference. For example, the moisture content in any two rolls of paper, opacity in any two spools of paper, or brightness of two lots of pulp will always vary. We aim to trace the sources of such variation as far back as possible to eliminate the source. The first step is to separate the common causes and the special causes of such sources of variation.

### 5.4.4.1 Common Causes (Random Causes)

Common causes or random causes refer to the various sources of variation within a process when it is in statistical control. They behave like a constant system of chance causes. While individual measured values may all be different, as a group, they tend to form a pattern that can be explained by a statistical distribution that can be characterized as follows:

- Location parameter
- Dispersion parameter
- Shape (the pattern of variation – symmetrical, right-skewed, left-skewed, etc.)

### 5.4.4.2 Special Causes (Assignable Causes)

Special causes or assignable causes refer to any source of variation that cannot be adequately explained by any statistical distribution of the process output, as would be the case if the process were in statistical control (Deming 1986). Unless all special causes of variation are identified and corrected, they will continue to affect the process output unpredictably.

Any process with assignable causes is considered unstable and hence not in statistical control. However, any process free of assignable causes is considered stable and therefore in statistical control. Assignable causes or special causes can be corrected by local actions, while common causes or chance causes can be corrected only by action on the system.

### 5.4.4.3 Local Actions and Actions on the System

*Local actions:*

- Are usually required to eliminate special causes of variation
- Can usually be performed by people close to the process
- Can correct about 15% of process problems

Deming (1982) believed that as much as 6% of all system variation is due to special or assignable causes, while no more than 94% is due to common causes.

*Actions on the system:*

- Are usually required to reduce variation due to common causes
- Almost always required management action for correction
- Are needed to correct about 85% of process problems

### 5.4.4.4 Relationship Between Two Types of Variation

Deming (1951) pointed out that there is an important relationship between the two types of variations and the two types of actions needed to reduce such variations. We now discuss this point in a bit more detail.

*Special causes* of variation can be detected by simple statistical techniques. These causes of variation are not the same in all operations involved. Detecting and removing special causes of variation is usually the responsibility of someone who is directly connected with the operation, although management is sometimes in a better position to make corrections. The resolution of a special cause of variation usually requires *local action.*

Simple statistical techniques can indicate the extent of *common causes* of variation, but the causes themselves need more exploration to isolate them. It is usually the responsibility of management to correct the common causes of variation, although other personnel directly involved with the operation are sometimes in a better position to identify such causes and pass them on to management for appropriate action. Overall, the resolution of common causes of variation usually requires action on the system.

As we noted earlier, about 15% (or, according to Deming, 6%) of industrial process troubles are correctable by local action taken by people directly involved with the operation, while 85% are correctable only by management's action on the system. Confusion regarding the type of action required is very costly to the organization in terms of wasted effort, delayed resolution of trouble, and other aggravating problems. It would be erroneous, for example, to take local action (e.g. change an operator or calibrate a machine) when, in fact, management action on the system was required (e.g. selecting a supplier that can provide better and more consistent raw material).

This reasoning shows that robust statistical analysis of any operation in any industrial production is necessary. Control charts are perhaps the best tool to separate the special causes from the common causes. The next section gives a general description of control charts, after which we discuss specific types of control charts.

### 5.4.5 Control Charts

As mentioned earlier, a control chart is

- A device for describing in concrete terms what a state of statistical control is
- A device for judging whether control has been attained and thus detecting whether assignable causes are present
- A device for attaining a stable process

Suppose that we take a sample of size $n$ from a process at regular intervals. For each sample, we compute a sample statistic, $X$. This statistic may be the sample mean, a fraction of nonconforming product, or any other appropriate measure. Now, since $X$ is a statistic, it is subject to fluctuations or variations. If no special causes are present, the variation in $X$ will have characteristics that can be described by some statistical distribution. By taking enough samples, we can estimate the desired characteristics of such a distribution. For instance, suppose that statistic $X$ is distributed normally, and we divide the vertical scale of a graph into units of $X$ and horizontal scale into units of time or any other characteristic. Then we may draw horizontal lines through the mean and the extreme values of $X$, called the *center line (CL)*, the *upper control limit (UCL),* and the *lower control limit (LCL),* which results in a *control chart*, as shown in Figure 5.7.

**Figure 5.7** A pictorial representation of a Shewhart control chart with UCL and LCL.

Suppose that we have a process that has a known mean $\mu$ and known standard deviation $\sigma$. Then a Shewhart control chart with three-sigma control limits is such that CL $= \mu$, LCL $= \mu - 3\sigma$, and UCL $= \mu + 3\sigma$.

The main goal in using the control charts is to identify abnormal variations in a process so that we may bring the process target value to the desired level. In other words, the goal of using a control chart is to bring the process into a state of statistical control.

If we plot data pertaining to a process on a control chart, and if the data conform to a pattern of random variation that falls within the upper and the lower limits, then we say that the process is *in statistical control*. If, however, the data fall outside these control limits and/or do not conform to a pattern of random variation, then the process is considered to be *out of control*. In the latter case, an investigation is launched to track down the special causes responsible for the process being out of control, and ultimately action is taken to correct them.

If any particular cause of variation is on the unfavorable side, an effort is made to eliminate it. If, however, the cause of variation is on the favorable side, an effort is made to perpetuate it, instead. In this manner, the process can eventually be brought into and kept in a state of statistical control.

Dr. Shewhart, the inventor of control charts, recommended very strongly that a process should not be judged to be in control unless the pattern of random variation has persisted for some time and for a sizeable volume of output.

> Note: If a control chart shows a process in statistical control, this does not mean all special causes have been eliminated completely. Rather, it means that for all practical purposes, it is reasonable to assume or adopt a hypothesis of common causes only.

### 5.4.5.1 Preparation for Using Control Charts

In order for control charts to serve their intended purpose, it is important to take a few preparatory steps prior to implementing them:

1) Establish an environment suitable for action. Any statistical method will fail unless management has prepared a responsive environment.
2) Define the process, and determine the characteristics to be studied. The process must be understood in terms of its relationship to other operations/users, as well as in terms of the process elements (e.g. people, equipment, materials, methods, and environment) that affect it at each stage. Some of the techniques discussed earlier, such as Pareto analysis and fishbone charts, can help make these relationships visible. Once the process is well understood, then the next step is to determine which characteristics are affecting the process, which characteristics should be studied in depth, and which characteristics should be controlled.
3) Determine correlation between characteristics. For an efficient and effective study, take advantage of the relationship between characteristics. If several characteristics of an item tend to vary together, it may be sufficient to chart only one of them. If some characteristics are negatively correlated, a deeper study is required before any corrective action can be taken.
4) Define the measurement system. Characteristics must be operationally defined so that the findings can be communicated to all concerned in ways that have the same meaning today as they did yesterday. This includes specifying what information is to be gathered, as well as where, how, and under what conditions. The operational definition is very important for collecting data since it can impact the control charts in many ways. Moreover, data analysis depends on how the data are collected. Thus it is extremely important that the data contain pertinent information

and be valid (e.g. appropriate sampling schemes are used) so that the data analysis and interpretation of the results are done appropriately. Always keep in mind that each measurement system has its own inherent variability. Thus the accuracy of any measurement system is as important as eliminating the special causes affecting the process.

5) Minimize unnecessary variation. Unnecessary external causes of variation should be reduced before the study begins. This includes over-controlling the process or avoiding obvious problems that could and should be corrected even without the use of control charts. Note that in all cases, a process log should be kept. It should include all relevant events (big or small), such as procedural changes, new raw materials, or change of operators. This will aid in subsequent problem analysis.

6) Consider the customer's needs. This includes any subsequent processes that use the product or service as an input. For example, a computer manufacturing company is a customer of the semiconductor industry, a car manufacturing company is a customer of tire manufacturing companies, and a paper-making unit in a paper mill is a customer of the pulp-making unit. The most important quality characteristic is to deliver a product that meets the needs of the customer. A company may produce excellent products, but if it cannot meet the needs of the customer, then the customer will not buy that product. For instance, if the pulp-making mill cannot deliver the kind of pulp the paper-making mill needs, the paper-making mill will seek another supplier.

### 5.4.5.2 Benefits of Control Charts

Properly used, control charts can

- Be used by operators for ongoing control of a process
- Help the process perform consistently and predictably
- Allow the process to achieve higher quality, higher effective capacity (since there will be no rejections or fewer rejections) and hence lower cost per unit
- Provide a common language for discussing process performance
- Help distinguish special causes from common causes for variation and hence serve as a guide for management to take local action or action on the system

### 5.4.5.3 Rational Samples for control Charts

It is very important to note that the samples used to prepare a control chart should represent subgroups of output that are as homogeneous as possible. In other words, the subgroups should be such that if special causes are present, they should appear between subgroups rather than within a subgroup. If there is any process shift, it should occur when sample is not taken. If such a shift occurs when a sample is taken, the sample average can hide the effect of the shift. One way to avoid this unfortunate scenario is to take small samples. Samples of size four or five are considered appropriate. Sometimes it is possible to take large samples of size 20 or 25 if the rate of the production process is high – say, 30,000–40,000 units per hour. Another appropriate precaution is to ensure a natural subgroup or sample. For example, for a sample taken from the output of a single shift, it is not appropriate to select a sample from an arbitrary period of time, especially if it overlaps two or more shifts. If a sample does come from two or more shifts, then any difference between the shifts will be "averaged out"; consequently, the plotted point won't indicate the presence of any special cause due to shifts. As another example, if a process uses six machines, it is better to take a separate sample from each machine rather than choose samples consisting of items from all six machines. If the difference between machines, for example, is a special cause of variation, then samples that include items from all six machines will obscure the effect of the special cause. Thus careful selection of a subgroup or sample is of paramount importance when setting up a control chart.

Note that the process of selecting samples consists of determining both the sample size and how often samples should be taken. Factors that are usually taken into consideration for determining sample size and frequency are the *average run length* (ARL) and the *operating characteristic curve* (OC curve).

#### 5.4.5.3.1  Average Run Length

> **Definition 5.2**  A *run* is a number of successive points plotted on a control chart.

> **Definition 5.3**  The *average run length* (ARL) is the average number of points plotted before a point falls outside the control limits, indicating the process is out-of-control.

In Shewhart control charts, the ARL can be determined by using the formula

$$ARL = \frac{1}{p} \tag{5.1}$$

where $p$ is the probability of any point falling outside the control limits. ARL is commonly used as a benchmark to check the performance of a control chart.

As an illustration, consider a process quality characteristic that is normally distributed. Then for an $\overline{X}$ control chart with three-sigma control limits, the probability that a point will fall outside the control limits when the process is stable is $p = 0.0027$. Note that $p = 0.0027$ is the probability that a normal random variable deviates from the mean, $\mu$, by at least $3\sigma$. Thus the ARL for the $\overline{X}$ control chart when the process is stable or in statistical control is

$$ARL_0 = \frac{1}{0.0027} = 370$$

So, when the process is stable, we should expect that, on average, an out-of-control signal or false alarm will occur once in every 370 samples. The ARL can also be used to determine how often a false alarm will occur by simply multiplying $ARL_0$ by the time $t$ between samples. For example, if samples are taken every 30 minutes, a false alarm will occur once every 185 hours, on average. On the other hand, the ARL can be used in the same manner to find how long it will be before a given shift in the process mean is detected. We illustrate this concept with the following example.

#### Example 5.2
Suppose a process quality characteristic that is normally distributed is plotted on a Shewhart $\overline{X}$ control chart with three-sigma control limits. Suppose that the process mean, $\mu_0$, experiences an upward shift of $1.5\sigma$. Determine how long, on average, it will take to detect this shift if samples of size four are taken every hour.

## Solution

Since the process mean has experienced an upward shift of $1.5\sigma$, the new process mean is $\mu_0 + 1.5\sigma$. Furthermore, since the sample size is 4, the UCL, in this case, is also $\mu_0 + 3\sigma_{\bar{x}} = \mu_0 + 3\frac{\sigma}{\sqrt{4}} = \mu_0 + 1.5\sigma$. In other words, the CL of the control chart coincides with the UCL. Thus the probability $p$ that a point will fall beyond the control limits is

$$p = P\left(\bar{X} \leq (\mu_0 - 1.5\sigma)\right) + P\left(\bar{X} \geq (\mu_0 + 1.5\sigma)\right)$$

$$= P\left(\frac{\bar{X} - (\mu_0 + 1.5\sigma)}{\sigma/2} \leq \frac{(\mu_0 - 1.5\sigma) - (\mu_0 + 1.5\sigma)}{\sigma/2}\right)$$

$$+ P\left(\frac{\bar{X} - (\mu_0 + 1.5\sigma)}{\sigma/2} \geq \frac{(\mu_0 + 1.5\sigma) - (\mu_0 + 1.5\sigma)}{\sigma/2}\right)$$

$$= P(z \leq -6) + P(z \geq 0)$$

$$= 0.00000 + 0.5 = 0.5$$

Thus the ARL in this case is given by

$$ARL = \frac{1}{0.5} = 2$$

This implies that it will take, on average, two hours to detect a shift of $1.5\sigma$ in the process mean.

In practice, the decision of how large samples should be and how frequently they should be taken is based on the cost of taking samples and how quickly we would like to detect the shift. Large samples taken more frequently gives better protection against shifts since it takes less time to detect any given shift. Thus, for instance, in Example 5.2, if the samples are taken every half hour instead of every hour, it will take only 1 hour (instead of 2 hours) to detect a shift of $1.5\sigma$. Also, it can easily be shown that if larger samples are taken, the shifts can be detected more quickly because as the sample size increases, the control limits become narrower. This means if large samples are taken more frequently, shifts in the process mean are detected more quickly, and the process will yield fewer nonconforming units. Therefore, when calculating the cost of taking samples, we must take into account how much money will be saved by detecting the shift more quickly and consequently producing fewer nonconforming units. On the other hand, by taking large samples, the shift may occur while sampling, and the sample mean can obscure the effect of the shift. This is particularly true when the shift occurs while we are collecting the last units of the sample. So, when taking samples, we should strike a balance. Taking large samples ($\geq 20$) is generally safe if the production rate is very high (several thousand units per hour, say) so that large samples can be taken very quickly and consequently avoid shifts occurring during sampling.

### 5.4.5.3.2 *Operating Characteristic Curve (OC Curve)*

**Definition 5.4** A type I error (usually denoted by $\alpha$) is the probability of a point falling outside the control limits when a process is in statistical control.

> **Definition 5.5**   A type II error (usually denoted by $\beta$) is the probability of a point falling within the control limits when a process is not in statistical control.

> **Definition 5.6**   An operating characteristic curve is a graph characterizing the relationship between the probability of a type II error, $\beta$, and the process mean shift.

Figure 5.8 shows a set of OC curves for an $\overline{X}$ control chart with three-sigma limits for different sample sizes, $n$, and various shifts. Note that the scale on the horizontal axis is in units of process standard deviation, $\sigma$.

By observing the OC curves carefully in Figure 5.8, we may note the following:

- For a given sample size $n$ and $\alpha$, a larger shift corresponds to a smaller probability $\beta$, where $\alpha$ is the probability of a point exceeding the control limits when the process is stable and $\beta$ is the probability of the point not exceeding the control limits when the process is not stable.
- With a larger sample size, there is a smaller probability $\beta$ for a given process mean shift.

As mentioned earlier, OC curves are very helpful in determining the sample size needed to detect a shift of size $r\sigma$ with a given probability $(1 - \beta)$, where $(1 - \beta)$ is the probability of detecting the shift when the process is not stable. For example, if we are interested in determining the sample size required to detect a one-sigma shift with probability 95%, then from Figure 5.8 with $r = 1$ and $\beta = 0.05$, we find the size $n$ will be slightly greater than 20.

Having studied some basic concepts of control charts, we next discuss special control charts that are very useful in any type of industry – manufacturing or non-manufacturing.

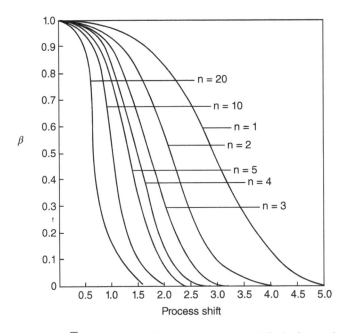

**Figure 5.8**   OC curves for an $\overline{X}$ control chart with three-sigma control limits for varying sample sizes $n$.

## 5.5 Shewhart $\overline{X}$ and $R$ Control Charts

Some common rules that are widely used in preparing Shewhart $\overline{X}$ and $R$ control charts are as follows:

- Take a series of samples from the process under investigation. Samples consisting of four or five items taken frequently are usually sufficient. This is because
  - Samples of size four or five are more cost-effective.
  - If samples are larger than 10, the estimate of the process standard deviation obtained using the range is not very efficient. As a result, the $R$ chart is also not very effective.
  - With samples of size four or five, there are fewer chances for special causes to occur while taking a sample. As Walter A. Shewhart states, "If the cause system is changing, the sample size should be as small as possible so that the average of samples does not mask the changes."
- Enough samples should be collected that the major source of variation has an opportunity to occur. Generally, at least 25 samples of size 4 or 5 are considered sufficient to give a good test for process stability.
- During the collection of data, a complete log of any changes in the process, such as changes in raw materials, changes in operators, changes in tools, calibration of tools or machines, training of workers, etc. must be maintained. Nothing is more important than maintaining the log for finding the special causes in a process.

### 5.5.1 Calculating Sample Statistics

The sample statistics needed to prepare a Shewhart $\overline{X}$ and $R$ control chart are the sample mean $(\overline{x})$ and sample range $(R)$. Thus for example, let $x_1, x_2, ..., x_n$ be a random sample from a process that is under investigation. Then we have

$$\overline{x} = \frac{x_1 + x_2 + ... + x_n}{n} \tag{5.2}$$

$$R = \text{Max}\ (x_i) - \text{Min}\ (x_i) \tag{5.3}$$

**Example 5.3**

Let 5, 6, 9, 7, 8, 6, 9, 7, 6, 5 be a random sample from a process under investigation. Find the sample mean and the sample range.

**Solution**

Using the given data, we have

$$\overline{x} = \frac{5 + 6 + 9 + 7 + 8 + 6 + 7 + 6 + 5}{10} = 5.9$$

$$R = \text{Max}\ (x_i) - \text{Min}\ (x_i) = 9 - 5 = 4$$

### 5.5.2 Calculating Control Limits

1) Calculate $\overline{x}_i$ and $R_i$ for the $i$th sample, for $i = 1, 2, 3, ..., m$, where $m$ is the number of samples collected during the study period.
2) Calculate

$$\bar{R} = \frac{R_1 + R_2 + \dots + R_m}{m} \tag{5.4}$$

$$\bar{\bar{x}} = \frac{\bar{x}_1 + \bar{x}_2 + \dots + \bar{x}_m}{m} \tag{5.5}$$

3) Calculate the three-sigma control limits for the $\bar{X}$ chart:

$$\text{UCL} = \bar{\bar{x}} + 3\hat{\sigma}_{\bar{x}}$$

$$= \bar{\bar{x}} + 3\frac{\hat{\sigma}}{\sqrt{n}}$$

$$= \bar{\bar{x}} + 3\frac{\bar{R}}{d_2\sqrt{n}}$$

$$= \bar{\bar{x}} + A_2\bar{R} \tag{5.6}$$

$$\text{CL} = \bar{\bar{x}} \tag{5.7}$$

$$\text{LCL} = \bar{\bar{x}} - 3\hat{\sigma}_{\bar{x}}$$

$$= \bar{\bar{x}} - 3\frac{\hat{\sigma}}{\sqrt{n}}$$

$$= \bar{\bar{x}} - 3\frac{\bar{R}}{d_2\sqrt{n}}$$

$$= \bar{\bar{x}} - A_2\bar{R} \tag{5.8}$$

where the values of $A_2$ and $d_2$ for various sample sizes are given in Table A.2 in the Appendix.

Note: Instead of calculating three-sigma limits, we can also calculate the probability control limits at the desired level of significance $\alpha$ simply by replacing 3 with $z_{\alpha/2}$ in Eqs. (5.6) and (5.8). The probability control limits at the level of significance $\alpha$ means the probability of a sample mean falling beyond the control limits when the process is stable / in statistical control is $\alpha$. For example, the probability control limits for an $\bar{X}$ control chart are defined as follows:

$$\text{UCL} = \bar{\bar{x}} + z_{\alpha/2}\frac{\hat{\sigma}}{\sqrt{n}}$$

$$= \bar{\bar{x}} + z_{\alpha/2}\frac{\bar{R}}{d_2\sqrt{n}} \tag{5.9}$$

$$\text{LCL} = \bar{\bar{x}} - z_{\alpha/2}\frac{\hat{\sigma}}{\sqrt{n}}$$

$$= \bar{\bar{x}} - z_{\alpha/2}\frac{\bar{R}}{d_2\sqrt{n}} \tag{5.10}$$

4) Calculate the control limits for the $R$ control chart:

$$\text{UCL} = \overline{R} + 3\hat{\sigma}_R$$

$$= \overline{R} + 3d_3\frac{\overline{R}}{d_2}$$

$$= \left(1 + 3\frac{d_3}{d_2}\right)\overline{R}$$

$$= D_4\overline{R} \tag{5.11}$$

$$\text{CL} = \overline{R} \tag{5.12}$$

$$\text{LCL} = \overline{R} - 3\hat{\sigma}_R$$

$$= \overline{R} - 3d_3\frac{\overline{R}}{d_2}$$

$$= \left(1 - 3\frac{d_3}{d_2}\right)\overline{R}$$

$$= D_3\overline{R} \tag{5.13}$$

where the values of $D_3$ and $D_4$ for various sample sizes are given in Table A.2 in the Appendix.

The first implementation of control charts is referred to as *phase I*. In phase I, it is important that we calculate the preliminary control limits to find the extent of variation in sample means and sample ranges affecting the process. In other words, at this point, only common causes are affecting the process. If all of the plotted points fall within the control limits – that is, there is no evidence of any special causes – then the control limits are suitable for current or future processes. However, if some points exceed the control limits, then every effort should be made to eliminate any evident special causes in the process, and the points falling beyond the control limits are ignored. New control limits are calculated by using the remaining sample data points, and the entire process is repeated with the new limits.

Remember that ignoring the points that exceed the control limits without eliminating the special causes may result in unnecessarily narrow control limits, which may cause some points to fall beyond the control limits when, in fact, they should not. Furthermore, it is highly recommended that we use at least 25 samples of size 4 or 5 for preliminary control limits. Otherwise, the control limits may not be suitable for current and future processes.

### Example 5.4

Table 5.4 provides data on bolt lengths used in the assembly of mid-size car engines. Twenty-five samples, each of size four, are taken directly from the production line. Samples come from all three shifts, and no sample contains data from two or more shifts. Use this data to construct $\overline{X}$ and R charts and to verify whether the process is stable.

### Solution

From Table A.2 in the Appendix, for a sample of size $n = 4$, we have $D_3 = 0$ and $D_4 = 2.282$. Thus the control limits for the $R$ chart are

$$\text{LCL} = D_3\overline{R} = 0 \times 0.03479 = 0$$

$$\text{UCL} = D_4\overline{R} = 2.282 \times 0.03479 = 0.07936$$

**Table 5.4** Lengths (cm) of bolts used in the assembly of mid-size car engines.

| Sample | Observations | $\bar{x}_i$ | $R_i$ |
|--------|-------------|-------------|-------|
| 1 | 6.155 6.195 6.145 6.125 | 6.155 | 0.070 |
| 2 | 6.095 6.162 6.168 6.163 | 6.147 | 0.073 |
| 3 | 6.115 6.126 6.176 6.183 | 6.150 | 0.068 |
| 4 | 6.122 6.135 6.148 6.155 | 6.140 | 0.033 |
| 5 | 6.148 6.152 6.192 6.148 | 6.160 | 0.044 |
| 6 | 6.169 6.159 6.173 6.175 | 6.169 | 0.016 |
| 7 | 6.163 6.147 6.137 6.145 | 6.148 | 0.026 |
| 8 | 6.150 6.164 6.156 6.170 | 6.160 | 0.020 |
| 9 | 6.148 6.162 6.163 6.147 | 6.155 | 0.016 |
| 10 | 6.152 6.138 6.167 6.155 | 6.153 | 0.029 |
| 11 | 6.147 6.158 6.175 6.160 | 6.160 | 0.028 |
| 12 | 6.158 6.172 6.142 6.120 | 6.148 | 0.052 |
| 13 | 6.133 6.177 6.145 6.165 | 6.155 | 0.044 |
| 14 | 6.148 6.174 6.155 6.175 | 6.155 | 0.027 |
| 15 | 6.143 6.137 6.164 6.156 | 6.150 | 0.027 |
| 16 | 6.142 6.150 6.168 6.152 | 6.153 | 0.026 |
| 17 | 6.132 6.168 6.154 6.146 | 6.150 | 0.036 |
| 18 | 6.172 6.188 6.178 6.194 | 6.183 | 0.022 |
| 19 | 6.174 6.166 6.186 6.194 | 6.180 | 0.028 |
| 20 | 6.166 6.178 6.192 6.184 | 6.180 | 0.026 |
| 21 | 6.172 6.187 6.193 6.180 | 6.183 | 0.021 |
| 22 | 6.182 6.198 6.185 6.195 | 6.190 | 0.016 |
| 23 | 6.170 6.150 6.192 6.180 | 6.173 | 0.042 |
| 24 | 6.186 6.194 6.175 6.185 | 6.185 | 0.019 |
| 25 | 6.178 6.192 6.168 6.182 | 6.180 | 0.024 |
| | | $\bar{\bar{x}} = 6.1628$ | $\bar{R} = 0.03479$ |

It is customary to prepare the $R$ chart first and verify that all of the plotted points fall within the control limits; only then do we proceed to construct the $\bar{X}$ control chart. The concept of bringing the process variability under control first and then proceeding to control the process average makes a lot of sense: without controlling process variability, it is almost impossible to bring the process average under control.

The $R$ chart for the example data is given in Figure 5.9, which shows that all the plotted points fall within the control limits and there is no evidence of any special pattern. Thus we may conclude that the only variation present in the process is due to common causes. We can now

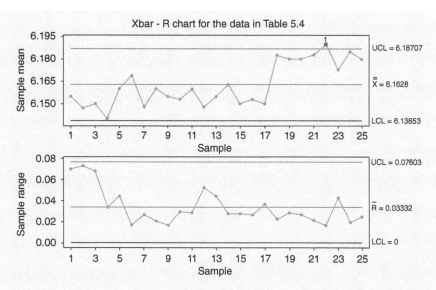

**Figure 5.9** Minitab printout of $\overline{X}$ and $R$ control charts for the data on bolts in Table 5.4.

proceed to construct the $\overline{X}$ chart. From Table A.2 in the Appendix, for a sample of size $n = 4$, we get

$$A_2 = 0.729$$

$$\text{LCL} = \overline{\overline{x}} - A_2\overline{R} = 5.1628 - 0.729 \times 0.03479 = 5.13746$$

$$\text{UCL} = \overline{\overline{x}} + A_2\overline{R} = 5.1628 + 0.729 \times 0.03479 = 5.18814$$

The $\overline{X}$ chart for the data is given in Figure 5.9, which shows that point 22 exceeds the UCL. Moreover, 10 consecutive points, starting from the seventh observation, fall below the CL. This indicates that the process is not in statistical control, or special causes are affecting the process average. Thus a thorough investigation should be launched to find these special causes, and appropriate actions should be taken to eliminate them before recalculating new control limits for an ongoing process.

We can construct $\overline{X}$ and $R$ charts using any of the statistical packages discussed in this book. Thus we can construct these charts using Minitab or R by taking the following steps.

### Minitab

1) Enter the data in column C1 (if the data is not divided into subgroups) or in columns C1–C$n$ (if the data is divided into subgroups), where $n$ is the subgroup size.
2) Choose **Stat > Control Charts > Variable Charts for Subgroups > Xbar-R**.
3) In the **Xbar-R Chart** dialog box that appears on the screen, choose the option **Observations for a subgroup in one row of columns**. Then in the next box, enter the columns C1, C2, ..., C$n$. If the data was not divided into subgroups, choose the option **All observations for a chart in one column** and enter C1 instead of C1, C2, ..., C$n$; in the box next to **Subgroup size**, enter the sample size.
4) In the **Xbar-R Chart** dialog box, there are several options such as Scale, Labels, etc. Select the options that are applicable, and complete all the required/applicable entries. Click **OK**. The Xbar-R chart will appear in the **Session Window**.

**R**

The built-in R function "qcc ()" can be used to generate many control charts. The chart type must be identified by assigning the type of chart we want to generate. For instance, in this example, we have to use the options "type = "Xbar"" and "type = "R"" to generate the $\overline{X}$ and R charts, respectively. The following R code, shown in the R console, can be used to generate charts similar to the one generated using Minitab (see Figure 5.9):

```
library (qcc)
# Read the data from the local directory. In this example, the data in Table 5.4 is saved in folder
"SQC-Data folder" on the desktop. The file name is Table 5.4.csv. For more details on this, see
Chapter 10, which is available for download on the book's website:
www.wiley.com/college/gupta/SQC.
# The data is read by using the "working directory" in R. One way to get the complete directory path
is to right-click the file and then click on "properties" or "get info". The working directory will
be shown.
setwd("C:/Users/person/Desktop/SQC-Data folder")
Bolt = read.csv("Table5.4.csv")
qcc(Bolt, type = "xbar", std.dev = "UWAVE-R")
qcc(Bolt, type = "R", std.dev = "UWAVE-R")
```

The construction of these control charts using JMP and Python is discussed in Chapter 10, which is available for download on the book's website: www.wiley.com/college/gupta/SQC.

### 5.5.3 Interpreting Shewhart $\overline{X}$ and $R$ Control Charts

A process is considered out of control not only when points exceed the control limits, but also when points show patterns of non-randomness. The Western Electric (1985) *Statistical Quality Control Handbook* gives a set of decision rules for determining nonrandom patterns in control charts. In particular, it suggests the patterns are nonrandom if one or more of the following conditions are met:

1) Two out of three successive points exceed the two-sigma warning limits
2) Four out of five successive points fall on or beyond the one-sigma control limits
3) Eight successive points fall on one side of the CL
4) Seven successive points run either upward or downward

We should investigate any out-of-control points – points on or beyond the UCL or LCL – or any patterns of non-randomness on the $R$ chart, before interpreting the $\overline{X}$ chart. As discussed earlier, the reason for doing this is simple: it is not possible to bring the process averages under control without first bringing process variability under control. Usually, the $\overline{X}$ chart is placed above the $R$ chart, and they are aligned with each other in such a manner that the average and the range for any sample are plotted on the same vertical line. Examine whether one, both, or neither chart indicates that the process is out of control for a given sample. If any point exceeds the control limits in one or both control charts, then the sample did not come from a stable process: special or assignable causes are present in the system. More precisely, if the plotted point exceeds the control limits in the $R$ chart, then it is evident that the variability of the process has changed. In such cases, before any full-blown investigation is launched, preliminary checks should be made. For example, we may check whether all the calculations are correct; the data were entered in the computer correctly; or there were any change in workers, machines, or suppliers of raw materials.

If the points exceed the control limits in the $\overline{X}$ chart, then the process mean has changed. Again, follow the preliminary checks before launching a full-blown investigation.

If points exceed the control limits in both the $\overline{X}$ and $R$ charts, then (usually) a sudden shift occurred in the lot from which the samples were taken. In such cases, after the preliminary checks, there should be a full investigation concentrating on the period during which that lot was produced. Depending on the process, the possibility of stopping production until the special causes are detected should be considered: if the process is not stable and the production process is not stopped, then it may produce a high percentage of defectives.

In addition to the points exceeding the control limits, nonrandom patterns should also be checked.

An upward run or a run above the CL in the R chart indicates one (or both) of the following:

- More significant variability or a tendency to perpetuate more significant variability in the output of the process is occurring. This may be due to new material of undesirable low quality or a difference between shifts. Immediate attention is warranted to detect special causes.
- The measurement system has changed.

A downward run or a run below the CL in the R chart indicates one (or both) of the following:

- A smaller variability or a tendency to perpetuate a smaller variability in the output of the process is occurring. This is usually a good sign for the process. However, a thorough investigation should be made so that similar conditions are maintained as long as possible and these conditions are implemented elsewhere in the process.
- The measurement system has changed.

A run relative to the $\overline{X}$ chart indicates one (or both) of the following:

- The process average has changed or is still changing.
- The measurement system has changed.

### 5.5.4 Extending the Current Control Limits for Future Control

If current data from at least 25 sample periods is contained within the current control limits, then we may use these limits to cover future periods. However, these limits are used for future periods with the condition that immediate action will be taken if any out-of-control indication appears or if the points consistently fall very close to the CL. The latter case indicates that the process has improved and that new control limits should be recalculated. Again, note that in any case, control limits for future use should be extended for only 25 to 30 sample periods at a time.

It is pertinent to note that a change in sample size would affect the control limits for both the $\overline{X}$ and R charts, and therefore, whenever there is a change in sample size, new control limits should be calculated. This situation may arise if we decide to take smaller samples more frequently, which usually is the case when we want to catch larger shifts (larger than 1.5 sigma) without increasing the total number of parts sampled over the whole sampling period. Another scenario is when we decide to increase the sample size, but samples are taken less frequently, which usually is the case when we want to catch smaller shifts (shifts of 1.5 sigma or smaller). To calculate the new control limits, proceed as follows:

1) Estimate the process standard deviation, using the existing sample size

$$\hat{\sigma} = \frac{\overline{R}}{d_2} \qquad (5.14)$$

where $\overline{R}$ is the sample range average for the period with ranges in control, and the value of $d_2$ is found from Table A.2 in the Appendix for the existing sample size.
2) Using the values for $d_2$, $D_3$, $D_4$, and $A_2$ from Table A.2 in the Appendix for the new sample size, calculate the new range and control limits as follows:

a) Estimate the new sample range average:

$$\overline{R}_{new} = d_2 \times \hat{\sigma} \qquad (5.15)$$

b) Then the new control limits for the $\overline{X}$ control chart are

$$LCL = \overline{\overline{x}} - A_2 \times \overline{R}_{new} \qquad (5.16)$$

$$CL = \overline{\overline{x}} \qquad (5.17)$$

$$UCL = \overline{\overline{x}} + A_2 \times \overline{R}_{new} \qquad (5.18)$$

and the new control limits for the $R$ control chart are

$$LCL = D_3 \times \overline{R}_{new} \qquad (5.19)$$

$$CL = \overline{R}_{new} \qquad (5.20)$$

$$UCL = D_4 \times \overline{R}_{new} \qquad (5.21)$$

**Example 5.5**
To illustrate the technique of calculating the new control limits, consider the $\overline{X}$ and R control charts developed for the data on bolt lengths in Example 5.4. The charts in Example 5.4 were based on a sample size of four. Since one point in the $\overline{X}$ chart exceeded the control limit, there may be some shift in the process mean. Thus the company's Six Sigma Green Belt engineer wants to increase the sample size to six. Determine the control limits for both the $\overline{X}$ and R control charts required for the new samples of size six.

**Solution**

From Table A.2 in the Appendix, for $n = 4$, $d_2 = 2.059$. Thus from Example 5.4, and using Eq. (5.14), we have

$$\hat{\sigma} = \frac{\overline{R}}{d_2} = \frac{0.03479}{2.059} = 0.0169$$

Again, from Table A.2 in the Appendix, for $n = 6$, $d_2 = 2.534$. Thus, using Eq. (5.15), we have

$$\overline{R}_{new} = 2.534 \times 0.0169 = 0.0428246$$

Therefore, from Eq. (5.16), the new control limits for the $\overline{X}$ control chart for samples of size six are

$$\text{LCL} = 15.1628 - 0.483 \times 0.0428246 = 15.1421$$

$$\text{CL} = 15.1628$$

$$\text{UCL} = 15.1628 + 0.483 \times 0.0428246 = 15.18348$$

and the new control limits for the $R$ control chart for samples of size six are

$$\text{LCL} = 0 \times 0.0428246 = 0$$

$$\text{CL} = 0.0428246$$

$$\text{UCL} = 2.004 \times 0.0428246 = 0.08582$$

Note that the net result of increasing the sample size is to narrow the control limits for the $\overline{X}$ chart and to move the CL and control limits for the $R$ chart higher. This is because the expected range value for larger sample increases. In this example, however, the LCL for the $R$ chart remains the same since the value of $D_3$ for sample sizes four and six is zero.

## 5.6 Shewhart $\overline{X}$ and $R$ Control Charts When the Process Mean and Standard Deviation are Known

If the process mean, $\mu$, and the process standard deviation, $\sigma$, are known, then $\overline{X}$ and $R$ control charts are developed as follows:

1) Calculate $\bar{x}_i$ and $R_i$ for the $i$th sample for $i = 1, 2, 3, ..., m$, where $m$ is the number of samples collected during the study period.
2) Calculate the control limits for the $\overline{X}$ control chart:

$$\text{UCL} = \mu + 3\frac{\sigma}{\sqrt{n}} \tag{5.22}$$

$$\text{CL} = \mu \tag{5.23}$$

$$\text{LCL} = \mu - 3\frac{\sigma}{\sqrt{n}} \tag{5.24}$$

Note: Instead of calculating three-sigma limits, we can also calculate the probability limits at the desired level of significance, $\alpha$, simply by replacing 3 with $z_{\alpha/2}$ in Eqs. (5.22) and (5.24).

3) Calculate the control limits for the $R$ control chart. Recalling that $\sigma = R/d_2$ and $\sigma_R = d_3\sigma$, we have

$$\text{UCL} = d_2\sigma + 3\sigma_R$$

$$= d_2\sigma + 3d_3\sigma$$

$$= (d_2 + 3d_3)\sigma$$

$$= D_2\sigma \tag{5.25}$$

$$\text{CL} = \overline{R} \tag{5.26}$$

$$
\begin{aligned}
\text{LCL} &= d_2\sigma - 3\sigma_R \\
&= d_2\sigma - 3d_3\sigma \\
&= (d_2 - 3d_3)\sigma \\
&= D_1\sigma
\end{aligned}
\tag{5.27}
$$

where the values of $d_3$, $D_1$ and $D_2$ for various sample sizes are given in Table A.2 in the Appendix.

Shewhart $\overline{X}$ and $R$ control charts, when the process mean, $\mu$, and process standard deviation, $\sigma$, are known, can be generated by using one of the statistical packages discussed in this book. For example, to use Minitab, use the same steps given earlier when $\mu$ and $\sigma$ were unknown. In this case, in step 4, when the **Xbar – R chart: Options** dialog box appears, enter the given value of $\mu$ and $\sigma$, and then click **OK**. The Xbar-R chart will appear in the **Session Window**.

**R.**

As in Example 5.4, we use the built-in R function "qcc ()" to generate Xbar-R charts. For example, using the data given in Example 5.4, and assuming that $\mu = 6.162$ and $\sigma = 0.016$, we use the following R code:

```
library (qcc)
#Read the data from a local directory.
setwd("C:/Users/person/Desktop/SQC-Data folder")
Bolt = read.csv("Table5.4.csv")
qcc(Bolt, type = "xbar", center = 6.162, std.dev = 0.016)
qcc(Bolt, type = "R", std.dev = 0.016)
```

Constructing control charts using JMP and Python is discussed in Chapter 10, which is available for download on the book's website: www.wiley.com/college/gupta/SQC.

## 5.7   Shewhart $\overline{X}$ and R Control Charts for Individual Observations

Sometimes, it is necessary to study SPC based on individual observations, since it may not be feasible to form rational subgroups of size greater than one. This type of scenario, for example, would arise when

- Sampling is very expensive, and it is not economical to take samples of size greater than one.
- Observations are collected through experimentation, and it may take several hours or days to make one observation.
- The circumstances warrant that each unit must be inspected or the process completely automated so that the measurement on each observation can be taken without much extra expense.
- Only a few items are produced by each unit per day, and any observed differences are attributed to being between the units and not within the units; thus one observation from each unit is sufficient.
- Sampling is destructive, and the units are costly. For example, certain bulbs for projectors are very expensive, and if we are interested in studying the life of such bulbs, collecting data will cause the bulbs to be destroyed.

Control charts for individual observations are very similar to $\overline{X}$ and $R$ control charts. However, in the case of individual observations, since the sample contains only one observation, it is not possible to find the sample range in the usual manner. Instead, we find the sample range as the absolute difference between the two successive observations. This type of sample range is usually known as the *moving range* (MR), and

$$MR_k = |x_k - x_{k-1}| \qquad (5.28)$$

where $k = 2, 3,..., m$; and $m$ is the total number of samples or observations. Instead of using the sample means, $\bar{x}_i$, the individual observations, $x_i$, are used.

Note that sometimes the control chart for individual observations is also known as an *individual moving range control chart or simply MR control chart*. To illustrate the procedure of constructing an MR control chart, we reproduce the observations of column 1 from Table 5.4 in Table 5.5 for the following example.

### Example 5.6
Using just the first data point of each sample from Table 5.4, construct the $X$ and $R$, MR control charts. These data are reproduced in Table 5.5.

**Table 5.5** Lengths (cm) of bolts used in the assembly of mid-size car engines.

| Sample | Observations | $MR_k$ |
|---|---|---|
| 1 | 6.155 | |
| 2 | 6.095 | 0.060 |
| 3 | 6.115 | 0.020 |
| 4 | 6.122 | 0.007 |
| 5 | 6.148 | 0.026 |
| 6 | 6.169 | 0.021 |
| 7 | 6.163 | 0.006 |
| 8 | 6.150 | 0.013 |
| 9 | 6.148 | 0.002 |
| 10 | 6.152 | 0.004 |
| 11 | 6.147 | 0.005 |
| 12 | 6.158 | 0.011 |
| 13 | 6.133 | 0.025 |
| 14 | 6.148 | 0.015 |
| 15 | 6.143 | 0.005 |
| 16 | 6.142 | 0.001 |
| 17 | 6.132 | 0.010 |
| 18 | 6.172 | 0.040 |
| 19 | 6.174 | 0.002 |
| 20 | 6.166 | 0.008 |
| 21 | 6.172 | 0.006 |
| 22 | 6.182 | 0.010 |
| 23 | 6.170 | 0.012 |
| 24 | 6.186 | 0.016 |
| 25 | 6.178 | 0.008 |
| | $\bar{x} = 6.1628$ | $\bar{R} = 0.01387$ |

**Solution**

Since the sample range is determined from two successive observations, the sample size for estimating the process standard deviation and constructing the $R$ chart is considered to be equal to two. Thus, from Table A.2 in the Appendix, for $n = 2$, we have $D_3 = 0$, $D_4 = 3.267$, and

$$\text{LCL} = D_3\bar{R} = 0 \times 0.01387 = 0$$
$$\text{UCL} = D_4\bar{R} = 3.267 \times 0.01387 = 0.0453$$

As in the case of the $\bar{X}$ and $R$ control charts, it is customary to prepare the $R$ control chart first and verify that all the plotted points fall within the control limits and only then proceed to construct the $X$ control chart. To construct the $X$ and $R$ control charts using Minitab and R, proceed as follows.

**Minitab**

1) Enter all the data in column C1 of the data (Worksheet) window.
2) Choose **Stat > Control Charts > Variable Charts for Individuals > I-MR**.
3) Enter **C1** in the box next to **Variables**, select the other options from the **Individuals-Moving-Range Chart** dialog box, and complete all the required entries. Then click **OK**. The output of the Shewhart control chart for individual observations will appear in the **Session Window** (see Figure 5.10).

**R**

As in Example 5.4, we can use the built-in R function "qcc ()" to obtain an $\bar{X}$ chart for individual observations. The following R code generates a control chart similar to that shown in Figure 5.10:

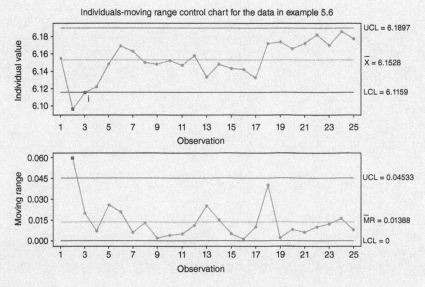

**Figure 5.10** The MR control chart constructed using Minitab for the data in Table 5.5.

```
library (qcc)
setwd("C:/Users/person/Desktop/SQC-Data folder")
Bolt = read.csv("Table5.5.csv")
qcc(Bolt, type ="xbar.one", std.dev = "MR")
```

Unfortunately, "qcc ()" does not automatically handle $R$ chart creation for individual observations. However, it can be made to do so by first creating a two-column spreadsheet for the given data. For instance, we can create a file with observations 1–24 in column 1 and observations 2–25 in column 2, and save the file as Table5.5.1.csv. The following R code generates an $R$ chart similar to the one shown in Figure 5.10:

```
library (qcc)
setwd("C:/Users/person/Desktop/SQC-Data folder")
Bolt = read.csv("Table5.5.1.csv")
qcc(Bolt, type = "R", std.dev = "UWAVE-R")
```

The construction of these control charts using JMP and Python is discussed in Chapter 10, which is available for download on the book's website: www.wiley.com/college/gupta/SQC.

The $R$ control chart for the example data in Figure 5.10 shows that process variability is not under control, since the first point exceeds the UCL and point 16 is almost on the LCL. Thus, as a matter of principle, we should first investigate the special causes of this variability and eliminate such causes before we construct the $X$ control chart.

We assume here, for illustration, that these special causes have been detected and eliminated, and therefore we proceed to calculate the control limits for the $X$ control chart. Note that in practice, observations corresponding to points found beyond the control limits should be ignored, but only after detecting and eliminating the special causes. Now, from Table A.2 in the Appendix, for a sample of size $n = 2$, we get $d_2 = 1.128$. Thus we have

$$\text{LCL} = \bar{x} - 3\hat{\sigma} = \bar{x} - 3\frac{\bar{R}}{d_2} = 6.1628 - 3\frac{0.01387}{1.128} = 6.1159$$

$$\text{CL} = 6.1628$$

$$\text{UCL} = \bar{x} + 3\hat{\sigma} = \bar{x} + 3\frac{\bar{R}}{d_2} = 6.1628 + 3\frac{0.01387}{1.128} = 6.1897$$

Notice, however, that there is a small difference between how we compute the control limits for the $\bar{X}$ control chart and the $X$ control chart. While calculating the control limits of the $X$ control chart, we always use sample size $n = 1$. In other words, we do not use sample size $n = 2$ or the control limits $\bar{x} \pm 3\hat{\sigma}/\sqrt{2}$; rather, we simply use the control limits $\bar{x} \pm 3\hat{\sigma}$, since n = 2 is used only for determining the control limits of the $R$ control chart and estimating $\sigma$. The $X$ control chart for the example data is given in Figure 5.10: it shows that points 2 and 3 fall beyond the LCL. Moreover, too many consecutive points fall above the CL. This indicates that the process in not under control and special causes in the system are affecting the process average. Thus a thorough investigation should be launched to find the special causes, and appropriate action should be taken to eliminate them.

Note: There are some limitations on using Shewhart charts for individual observations. If a process characteristic is non-normal, the rules applicable for $\bar{X}$ control charts may not hold for $X$ control charts, since $\bar{X}$ usually behaves as normal even if the process distribution is not normal. This is not true for individual observations. Therefore, in such cases, $X$ charts may signal special causes when such causes are not actually present. Also, note that the ranges are not independent since the two adjacent ranges have a common point. Hence a run of successive points falling near or beyond the control limits does not have the same significance as in control charts with a sample size greater than one. For a more detailed discussion of this topic, consult Montgomery (2013).

## 5.8 Shewhart $\overline{X}$ and $S$ Control Charts with Equal Sample Sizes

$\overline{X}$ and $S$ control charts, like $\overline{X}$ and $R$ control charts, are developed from measured process output data, and $\overline{X}$ and $S$ control charts are used together. The standard deviation $s$ is usually a more efficient indicator of process variability than the range, particularly when sample sizes are large (10 or greater). When the sample size is 10 or greater, $R$ is no longer a very efficient measure to estimate the standard deviation. The sample standard deviation $s$ for the $S$ control chart is calculated using all the data points in each subgroup, rather than just the maximum and minimum values as is done for the $R$ control chart. $S$ control charts are usually preferred over $R$ control charts when

- Samples are of size 10 or greater than 10.
- Sample size is variable.
- The process is automated, so $s$ for each sample can be calculated easily.

The sample standard deviation $s$ is determined using the formula

$$s = \sqrt{\frac{1}{n-1}\left(\sum_{i=1}^{n}x_i^2 - \frac{1}{n}\left(\sum_{i=1}^{n}x_i\right)^2\right)} \qquad (5.29)$$

Then the control limits for the $\overline{X}$ and $S$ control charts are determined as follows:

1) Calculate $\overline{x}_i$ and $s_i$ for the $i$th sample, for $i = 1, 2, 3, ..., m$, where $m$ is the number of samples collected during the study period.
2) Calculate

$$\overline{s} = \frac{s_1 + s_2 + ... + s_m}{m} \qquad (5.30)$$

$$\overline{\overline{x}} = \frac{\overline{x}_1 + \overline{x}_2 + ... + \overline{x}_m}{m} \qquad (5.31)$$

3) Calculate the control limits for the $\overline{X}$ chart:

$$\text{UCL} = \overline{\overline{x}} + 3\frac{\hat{\sigma}}{\sqrt{n}}$$

$$= \overline{\overline{x}} + 3\frac{\overline{s}}{c_4\sqrt{n}}$$

$$= \overline{\overline{x}} + A_3\overline{s} \qquad (5.32)$$

$$\text{CL} = \overline{\overline{x}} \qquad (5.33)$$

$$\text{LCL} = \bar{\bar{x}} - 3\frac{\hat{\sigma}}{\sqrt{n}}$$

$$= \bar{\bar{x}} - 3\frac{\bar{s}}{c_4\sqrt{n}}$$

$$= \bar{\bar{x}} - A_3\bar{s} \tag{5.34}$$

where the values of $A_3$ and $c_4$ for various sample sizes are given in Table A.2 in the Appendix.

Note: Instead of calculating three-sigma limits for the $\overline{X}$ chart, we can also calculate the probability limits at the desired level of significance, $\alpha$, simply by replacing 3 with $z_{\alpha/2}$ in Eqs. (5.32) and (5.34). Thus the control limits will be

$$\text{UCL} = \bar{\bar{x}} + z_{\alpha/2}\frac{\hat{\sigma}}{\sqrt{n}}$$

$$= \bar{\bar{x}} + z_{\alpha/2}\frac{\bar{s}}{c_4\sqrt{n}} \tag{5.35}$$

$$\text{CL} = \bar{\bar{x}} \tag{5.36}$$

$$\text{LCL} = \bar{\bar{x}} - z_{\alpha/2}\frac{\hat{\sigma}}{\sqrt{n}}$$

$$= \bar{\bar{x}} - z_{\alpha/2}\frac{\bar{s}}{c_4\sqrt{n}} \tag{5.37}$$

4) Calculate the control limits for the $S$ chart:

$$\text{UCL} = \bar{s} + 3\hat{\sigma}_s$$

$$= \bar{s} + 3\frac{\bar{s}}{c_4}\sqrt{1-c_4^2}$$

$$= \left(1 + 3\frac{1}{c_4}\sqrt{1-c_4^2}\right)\bar{s}$$

$$= B_4\bar{s} \tag{5.38}$$

$$\text{CL} = \bar{s} \tag{5.39}$$

$$\text{LCL} = \bar{s} - 3\hat{\sigma}_s$$

$$= \bar{s} - 3\frac{\bar{s}}{c_4}\sqrt{1-c_4^2}$$

$$= \left(1 - 3\frac{1}{c_4}\sqrt{1-c_4^2}\right)\bar{s}$$

$$= B_3\bar{s} \tag{5.40}$$

where the values of $B_3$ and $B_4$ for various sample sizes are given in Table A.2 in the Appendix.

We illustrate the development of the $\overline{X}$ and $S$ control charts with the following example.

**Example 5.7**

Use the bolt lengths data in Table 5.4 to construct the $\overline{X}$ and S control charts.

**Solution**

From Table A.2 in the Appendix, for a sample of size $n = 4$, we have $B_3 = 0$ and $B_4 = 2.266$, and the control limits for the $S$ control chart are

$$\text{LCL} = B_3\overline{s} = 0 \times 0.01557 = 0$$

$$\text{UCL} = B_4\overline{s} = 2.266 \times 0.01492 = 0.03381$$

It is customary to prepare the $S$ control chart first and verify that all the plotted points fall within the control limits, and only then proceed to construct the $\overline{X}$ control chart. As described earlier, the concept of first bringing process variability under control and only then proceeding to control the average makes a lot of sense. That is, without controlling process variability, it is practically impossible to bring the process average under control.

The $S$ control chart for the example data is given in Figure 5.11 and shows that points 2 and 3 exceed the UCL. Moreover, points 14–22 fall below the CL, so we have a run of 9 points below the CL. Note that point 17 seems to fall on the CL, but from the calculation of the $s$ value for sample 17, we can see that it is less than the CL limit. These observations show that the process variability is not under control, and therefore the process should be carefully investigated and any special cause present in the system should be eliminated.

After eliminating the special cause(s), we can assume that process variability is under control, and then we can evaluate the $\overline{X}$ control chart by ignoring the samples corresponding to points 2 and 3. From Table A.2 in the Appendix, for a sample of size $n = 4$, we have $A_3 = 1.628$. Thus, for the example data, the control limits for the $\overline{X}$ chart are

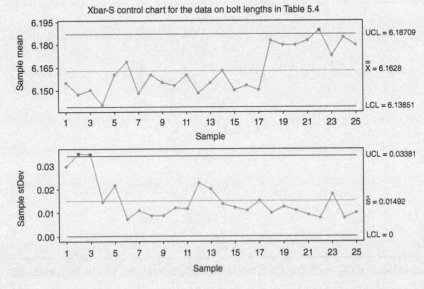

**Figure 5.11** The $\overline{X}$ and $S$ control charts, constructed using Minitab, for the data on bolt lengths in Table 5.4.

$$\text{LCL} = \overline{\overline{x}} - A_3\overline{s} = 6.1628 - 1.628 \times 0.01492 = 6.13851$$

$$\text{UCL} = \overline{\overline{x}} + A_3\overline{s} = 6.1628 + 1.628 \times 0.01492 = 6.18709$$

To construct the $\overline{X}$ and $S$ charts by using Minitab and R, we proceed as follows.

**Minitab**

To construct the $\overline{X}$ and $S$ charts using Minitab, all the steps are the same as discussed in Example 5.4 to construct $\overline{X}$ and $R$ charts, except step 2. Replace it with the following:

2) Choose **Stat > Control Charts > Variable Charts for Subgroups > Xbar-S**.

**R**

As in Example 5.4, we can use the built-in R function "qcc ()" to obtain $\overline{X}$ and $S$ charts. The following R code generates charts similar to those shown in Figure 5.11:

```
library (qcc)
setwd("C:/Users/Person/desktop/SQC-Data folder")
Bolt = read.csv("Table5.4.csv")
qcc(Bolt, type = "xbar", size = 4, std.dev = "UWAVE-SD", nsigmas = 3)
qcc(Bolt, type = "S", size = 4, std.dev = "UWAVE-SD", nsigmas = 3)
```

The construction of these control charts using JMP and Python is discussed in Chapter 10, which is available for download on the book's website: www.wiley.com/college/gupta/SQC.

## 5.9  Shewhart $\overline{X}$ and $S$ Control Charts with Variable Sample Sizes

Sometimes it is not possible to select samples of the same size. In such cases, we construct $\overline{X}$ and $S$ charts by using the weighted mean, $\overline{\overline{x}}$, and the square root of the pooled estimator, $s^2$, of $\sigma^2$ as the CLs for the $\overline{X}$ and $S$ charts, respectively. The UCL and LCL are calculated for individual samples. For example, if the sample sizes are $n_1, n_2, n_3, ..., n_m$, then the weighted mean, $\overline{\overline{x}}$, and the pooled estimator, $s^2$, are given by

$$\overline{\overline{x}} = \frac{\sum\limits_{i=1}^{m} n_i\overline{x}_i}{\sum\limits_{i=1}^{m} n_i} = \frac{\sum\limits_{i=1}^{m} n_i\overline{x}_i}{N} \tag{5.41}$$

$$\overline{s} = \sqrt{s^2} = \sqrt{\frac{\sum\limits_{i=1}^{m} (n_i-1)s_i^2}{\sum\limits_{i=1}^{m} (n_i-1)}} = \sqrt{\frac{\sum\limits_{i=1}^{m} (n_i-1)s_i^2}{N-m}} \tag{5.42}$$

where $N = n_1 + n_2 + n_3 + ... + n_m$; $m$ is the number of samples selected during the study period. The $i$th sample control limits for the $\overline{X}$ and $S$ charts, respectively, are given by

$$UCL_{\bar{x}} = \bar{\bar{x}} + A_3 s_i \tag{5.43}$$

$$LCL_{\bar{x}} = \bar{\bar{x}} - A_3 s_i \tag{5.44}$$

$$UCL_{s(i)} = \bar{s} + 3(C_5(n_i)/C_4(n_i))\bar{s}, i = 1, 2, \cdots, m \tag{5.45}$$

$$LCL_{s(i)} = \bar{s} - 3(C_5(n_i)/C_4(n_i))\bar{s}, i = 1, 2, \cdots, m \tag{5.46}$$

The values of $A_3$, $C_4(n_i)$, and $C_5(n_i)$ depend on the individual sample sizes, $n_i$, and they are given in Table A.2 in the Appendix. To illustrate the development of these charts, we use the data presented in Example 5.8.

**Example 5.8  Car Manufacturing Data**
A manufacturing process is designed to produce piston rings for car engines with a finished inside diameter of 9 cm and a standard deviation of 0.03 cm. From this process, 25 samples of variable sizes are carefully selected so that no sample comes from two machines or two shifts. In other words, every effort is made to ensure that no sample masks any special causes. The data is shown in Table 5.6. Construct $\overline{X}$ and S control charts for these data, and plot all the points on them to determine whether the process is stable.

**Solution**

Using Eqs. (5.41) and (5.42), and the data given in Table 5.6, we get

$$\bar{\bar{x}} = 9.0002$$

$$\bar{s} = 0.02812$$

Now, using Eqs. (5.43) – (5.46), we calculate the control limits for both the $\overline{X}$ and $S$ charts, shown in Figure 5.12. From the figure, we see that process variability is under control. But point 19 in the $\overline{X}$ chart falls below the LCL, which implies that special causes in the process are affecting the process mean.

To construct $\overline{X}$ and $S$ charts with unequal sample sizes using Minitab and R, proceed as follows.

**Minitab**

1) Enter the data in columns C1–C10 of the Worksheet window. Note that the largest sample size in this example is 10.
2) Choose **Stat > Control Charts > Variable Charts for Subgroups > Xbar-S**.
3) In the **Xbar-S Chart** dialog box, several options are available, such as **Scale** and **Labels**. Select the desired options, and complete all the applicable entries. For example, we select the option **Xbar-S** option. A **Xbar-S Options** dialog box appears. If the parameters (process mean and standard deviation) are known, as in this example, select the **Parameters** entry and enter the values of the mean and standard deviation. If the parameters are unknown, select the **Estimate** entry and then complete the applicable entries. For example, in the box next to the entry **Omit the following subgroups when estimating parameters**, enter any row numbers that you do not want to include for estimating the parameters. Normally you will not omit any rows, in which case you leave that box empty. Next, under the entry **Method for estimating standard deviation**, check the appropriate circle next to **Sbar** or **Pooled standard deviation**, and then click **OK**.

**Table 5.6** Inside diameter measurements (cm) of piston rings for car engines.

| Sample | Observations | | | | | | | | | |
|---|---|---|---|---|---|---|---|---|---|---|
| 1 | 9.01759 | 8.97730 | 9.02472 | 9.03616 | 9.05257 | 9.00121 | 9.00361 | 8.98826 | 9.05866 | 9.05982 |
| 2 | 9.00974 | 8.95902 | 9.02650 | 8.97163 | 9.04754 | 9.05096 | * | * | * | * |
| 3 | 8.97809 | 8.98972 | 8.97877 | 9.01464 | 9.00623 | 8.98055 | 9.00877 | 9.03749 | 8.96140 | 9.00595 |
| 4 | 9.04504 | 8.99329 | 8.99112 | 9.01239 | 8.97850 | 8.97543 | 8.97958 | 9.02115 | * | * |
| 5 | 9.04544 | 8.95615 | 9.00747 | 8.99861 | 8.99077 | 9.03892 | 8.98887 | 8.98418 | 9.01795 | 9.01016 |
| 6 | 8.98229 | 9.02346 | 8.98103 | 8.97341 | 8.96475 | 9.00518 | 8.99561 | * | * | * |
| 7 | 9.02982 | 9.00707 | 9.00735 | 8.96042 | 8.97471 | 8.94890 | 8.99009 | 9.02050 | * | * |
| 8 | 9.01322 | 8.99716 | 8.98482 | 8.92890 | 9.03727 | 8.96801 | 9.02173 | 8.98401 | 8.95438 | * |
| 9 | 8.99775 | 9.02773 | 8.99505 | 9.00820 | 9.03221 | 8.94485 | 8.98242 | * | * | * |
| 10 | 8.98140 | 8.99443 | 8.99374 | 9.05063 | 9.00155 | 9.03501 | 9.04670 | 9.03164 | 9.04678 | 9.00865 |
| 11 | 8.94018 | 9.06380 | 8.97868 | 9.01246 | 8.96870 | 8.99593 | * | * | * | * |
| 12 | 8.96117 | 9.00420 | 9.02410 | 9.06514 | 8.97921 | 9.02223 | 8.98201 | 9.02577 | 8.94999 | * |
| 13 | 9.02374 | 8.99673 | 8.96257 | 8.99945 | 9.01848 | 9.02770 | 9.02901 | 9.00383 | * | * |
| 14 | 8.94363 | 9.00214 | 8.97526 | 8.98661 | 9.02040 | 9.02238 | 8.96735 | 8.99283 | 9.02029 | 9.01754 |
| 15 | 8.99341 | 9.02318 | 8.96106 | 9.00023 | 9.01309 | 8.95582 | * | * | * | * |
| 16 | 9.02781 | 9.02093 | 8.97963 | 8.98019 | 9.01428 | 9.03871 | 9.03058 | 9.01304 | 9.06247 | 9.00215 |
| 17 | 8.96783 | 8.94662 | 9.02595 | 9.07156 | 9.02688 | 8.96584 | 8.99007 | * | * | * |
| 18 | 8.98993 | 8.97726 | 8.97231 | 8.93709 | 8.97035 | 8.96499 | * | * | * | * |
| 19 | 8.95677 | 9.00000 | 8.98083 | 8.97502 | 8.93080 | 8.94690 | 8.98951 | 8.96032 | * | * |
| 20 | 9.01965 | 9.00524 | 9.02506 | 8.99299 | 8.95167 | 8.98578 | 9.03979 | 9.00742 | 8.99202 | 9.01492 |
| 21 | 8.97892 | 8.98553 | 9.01042 | 8.97291 | 9.01599 | * | * | * | * | * |
| 22 | 8.99523 | 9.00044 | 9.02239 | 8.93990 | 8.96644 | 8.99666 | 8.96103 | * | * | * |
| 23 | 9.05596 | 9.02182 | 8.94953 | 9.03914 | 8.97235 | 9.00869 | 9.01031 | 9.01371 | 8.99326 | 9.03646 |
| 24 | 9.03412 | 8.97335 | 9.00136 | 9.08037 | 9.04301 | 8.97701 | 9.02727 | 9.03449 | * | * |
| 25 | 9.00277 | 9.00651 | 9.02906 | 8.97863 | 8.99956 | 8.99291 | 8.97211 | 9.02725 | 8.97847 | 9.03710 |

## R

As in Example 5.4, we can use the built-in R function "qcc ()" to obtain $\overline{X}$ and S charts with variable sample sizes. The following R code generates $\overline{X}$ and S charts similar to the ones shown in Figure 5.12:

```
library (qcc)
setwd("C:/Users/Person/desktop/SQC-Data folder")
Bolt = read.csv("Table5.6.csv")
qcc(Bolt, type = "xbar" , std.dev = 0.01492, nsigmas = 3)
qcc(Bolt[,2:10], sizes = Bolt[,1], type ="xbar", nsigmas = 3)
qcc(Bolt[,2:10], sizes = Bolt[,1], type ="S", nsigmas = 3)
```

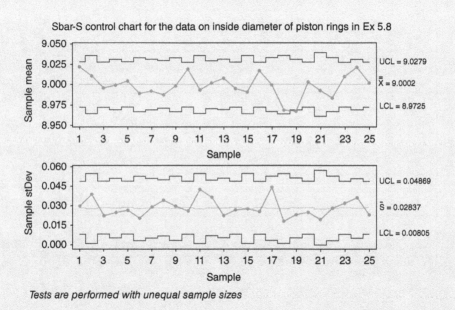

**Figure 5.12** $\overline{X}$ and S control charts for variable sample sizes, constructed using Minitab, for the data in Table 5.6.

The construction of these control charts using JMP and Python is discussed in Chapter 10, which is available for download on the book's website: www.wiley.com/college/gupta/SQC.

Note: If the sample sizes do not vary too much, then it is quite common to use the average sample size, $\overline{n}$ ($\overline{n} = (n_1 + n_2 + n_3 + ... + n_m)/m$), instead of variable sample sizes to calculate control limits. As a rule of thumb, if all the samples are within 20% of the average sample size, $\overline{n}$, it is reasonable to use the average sample size instead of variable sample sizes. However, if any point(s) in the $\overline{X}$ or S control chart falls on or very close to the control limits, it is prudent to recalculate the control limits, at least for that sample(s), using the actual sample size(s), and determine whether that point(s) falls within the control limits. If the point falls on or exceeds the recalculated control limits using the actual sample size, then the process is deemed unstable; otherwise, the process is considered stable.

## 5.10  Process Capability

In this section, we briefly discuss the concept of process capability. Later, in Chapter 8, we will study how to quantify this concept. So far, our emphasis has been on how to control a quality characteristic in a given product, and we have measured everything in terms of the control limits. But it is a well-known fact that a quality characteristic of a product is usually evaluated in terms of the specification limits, which are very often determined by the customer. It is of paramount importance to understand, however, that control limits and specification limits are two entirely different entities. Control limits, as we saw earlier in this chapter, are determined by the natural variability of the process, whereas specification limits are determined by the customer, management, or by a team from the design department. Specification limits are usually defined in terms of the expectations of how a product should perform, whereas control limits are the means to achieve these expectations. A process that can produce a product that can meet the expectations is called a *capable process*.

The expected value of a quality characteristic is called the *target* value. The largest and smallest acceptable values of a quality characteristic are known as the *upper specification limit* (USL) and *lower specification limit* (LSL), respectively. The upper and lower control limits used in $\overline{X}$ and $R$ control charts for individual values are usually known as the *upper natural tolerance limit* (UNTL) and *lower natural tolerance limit* (LNTL) – that is, $\mu + 3\sigma$ and $\mu - 3\sigma$, respectively.

It is very common to examine the process capability of a stable process. However, it is also important to note that a stable process is not necessarily a capable process. In other words, a stable process may or may not be capable. A visual representation of this scenario is given in Figure 5.13.

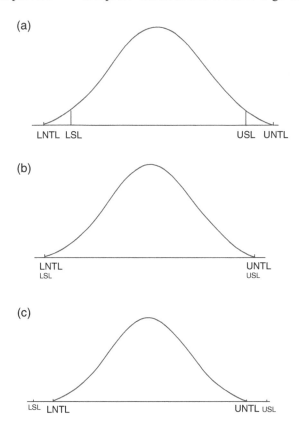

**Figure 5.13**   (a) Process is stable but not capable; (b) process is stable and barely capable; (c) process is stable and capable.

To further illustrate the concept of process capability, we use the following numerical example.

**Example 5.9**
Consider a quality characteristic of a process that is normally distributed and stable with respect to the three-sigma control limits. Furthermore, suppose that 30 samples of size 5 from this process provide the following summary statistics:

$$\bar{\bar{x}} = 0.740, \bar{R} = 0.175, n = 5$$

We are also given that LSL = 0.520 and USL = 0.950. Determine whether the process is capable.

**Solution**
From the given information, we have

$$\hat{\sigma} = \frac{\bar{R}}{d_2} = \frac{0.175}{2.326} = 0.075$$

$$z_{LSL} = \frac{0.520 - 0.740}{0.075} = -2.93$$

$$z_{USL} = \frac{0.950 - 0.740}{0.075} = 2.80$$

Therefore, the percentage of nonconforming products is

$$P(z \le -2.93) + P(z \ge 2.80) = 0.0017 + 0.0026$$
$$= 0.0043$$

That is, less than 0.5% of the product is outside of the specification limits, and therefore the process is almost capable.

From this example, it can easily be seen that if the specification limits are narrower, the percentage of nonconforming products will increase. On the other hand, if the specification limits remain the same but the process standard deviation becomes smaller, then the percentage of nonconforming products will be reduced (or the percentage of conforming products will be increased). So, to make the process more capable and improve it continuously, we should not only focus on eliminating special causes, but also work to eliminate common causes. These efforts will consequently reduce the process standard deviation and make the process more capable.

## Review Practice Problems

1  What is the difference between common (random) causes and assignable (special) causes?

2  Define *action on the process* and *action on the output*.

3  In your own words, give a brief answer to this question: "What is a control chart?"

4  What is the difference between control limits and specification limits?

**5** Define the average run length (ARL). What are the advantages and disadvantages of having large or small values for the ARL?

**6** Suppose sample means, $\overline{X}$, are calculated for 25 samples of size $n = 4$, for a process that is considered under statistical control. Also suppose that the process mean experiences shifts. Determine the type II errors if the process mean shifts by 0.5, 0.75, 1.0, 1.5, and 2.0 standard deviations when three-sigma control limits are used. How will the value of the probability of type II errors ($\beta$) change if the sample size is increased? Finally, calculate the value of ARL for each mean shift.

**7** Suppose sample means, $\overline{X}$, are calculated for 30 samples of size $n = 9$ for a process that is considered under statistical control. Also suppose that the process mean experiences shifts. Determine the probability of type II errors($\beta$) if the process mean shifts by 0.5, 0.75, 1.0, 1.5, and 2.0 standard deviations and three-sigma control limits are used. How does the value of $\beta$ change if the sample size is increased? Finally, calculate the value of ARL for each mean shift.

**8** Problem 7 is the same as Problem 6, except for the sample size. Compare the results you obtained in Problems 6 and 7, and comment on what you observe.

**9** Suppose that in Problem 6, samples are taken every hour. Determine how long it will take on average to detect each shift.

**10** Suppose that in Problem 7, samples are taken every half hour. Determine how long it will take on average to detect each shift.

**11** Suppose the sample means, $\overline{X}$, and ranges, $R$, are calculated for 30 samples of size $n = 5$ for a process that is considered to be under statistical control. Also suppose that the means of the 30 sample means and ranges are calculated, yielding $\overline{\overline{X}} = 40.50$ and $\overline{R} = 12.60$.
a) Estimate the process standard deviation using the mean of the sample ranges.
b) Determine the control limits for the $\overline{X}$ chart.
c) Determine the control limits for the $R$ chart.

**12** What is the significance of the $R$ chart? Explain its importance in the development of a process.

**13** Suppose sample means, $\overline{X}$, and ranges, $R$, are calculated for 25 samples of size $n = 4$ for a process that is considered to be under statistical control. Also suppose that the 25 sample means and ranges average $\overline{\overline{X}} = 60.25$ and $\overline{R} = 14.30$.
a) Estimate the process standard deviation using the mean of the sample ranges.
b) Determine the control limits for the $\overline{X}$ chart.
c) Determine the control limits for the $R$ chart.

**14** The following data are sample means (rows 1 and 3) and sample ranges (rows 2 and 4) of 25 samples of size 5 taken from a production process for rods used in small-car engines. The measurements are the lengths of the rods in centimeters.

| 20.4 | 19.5 | 19.6 | 19.8 | 20.6 | 19.9 | 20.52 | 19.55 | 19.39 | 20.78 | 20.29 | 19.82 | 19.90 |
|------|------|------|------|------|------|-------|-------|-------|-------|-------|-------|-------|
| 1.58 | 1.56 | 1.55 | 1.66 | 1.38 | 1.57 | 1.38 | 1.45 | 1.18 | 1.32 | 1.45 | 1.59 | 1.68 |
| 20.54 | 20.6 | 19.3 | 20.10 | 19.4 | 20.30 | 20.78 | 20.13 | 20.75 | 20.29 | 19.95 | 19.47 | 1.78 |
| 1.83 | 1.78 | 1.59 | 1.78 | 1.82 | 1.83 | 1.87 | 1.79 | 1.82 | 1.76 | 1.83 | | |

a) Set up the trial limits on $\overline{X}$ and R charts for this production process. Is this process in statistical control?

b) If, in part (a), points fall beyond the control limits, find the revised control limits by eliminating the points that fall beyond the control limits.

**15** a) Given the data from Problem 14, estimate the process mean, $\mu$, and standard deviation, $\sigma$.

b) Suppose that the specification limits for the rod lengths are $20 \pm 0.95$. Find the percentage of nonconforming rods manufactured by this process. Assume that the rod lengths are normally distributed.

**16** Suppose in Problem 15 that the manufacturer renegotiates new specification limits on the rod lengths: $20 \pm 1.05$. Under the new specification limits, determine the percentage of nonconforming rods produced by the process. Again, assume that the lengths are normally distributed.

**17** In Problem 14, a quality control engineer collects 25 new samples of size $n = 5$ and then sets up $\overline{X}$ and S control charts. The data collected are shown in the following table. Determine the control limits of the $\overline{X}$ and S control charts, and plot all the points. Compare the $\overline{X}$ and S control charts with the $\overline{X}$ and R control charts from Problem 14, and check whether the conclusions change.

| Sample # | $x_1$ | $x_2$ | $x_3$ | $x_4$ | $x_5$ |
|----------|-------|-------|-------|-------|-------|
| 1 | 19.25 | 20.48 | 20.66 | 19.86 | 19.95 |
| 2 | 19.62 | 20.19 | 20.38 | 20.52 | 20.50 |
| 3 | 19.69 | 20.45 | 19.76 | 19.66 | 19.74 |
| 4 | 19.57 | 20.94 | 19.41 | 21.10 | 19.82 |
| 5 | 19.84 | 20.87 | 19.86 | 19.96 | 20.68 |
| 6 | 20.95 | 19.87 | 19.91 | 19.26 | 19.95 |
| 7 | 19.84 | 20.59 | 20.77 | 19.67 | 21.77 |
| 8 | 20.49 | 20.28 | 19.57 | 20.90 | 19.79 |
| 9 | 20.30 | 20.26 | 20.66 | 21.04 | 19.81 |
| 10 | 20.59 | 19.26 | 19.68 | 20.96 | 19.93 |
| 11 | 20.40 | 19.55 | 21.11 | 20.42 | 20.57 |
| 12 | 20.10 | 20.27 | 19.69 | 19.76 | 20.43 |
| 13 | 19.70 | 20.14 | 21.06 | 20.23 | 19.68 |
| 14 | 20.10 | 21.02 | 19.99 | 19.71 | 19.75 |
| 15 | 20.30 | 20.15 | 19.95 | 19.76 | 20.49 |
| 16 | 19.88 | 21.10 | 19.95 | 21.13 | 19.58 |
| 17 | 19.90 | 19.66 | 19.97 | 19.74 | 20.36 |
| 18 | 20.43 | 19.84 | 20.42 | 21.05 | 20.35 |
| 19 | 19.83 | 20.30 | 19.95 | 20.44 | 20.53 |
| 20 | 20.52 | 20.66 | 19.66 | 21.05 | 19.91 |
| 21 | 20.32 | 19.91 | 20.40 | 20.38 | 20.27 |
| 22 | 21.02 | 21.04 | 20.49 | 19.75 | 19.90 |

| Sample # | $x_1$ | $x_2$ | $x_3$ | $x_4$ | $x_5$ |
|----------|-------|-------|-------|-------|-------|
| 23 | 19.60 | 20.22 | 20.46 | 21.07 | 21.13 |
| 24 | 19.86 | 19.97 | 20.61 | 20.49 | 20.68 |
| 25 | 20.61 | 19.74 | 20.19 | 21.02 | 19.89 |

**18** a) Use the data in Problem 17 to estimate the process mean, $\mu$, and standard deviation, $\sigma$.

b) What should the specification limits be so there are no more than 5% nonconforming rods?

**19** After establishing the trial control limits in Problem 17, the engineer decides to collect another set of 20 samples of size $n = 5$. The data are shown in the following table. Use the control charts in Problem 17 to examine whether the process continues to be in statistical control.

| Sample # | $x_1$ | $x_2$ | $x_3$ | $x_4$ | $x_5$ |
|----------|-------|-------|-------|-------|-------|
| 1 | 20.71 | 19.80 | 19.93 | 20.72 | 20.37 |
| 2 | 20.73 | 19.89 | 19.87 | 19.97 | 20.45 |
| 3 | 20.15 | 20.50 | 19.91 | 19.79 | 20.23 |
| 4 | 21.14 | 21.20 | 20.66 | 20.72 | 20.77 |
| 5 | 19.68 | 19.85 | 19.71 | 19.95 | 20.45 |
| 6 | 20.55 | 21.06 | 20.65 | 20.48 | 21.11 |
| 7 | 20.77 | 19.93 | 20.17 | 19.88 | 20.47 |
| 8 | 19.81 | 20.32 | 20.18 | 21.06 | 19.99 |
| 9 | 20.86 | 19.05 | 19.76 | 20.24 | 19.19 |
| 10 | 19.58 | 19.81 | 20.37 | 20.45 | 19.46 |
| 11 | 21.01 | 21.05 | 20.86 | 19.87 | 19.87 |
| 12 | 20.45 | 19.87 | 19.95 | 19.94 | 20.44 |
| 13 | 20.57 | 20.58 | 19.89 | 19.70 | 19.95 |
| 14 | 20.21 | 20.60 | 20.80 | 21.08 | 21.12 |
| 15 | 20.86 | 19.87 | 19.89 | 20.81 | 20.30 |
| 16 | 19.98 | 20.67 | 20.18 | 21.04 | 20.78 |
| 17 | 19.89 | 19.81 | 20.15 | 20.29 | 20.54 |
| 18 | 19.96 | 20.45 | 20.31 | 21.09 | 20.85 |
| 19 | 19.88 | 20.45 | 20.54 | 20.19 | 19.76 |
| 20 | 21.12 | 20.78 | 20.88 | 20.66 | 20.98 |

**20** If, after plotting all the points in Problem 19, you observe that the process is under control, find the average run length for the $\overline{X}$ control chart. Otherwise, find the new control limits by ignoring the points that fall outside the control limits.

**21** Twenty-four samples of four insecticide dispensers were taken during production. The averages and ranges of the charge weight (in grams) in the 24 samples are shown in the following table. The charging machine's specifications say that it should produce charges having a mean of 454 gm and a standard deviation of 9 gm. Set up $\overline{X}$ and $R_i$ control charts with three-sigma control limits for these data, and determine whether the process is in statistical control. If the process is in statistical control, determine whether the sample averages and sample ranges are behaving per the specifications.

| Sample no. | Sample average, $\bar{x}$ | Sample range, $R$ | Sample no. | Sample average, $\bar{x}$ | Sample range, $R$ |
|---|---|---|---|---|---|
| 1 | 471.5 | 19 | 13 | 457.5 | 4 |
| 2 | 462.2 | 31 | 14 | 431.0 | 14 |
| 3 | 458.5 | 22 | 15 | 454.2 | 19 |
| 4 | 476.5 | 27 | 16 | 474.5 | 18 |
| 5 | 461.8 | 17 | 17 | 475.8 | 18 |
| 6 | 462.8 | 38 | 18 | 455.8 | 24 |
| 7 | 464.0 | 38 | 19 | 497.8 | 34 |
| 8 | 461.0 | 37 | 20 | 448.5 | 21 |
| 9 | 450.2 | 19 | 21 | 453.0 | 58 |
| 10 | 479.0 | 41 | 22 | 469.8 | 32 |
| 11 | 452.8 | 15 | 23 | 474.0 | 10 |
| 12 | 467.2 | 28 | 24 | 461.0 | 36 |

**22** $\bar{X}$ and $R$ control charts with three-sigma control limits are desired for a certain dimension of rheostat knobs. During 3 days, samples of 5 knobs were taken every hour during production (a production day is 8 hours), and the averages and ranges of each of the 24 samples obtained are shown in the following table. Using the information from these samples, set up $\bar{X}$ and $R$ control charts with three-sigma control limits, and graph the sample averages. Is the rheostat knob dimension under statistical control?

| Sample no. | Sample average, $\bar{x}$ | Sample range, $R$ | Sample no. | Sample average, $\bar{x}$ | Sample range, $R$ |
|---|---|---|---|---|---|
| 1 | .1396 | .012 | 13 | .1394 | .010 |
| 2 | .1414 | .009 | 14 | .1422 | .010 |
| 3 | .1415 | .003 | 15 | .1412 | .009 |
| 4 | .1406 | .008 | 16 | .1417 | .005 |
| 5 | .1416 | .009 | 17 | .1402 | .006 |
| 6 | .1404 | .013 | 18 | .1425 | .006 |
| 7 | .1400 | .006 | 19 | .1407 | .011 |
| 8 | .1418 | .008 | 20 | .1408 | .010 |
| 9 | .1410 | .008 | 21 | .1430 | .006 |
| 10 | .1432 | .005 | 22 | .1398 | .007 |
| 11 | .1448 | .006 | 23 | .1415 | .008 |
| 12 | .1428 | .008 | 24 | .1409 | .004 |

**23** Measurements (in inches) of a particular dimension of fragmentation bomb heads manufactured during World War II by the American Stove Company were taken for successive samples of five bomb heads, each giving the results in the following table (data from Duncan (1958)). From the first 20 samples, set up $\bar{X}$ and $R$ control charts with three-sigma control limits. Plot the values of $\bar{X}$ and $R$ on these charts to see if the process is in a state of statistical control.

| Sample no. | Sample average | Sample range | Sample no. | Sample average | Sample range |
|---|---|---|---|---|---|
| 1 | .4402 | .015 | 16 | .4362 | .015 |
| 2 | .4390 | .018 | 17 | .4380 | .019 |
| 3 | .4448 | .018 | 18 | .4350 | .008 |
| 4 | .4432 | .006 | 19 | .4378 | .011 |

| Sample no. | Sample average | Sample range | Sample no. | Sample average | Sample range |
|---|---|---|---|---|---|
| 5 | .4428 | .008 | 20 | .4384 | .009 |
| 6 | .4382 | .010 | 21 | .4392 | .006 |
| 7 | .4358 | .011 | 22 | .4378 | .008 |
| 8 | .4440 | .019 | 23 | .4362 | .016 |
| 9 | .4366 | .010 | 24 | .4348 | .009 |
| 10 | .4368 | .011 | 25 | .4338 | .005 |
| 11 | .4360 | .011 | 26 | .4366 | .014 |
| 12 | .4402 | .007 | 27 | .4346 | .009 |
| 13 | .4332 | .008 | 28 | .4374 | .015 |
| 14 | .4356 | .017 | 29 | .4339 | .024 |
| 15 | .4314 | .010 | 30 | .4368 | .014 |

**24** Twenty samples of five mechanical components each were taken randomly from a production process, and the inside diameter of a hole in each component was measured to the nearest 0.00025 cm and rounded to whole numbers. The values of $x_1$, $x_2$, $x_3$, $x_4$, $x_5$ obtained for each of the 20 samples rounded to whole numbers are shown in the following table. Construct $\overline{X}$ and $S$ control charts for controlling the inside diameter in future production.

| | | | | | |
|---|---|---|---|---|---|
| 1 | 30 | 22 | 16 | 33 | 12 |
| 2 | 28 | 32 | 22 | 30 | 14 |
| 3 | 26 | 12 | 18 | 11 | 20 |
| 4 | 30 | 30 | 18 | 33 | 14 |
| 5 | 18 | 24 | 18 | 18 | 16 |
| 6 | 22 | 28 | 22 | 26 | 10 |
| 7 | 26 | 24 | 18 | 13 | 20 |
| 8 | 20 | 30 | 24 | 9 | 12 |
| 9 | 16 | 24 | 28 | 20 | 20 |
| 10 | 20 | 20 | 18 | 30 | 28 |
| 11 | 26 | 32 | 24 | 33 | 36 |
| 12 | 14 | 20 | 18 | 24 | 32 |
| 13 | 22 | 14 | 32 | 22 | 28 |
| 14 | 22 | 14 | 20 | 22 | 14 |
| 15 | 26 | 18 | 24 | 29 | 34 |
| 16 | 34 | 20 | 22 | 20 | 16 |
| 17 | 8 | 28 | 10 | 24 | 22 |
| 18 | 16 | 18 | 12 | 29 | 18 |
| 19 | 18 | 20 | 14 | 22 | 26 |
| 20 | 30 | 20 | 24 | 26 | 32 |

**25** The following data are sample means (rows 1 and 3) and sample ranges (rows 2 and 4) of 25 samples of size 5 taken from a production process for bolts. The bolts are used to mount wheels on a car, and the measurements are the lengths of the bolts in cm.

| | | | | | | | | | | | | |
|---|---|---|---|---|---|---|---|---|---|---|---|---|
| 10.37 | 9.71 | 9.58 | 9.71 | 10.41 | 9.88 | 10.52 | 9.55 | 9.39 | 10.68 | 10.09 | 9.82 | 9.90 |
| 1.48 | 1.51 | 1.35 | 1.46 | 1.27 | 1.57 | 1.38 | 1.26 | 1.05 | 1.48 | 1.34 | 1.39 | 1.41 |
| 10.24 | 9.42 | 9.26 | 10.10 | 9.45 | 10.10 | 10.48 | 10.53 | 10.65 | 10.26 | 9.90 | 10.47 | |
| 1.43 | 1.41 | 1.48 | 1.33 | 1.04 | 1.34 | 1.29 | 1.35 | 1.05 | 1.45 | 1.23 | 1.47 | |

a) Set up $\overline{X}$ and $R$ control charts with three-sigma control limits for this production process. Is this process in statistical control?

b) In (a), if some points fall beyond the control limits, find the revised control limits by eliminating the points beyond the control limits.

26 a) For the data in Problem 25, estimate the process mean, $\mu$, and standard deviation, $\sigma$.

b) Suppose that, in this problem, the specification limits on the length of bolts are $10 \pm 1.55$. Find the percentage of nonconforming bolts manufactured by this process. Assume that the bolt lengths are normally distributed.

27 Suppose that, in Problem 26, the manufacturer negotiates new specification limits on the bolt length: $10 \pm 1.60$ cm. Under the new specification limits, determine the percentage of nonconforming bolts produced by the process, assuming that the bolt lengths are normally distributed.

28 Referring to Problem 25, suppose that a quality control engineer collects 25 new samples of size $n = 5$ and sets up $\overline{X}$ and $S$ control charts with three-sigma control limits. The data collected are shown in the following table. Determine the control limits of the $\overline{X}$ and $S$ control charts, and plot all the points. Compare these $\overline{X}$ and $S$ control charts with the $\overline{X}$ and $R$ control charts in Problem 25 to check how the control limits have changed and also whether the conclusions have changed.

| Sample # | $x_1$ | $x_2$ | $x_3$ | $x_4$ | $x_5$ |
|----------|-------|-------|-------|-------|-------|
| 1 | 9.05 | 10.47 | 10.74 | 9.67 | 9.92 |
| 2 | 9.60 | 10.14 | 10.38 | 10.22 | 10.30 |
| 3 | 9.79 | 10.40 | 9.70 | 9.35 | 9.24 |
| 4 | 10.57 | 8.94 | 11.41 | 10.60 | 9.80 |
| 5 | 9.84 | 10.83 | 9.96 | 9.92 | 10.68 |
| 6 | 10.93 | 9.37 | 9.51 | 10.22 | 9.75 |
| 7 | 10.84 | 9.59 | 10.77 | 10.67 | 8.97 |
| 8 | 10.49 | 10.27 | 9.53 | 10.40 | 9.79 |
| 9 | 10.30 | 10.26 | 10.66 | 11.04 | 10.11 |
| 10 | 10.59 | 10.26 | 9.68 | 8.96 | 9.90 |
| 11 | 10.40 | 10.55 | 11.11 | 10.42 | 10.54 |
| 12 | 10.10 | 10.27 | 9.66 | 9.56 | 10.43 |
| 13 | 9.70 | 10.14 | 11.16 | 10.23 | 9.68 |
| 14 | 9.10 | 9.02 | 9.91 | 9.71 | 9.77 |
| 15 | 10.32 | 10.10 | 9.95 | 9.56 | 10.49 |
| 16 | 9.88 | 10.10 | 9.95 | 10.13 | 9.58 |
| 17 | 9.40 | 9.66 | 9.77 | 9.74 | 9.36 |
| 18 | 9.43 | 10.84 | 10.42 | 10.05 | 9.35 |
| 19 | 9.83 | 10.30 | 9.95 | 9.44 | 10.23 |
| 20 | 9.52 | 10.66 | 9.66 | 10.05 | 9.91 |
| 21 | 9.32 | 10.91 | 10.40 | 10.36 | 10.29 |
| 22 | 9.02 | 10.04 | 11.19 | 9.45 | 9.60 |
| 23 | 9.60 | 9.22 | 10.56 | 9.07 | 10.13 |
| 24 | 9.76 | 10.97 | 9.61 | 10.46 | 10.64 |
| 25 | 9.61 | 9.74 | 9.19 | 10.02 | 9.77 |

**29**  a) Use the data in Problem 25 to estimate the process mean, $\mu$, and standard deviation, $\sigma$.

b) What should the specification limits be so that the fraction of nonconforming bolts is no more than 5%?

**30**  A quality control engineer collected a set of 15 samples of size $n = 5$. The data are shown in the following table. Based on these data, determine control limits for the $\overline{X}$ and $S$ control charts. Plot the data on the control charts, and examine whether the process is in statistical control.

| Sample # | $x_1$ | $x_2$ | $x_3$ | $x_4$ | $x_5$ |
|----------|-------|-------|-------|-------|-------|
| 1 | 10.51 | 9.30 | 9.73 | 9.72 | 10.27 |
| 2 | 10.83 | 9.79 | 9.27 | 9.84 | 9.48 |
| 3 | 9.91 | 10.52 | 9.41 | 9.75 | 10.23 |
| 4 | 10.34 | 10.20 | 9.66 | 9.72 | 10.57 |
| 5 | 9.68 | 10.35 | 9.31 | 9.91 | 10.47 |
| 6 | 10.57 | 11.36 | 9.65 | 9.48 | 10.11 |
| 7 | 10.77 | 9.93 | 9.17 | 9.88 | 9.47 |
| 8 | 9.81 | 10.00 | 9.62 | 9.06 | 8.99 |
| 9 | 10.86 | 10.05 | 8.76 | 10.24 | 10.19 |
| 10 | 9.38 | 9.61 | 10.37 | 9.45 | 8.46 |
| 11 | 10.01 | 10.05 | 10.86 | 8.62 | 9.80 |
| 12 | 10.29 | 9.82 | 9.65 | 9.64 | 10.54 |
| 13 | 10.51 | 10.38 | 8.89 | 9.77 | 9.85 |
| 14 | 9.21 | 9.60 | 9.60 | 11.08 | 10.12 |
| 15 | 10.86 | 9.87 | 9.84 | 10.88 | 9.30 |

**31**  In Problem 28, if you observe that the process is under statistical control, find the average run length for the $\overline{X}$ control chart. Otherwise, find the new control limits by ignoring points that fall beyond the control limits.

**32**  The data shown in the following table provide 25 samples of size 5 of measurements of throttle-adjustment screws used in a car's fuel system. Using Minitab or R, construct three-sigma $\overline{X}$ and $R$ control charts for these data. Is the process in statistical control?

| | | | | |
|------|------|------|------|------|
| 1.66 | 1.93 | 1.76 | 2.10 | 2.10 |
| 1.93 | 2.17 | 1.82 | 1.88 | 1.98 |
| 1.94 | 1.88 | 1.88 | 1.66 | 1.71 |
| 2.37 | 1.94 | 2.09 | 2.03 | 2.02 |
| 2.40 | 2.14 | 2.08 | 2.36 | 2.14 |
| 1.77 | 2.25 | 2.25 | 1.83 | 1.83 |
| 2.03 | 2.16 | 1.69 | 2.19 | 1.91 |
| 2.03 | 1.85 | 2.03 | 2.02 | 1.81 |
| 1.93 | 2.24 | 1.62 | 2.02 | 2.07 |
| 2.36 | 2.27 | 2.18 | 2.08 | 2.32 |
| 2.07 | 1.86 | 1.75 | 2.08 | 1.78 |
| 1.70 | 2.01 | 2.02 | 1.75 | 2.05 |
| 1.97 | 2.18 | 1.69 | 2.24 | 1.97 |
| 1.75 | 1.99 | 2.07 | 1.57 | 1.77 |
| 1.78 | 1.67 | 2.17 | 1.91 | 2.16 |
| 1.91 | 2.13 | 1.74 | 1.95 | 1.77 |
| 2.04 | 1.94 | 2.14 | 1.75 | 2.16 |

*(Continued)*

| 2.06 | 2.19 | 2.03 | 1.93 | 2.13 |
|------|------|------|------|------|
| 1.85 | 2.07 | 2.25 | 2.09 | 1.78 |
| 1.94 | 2.07 | 1.98 | 1.98 | 1.90 |
| 2.02 | 2.38 | 2.09 | 1.72 | 1.93 |
| 1.70 | 2.17 | 2.31 | 1.98 | 1.87 |
| 1.90 | 2.03 | 2.26 | 1.79 | 2.28 |
| 2.09 | 2.18 | 2.13 | 2.00 | 1.83 |
| 2.04 | 2.22 | 1.77 | 1.94 | 2.05 |

**33** Refer to Problem 32.
  a) If the specifications are $2 \pm 0.5$, what percentage of the screws are nonconforming?
  b) Determine the specification limits so that the percentage of nonconforming screws is no more than 5%.

**34** If the process mean in Problem 32 shifts upward by one standard deviation, find the probability that the shift is detected in the first subsequent sample. Assume that the process is normally distributed.

**35** Set up 99% probability limits for the $\overline{X}$ and R control charts using the data from Problem 32 using Minitab, R, or JMP, and determine whether the process is in statistical control.

**36** A quality engineer in the semiconductor industry wants to maintain control charts for a critical characteristic. After collecting 30 samples of size 5, the engineer computes $\overline{X}$ and $R$ for each sample. Suppose that the samples produce the following statistics:

$$\sum_{i=1}^{30} \overline{X}_i = 450 \text{ and } \sum_{i=1}^{30} R_i = 80$$

  a) Set up $\overline{X}$ and $R$ control charts for these data using three-sigma control limits.
  b) Assuming that the process is in statistical control, estimate the process mean, $\mu$, and process standard deviation, $\sigma$.

**37** In Problem 36, the quality control engineer also computes the sample standard deviation for each sample, obtaining the following statistics:

$$\sum_{i=1}^{30} \overline{X}_i = 450, \quad \sum_{i=1}^{30} S_i = 35.5$$

  Set up $\overline{X}$ and S control charts for these data using three-sigma control limits.

**38** Using Minitab, R, or JMP, set up $\overline{X}$ and R control charts for the data in Problem 28 using 99% probability control limits, and then examine whether the process is in statistical control.

**39** Set up $\overline{X}$ and S control charts for the data in Problem 32 using 99% probability control limits, and reexamine whether the process is in statistical control.

**40** Suppose that the process mean in Problem 32 experiences a downward shift of one standard deviation immediately after the engineer has finished collecting the data. Determine the probability that the shift will be detected in the first subsequent sample.

**41** During a trial period, 25 samples of size 5 from a process provide the following statistics:

$$\overline{\overline{X}} = 25.5 \text{ oz}, \quad \overline{R} = 2.65 \text{ oz}$$

Construct the OC curve of the control chart using these values and three-sigma control limits. Assume that the process mean has suffered downward shifts of 0.1, 0.2, 0.6, 0.8, 1.0, 1.2, 1.6, 1.8, 2.0, and 2.5 standard deviations.

**42** Consider a three-sigma $\overline{X}$ control chart. Draw an OC curve for a process whose process mean has suffered upward shifts of 0.1, 0.3, 0.6, 0.8, 1.0, 1.5, 1.8, 2.0, and 2.5 standard deviations. Assume that the sample size is 4.

# 6

# Phase I Control Charts for Attributes

## 6.1   Introduction

In Chapter 5, we studied control charts for quality characteristics that are measured numerically. But not all quality characteristics are measured this way. For example, we may be interested in finding whether a new type of car paint meets certain specifications in terms of shine, uniformity, and scratches. In this case, we cannot quantify shine, blemishes, or scratches numerically, and instead classify these characteristics simply as conforming or nonconforming to specification.

Sometimes, even when a quality characteristic is measured numerically, it may be appropriate to classify the characteristic as conforming or nonconforming to specification. This may happen when the cost of taking measurements is too high, there is not enough time to take proper measurements, or both. In such scenarios, we prefer to use more economical and time-saving methods such as a *go or no-go* gauge.

In this chapter, we study control charts appropriate for conforming/nonconforming quality characteristics, called *control charts for attributes*. Specifically, we discuss control charts for detecting significant shifts in attributes that usually occur in phase I of implementing statistical process control (SPC).

## 6.2   Control Charts for Attributes

In day-to-day life, when a quality characteristic of a product is not measurable numerically, we usually classify the product as non-defective or defective. In SPC, however, it has become more common to use the terms *conforming or nonconforming,* respectively.

> **Definition 6.1**   A quality characteristic that classifies any product as conforming or nonconforming is called an *attribute*.

Examples of attributes include quality characteristics that identify whether a soft drink can is leaking, a stud has regular edges, a rod fits into a slot, a 100-watt light bulb meets the desired standard, or a steel rivet meets the manufacturer's specifications. Note that the data collected on a quality characteristic, which is an attribute, are simply a count data. A drawback of control charts for attributes is that larger sample sizes (sometimes in the hundreds) are required than those for variables

*Statistical Quality Control: Using Minitab, R, JMP, and Python*, First Edition. Bhisham C. Gupta.
© 2021 John Wiley & Sons, Inc. Published 2021 by John Wiley & Sons, Inc.
Companion website: www.wiley.com/go/college/gupta/SQC

**Table 6.1** Control charts for attributes.

| Control chart | Quality characteristic under investigation |
| --- | --- |
| $p$ control chart | Percent or fraction of nonconforming units in a subgroup or sample, where sample size may or may not be constant |
| $np$ control chart | Number of nonconforming units in a sample where sample size is constant |
| $c$ control chart | Number of nonconformities in a sample or in one or more inspection units |
| $u$ control chart | Number of nonconformities per unit, where sample size can be variable. |

(four or five). Furthermore, attribute control charts, which are used only after defects have occurred, are generally less informative than variable control charts, which are useful in detecting defects before they occur.

There are cases, however, when variable control charts have limitations. For example, consider a product that is nonconforming because 1 of 10 quality characteristics does not conform to specifications. In such a case, we cannot control all 10 quality characteristics with one variable control chart, since a variable control chart can control only one quality characteristic at a time. Thus, to control all 10 quality characteristics, we would need 10 different variable control charts. On the other hand, by identifying units as simply conforming or nonconforming, one attribute control chart can simultaneously and suitably analyze all 10 quality characteristics. As you can see, both variable and attribute control charts have pros and cons.

In some cases, the quality characteristic is such that instead of classifying a unit as conforming or nonconforming, it records the number of nonconformities per manufactured unit: the number of holes in a roll of paper, irregularities per unit area of a spool of cloth, blemishes on a painted surface, loose ends in a circuit board, nonconformities per unit length of a cable, and so on. For such cases, despite having a numerical measure for each unit, we still use control charts for attributes. Such control charts are used to reduce the number of nonconformities per unit length, unit area, or unit volume of a single manufactured product.

These control charts are similar to the control charts for variables in that the center line and control limits are set in the same manner as for variables. However, it is important to note that the purposes of using control charts for variables and control charts for attributes are distinct. As noted in Chapter 5, the purpose of using control charts for variables in any process is to reduce variability due to special or assignable causes, whereas control charts for attributes are used to reduce the number of nonconforming units, nonconformities per manufactured or assembled unit, or nonconformities per unit length, unit area, or unit volume of a single manufactured product.

In this chapter, we study four types of control charts for attributes: the $p$ control chart, $np$ control chart, $c$ control chart, and $u$ control chart. Table 6.1 give a brief description of these control charts, which may be helpful to determine the appropriate chart for the quality characteristic under investigation.

## 6.3 The $p$ Chart: Control Charts for Nonconforming Fractions with Constant Sample Sizes

The most frequently used attribute control chart is the $p$ control chart. It is used whenever we are interested in finding the fraction or percent of nonconforming units when the quality characteristic is an attribute or variable measured by a *go* or *no-go gauge*. A $p$ control chart can be used to study one or more quality characteristics simultaneously since each inspected unit is classified as conforming or nonconforming irrespective of the number or types of defects. Furthermore, it is assumed that the conforming or nonconforming status of each unit is defined independently, which is true only if the process is stable or the probability of occurrence of a nonconforming unit at any given time is the same.

The basic rules of the $p$ control chart are governed by the binomial probability distribution with parameters $n$ and $p$, where $n$ is the sample size and $p$ is the fraction of nonconforming units produced by the process under investigation. As discussed in Chapter 3, the binomial probability distribution function of a random variable X with parameters $n$ and $p$ is defined as

$$P(X = x) = \binom{n}{x} p^x (1-p)^{n-x}; \; x = 0, 1, ..., n \tag{6.1}$$

The mean and standard deviation of the random variable $X$ are given by $np$ and $\sqrt{np(1-p)}$, respectively. For more details on the binomial distribution, see Section 3.5.

### 6.3.1 Control Limits for the $p$ Control Chart

To develop a $p$ control chart, proceed as follows:

1) From the process under investigation, select $m$ ($m \geq 25$) samples of size $n(n \geq 50)$ units. Note that if we have some prior information or clue that the process is producing a very small fraction of nonconforming units, the sample size should be large enough that the probability that it contains nonconforming units is relatively high.
2) Find the number of nonconforming units in each sample.
3) For each sample, find the fraction $p_i$ of nonconforming units

$$\hat{p}_i = \frac{x}{n} \tag{6.2}$$

where $x$ is the number of nonconforming units in the $i$th ($i = 1, 2, ..., m$) sample.
4) Find the average nonconforming, $\bar{p}$, over the $m$ samples, that is,

$$\bar{p} = \frac{\hat{p}_1 + \hat{p}_2 + ... + \hat{p}_m}{m} \tag{6.3}$$

The value of $\bar{p}$ determines the center line for the $p$ control chart and is an estimate of $p$, the process fraction of nonconforming units.

Using the well-known result that the binomial distribution with parameters $n$ and $p$ for large $n$ can be approximated by the normal distribution with mean $np$ and variance $np(1-p)$, it can easily be seen that $\bar{p}$ will be approximately normally distributed with mean $p$ and variance $\frac{p(1-p)}{n}$ or standard deviation $\sqrt{\frac{p(1-p)}{n}}$. Hence, the upper and lower three-sigma control limits (UCL and LCL) and the center line (CL) for the $p$ control chart are as follows:

$$UCL = \bar{p} + 3\sqrt{\frac{\bar{p}(1-\bar{p})}{n}} \tag{6.4}$$

$$CL = \bar{p} \tag{6.5}$$

$$LCL = \bar{p} - 3\sqrt{\frac{\bar{p}(1-\bar{p})}{n}} \tag{6.6}$$

Note that in Eqs. (6.4) and (6.6), $\sqrt{\frac{\bar{p}(1-\bar{p})}{n}}$ is an estimator of $\sqrt{\frac{p(1-p)}{n}}$, the standard deviation of $\bar{p}$. Furthermore, note that if control charts are being implemented for the first time, the control limits given by Eqs. (6.4–6.6) should be treated as the trial limits. In other words, before using these control limits any further, the points corresponding to all the samples used to determine these limits should be plotted to verify that all points fall within the control limits and that there is no evident pattern. If any sample points exceed the control limits, or if there is any pattern, then possible special causes should be investigated, detected, and eliminated before using or recalculating the control limits for future use. When recalculating the control limits, points that exceeded the trial control limits should be ignored, provided the special causes related to such points have been detected and eliminated.

Note: Sometimes, for small values of $\bar{p}$, $n$, or both, the value of the LCL may turn out to be negative. In such cases, we always set the LCL to zero. This is because the fraction of nonconforming units can never go below zero.

### 6.3.2 Interpreting the Control Chart for Nonconforming Fractions

If any point(s) exceed the UCL or LCL, we conclude that the process is not stable and special causes are present in the process. A point above the UCL is generally an indication of one of these:

- The control limit or the plotted point is in error.
- The process performance has deteriorated or is deteriorating.
- The measurement system has changed.

A point below the LCL is generally an indication of one of these:

- The control limit or the plotted point is in error.
- The process performance has improved or is improving. This condition of the process should be investigated very carefully so that the conditions of improvement are implemented on a permanent basis at this location and elsewhere in the company.
- The measurement system has changed.

The presence of special causes, which may be *favorable* (points fall below the CL) or *unfavorable* (point(s) exceed the UCL or consistently fall above the CL) must be investigated, and appropriate action(s) should be taken.

As is the case for the $\bar{X}$ and $R$ or $\bar{X}$ and $S$ control charts, the presence of any unusual patterns or trends is either an indication of an unstable process or an advance warning of conditions that, if left unattended or without any appropriate action, could make the process unstable.

If $\bar{p}$ is moderately high ($n\bar{p} \geq 5$), then an approximately equal number of points should fall on either side of the CL. Therefore, either of the following conditions could indicate that the process has shifted or a trend of shifting has started:

- A run of seven or more points going up or going down
- A run of seven or more points falling either below or above the CL

A run above the CL or a run going up generally indicates

- Process performance has deteriorated and may still be deteriorating.
- The measurement system has changed.

A run below the CL or a run going down generally indicates

- Process performance has improved and may still be improving.
- The measurement system has changed.

To illustrate the construction of a $p$ chart, we consider the data in Example 6.1, shown in Table 6.2.

### Example 6.1   Fuel Pump Data

A manufacturer of auto parts decides to evaluate the quality of the accelerating pump lever (APL) used in the fuel system. A team of engineers is assigned to study this problem to reduce the fraction of nonconforming units. To achieve this goal, the team decides to set up a $p$ control chart based on daily inspections of 900 units of APL over 30 days. Table 6.2 gives the number of nonconforming APLs out of 900 inspected units each day during the study period.

**Table 6.2**   Number of nonconforming APLs inspected each day during the study period.

| Day | Number of nonconforming units, $x$ | Sample fraction of nonconforming units, $p_i$ | Day | Number of Nonconforming units, $x$ | Sample fraction of nonconforming units, $p_i$ |
|---|---|---|---|---|---|
| 1 | 9 | 0.01 | 16 | 11 | 0.012 |
| 2 | 8 | 0.009 | 17 | 9 | 0.01 |
| 3 | 6 | 0.007 | 18 | 8 | 0.009 |
| 4 | 10 | 0.011 | 19 | 13 | 0.014 |
| 5 | 12 | 0.013 | 20 | 11 | 0.012 |
| 6 | 9 | 0.01 | 21 | 8 | 0.009 |
| 7 | 8 | 0.009 | 22 | 7 | 0.008 |
| 8 | 8 | 0.009 | 23 | 10 | 0.011 |
| 9 | 7 | 0.008 | 24 | 12 | 0.013 |
| 10 | 5 | 0.006 | 25 | 7 | 0.008 |
| 11 | 8 | 0.009 | 26 | 14 | 0.016 |
| 12 | 8 | 0.009 | 27 | 11 | 0.012 |
| 13 | 7 | 0.008 | 28 | 9 | 0.01 |
| 14 | 5 | 0.006 | 29 | 12 | 0.013 |
| 15 | 6 | 0.007 | 30 | 8 | 0.009 |

## Solution

Using the data in Table 6.2, we construct the $p$ chart as follows. First we calculate the sample fraction of nonconforming values ($p_i$), listed in columns 3 and 6 of the table. Substituting the sample fraction of nonconforming values in Eq. (6.3), we get

$$\bar{p} = \frac{9 + 8 + 6 + \cdots + 9 + 12 + 8}{30 \times 900} \cong 0.00985$$

Plugging the value of $\bar{p} = 0.00985$ and $n = 900$ in Eqs. (6.4) and (6.6), we get the control limits for the $p$ control chart:

$$UCL = 0.00985 + 3\sqrt{\frac{0.00985(1-0.00985)}{900}} = 0.01973$$

$$CL = 0.00985$$

$$LCL = 0.00985 - 3\sqrt{\frac{0.00985(1-0.00985)}{900}} = -0.00003 = 0$$

Note that since the LCL turns out to be negative, we set it to zero.

The $p$ control chart for the data in Table 6.2 is shown in Figure 6.1.

From the control chart in Figure 6.1, we observe that all the points are well within the control limits. We should note, however, that starting from point 7, nine successive points fall below the CL. This indicates that from days 7–15, the number of nonconforming pump levers was relatively low. Investigation to determine the process conditions on these days should be launched so that similar conditions can be implemented for future use. Other than that, since all the points of the current data fall within the control limits, these control limits can be extended for use over the next 30 days, after which the control chart should be re-evaluated.

We can construct a $p$ chart using any of the statistical packages discussed in this book. To do so using Minitab or R, we proceed as follows.

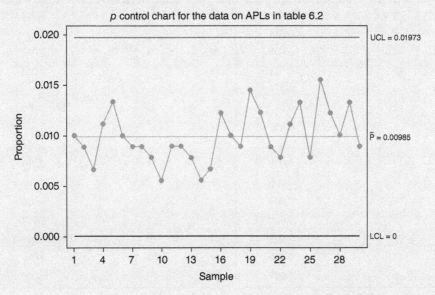

**Figure 6.1** Minitab printout of the $p$ control chart for the data on nonconforming APLs in Table 6.2.

**Minitab**

1) Enter the nonconforming data in column C1.
2) From the menu bar select **Stat > Control Charts > Attribute Chart > p** to open the **P Chart** dialog box. In this dialog box, enter **C1** in the box under **Variables**; and in the box next to **Subgroup size**, enter the sample size (e.g. **900** for this example). Then select any other desired option(s) and complete all the required entries for each selected one. Click **OK**. The *p* chart shown in Figure 6.1 will appear in the **Session Window**.

**R**

As in Chapter 5, we can use the built-in R function "qcc()" to construct a *p* chart. The following R code generates the *p* chart for the data in Table 6.2:

```
library (qcc)
setwd("C:/Users/person/Desktop/SQC-Data folder")
Nonconforming = read.csv("Table6.2.csv")
qcc(Nonconforming, type = "p", size = 900)
```

The *p* control chart generated by using R is exactly the same as shown in Figure 6.1.

Constructing the *p* control chart using JMP and Python is discussed in Chapter 10, which is available for download on the book's website: www.wiley.com/college/gupta/SQC.

## 6.4 The *p* Chart: Control Chart for Nonconforming Fractions with Variable Samples Sizes

There are times when, for various reasons, it is not possible to select samples of equal sizes. This is particularly true when the samples consist of 100% inspection of the process output during a fixed period of time on each day of the study period. The procedure to generate a *p* chart with variable sample sizes is very similar to the one used for generating a *p* chart with a constant sample size. For example, suppose we have $m$ samples of sizes $n_1, n_2, n_3, ..., n_m$. To generate a *p* chart with variable sample sizes, we proceed as follows:

1) From the process under investigation, select $m$ ($m \geq 25$) samples of sizes $n_1, n_2, n_3, ..., n_m$ ($n_i \geq 50$) units.
2) Find the number of nonconforming units in each sample.
3) For each sample, find the fraction $p_i$ of nonconforming units:

$$p_i = \frac{x}{n_i} \tag{6.7}$$

where $x$ is the number of nonconforming units in the $i$th ($i = 1, 2, ..., m$) sample.

4) Find the average fraction $\bar{p}$ of nonconforming units over the $m$ samples:

$$\bar{p} = \frac{n_1 p_1 + n_2 p_2 + \dots + n_m p_m}{n_1 + n_2 + n_3 + \dots + n_m} \qquad (6.8)$$

The value of $\bar{p}$ determines the CL for the $p$ chart and is an estimate of $p$, the process fraction of nonconforming units.

5) The control limits for $p$ with variable sample sizes are determined for each sample separately. Thus, for example, the three-sigma UCL and LCL for the $i$th sample are

$$\text{UCL} = \bar{p} + 3\sqrt{\frac{\bar{p}(1-\bar{p})}{n_i}} \qquad (6.9)$$

$$\text{CL} = \bar{p} \qquad (6.10)$$

$$\text{LCL} = \bar{p} - 3\sqrt{\frac{\bar{p}(1-\bar{p})}{n_i}} \qquad (6.11)$$

Note that the CL is the same for all samples, whereas the control limits are different for different samples.

To illustrate the construction of a $p$ chart with variable sample sizes, we consider the data in Table 6.3 of Example 6.2.

### Example 6.2 Fuel Pump Data

Suppose that in Example 6.1, all the APLs manufactured on each day during the 30-day study period are inspected. However, the number of APLs manufactured varies from day to day. The data collected during the study period are shown in Table 6.3. Construct the $p$ chart for these data, and determine whether the process is stable.

### Solution

Using the data in Table 6.3 and Eqs. (6.8–6.11), we can determine the control limits as

$$\text{UCL} = \bar{p} + 3\sqrt{\frac{\bar{p}(1-\bar{p})}{n_i}} = 0.00795 + 0.2664/\sqrt{n_i}$$

$$\text{CL} = 0.00795$$

$$\text{LCL} = \bar{p} - 3\sqrt{\frac{\bar{p}(1-\bar{p})}{n_i}} = 0.00795 - 0.2664/\sqrt{n_i} = 0$$

Note that since the LCL for each sample size turns out to be negative, we set them to zero. (So, the LCL is fixed at zero.) The $p$ chart for the data in Table 6.3 obtained using Minitab is shown in Figure 6.2.

This control chart shows that all of the points are well within the control limits and that there is no apparent pattern or trend in the chart. Thus, the process is stable. Also, note that in Figure 6.1 we had a run of nine points below the CL, whereas in Figure 6.2 there is no such run, even though we are dealing with the same process. Such differences are normal when samples are taken at different times.

**Table 6.3**  Number of nonconforming APLs with 100% inspection each day.

| Day | Number of nonconforming units, x | Sample size | Day | Number of nonconforming units, x | Sample size |
|---|---|---|---|---|---|
| 1 | 7 | 908 | 16 | 7 | 962 |
| 2 | 11 | 986 | 17 | 11 | 926 |
| 3 | 8 | 976 | 18 | 7 | 917 |
| 4 | 10 | 991 | 19 | 9 | 978 |
| 5 | 7 | 944 | 20 | 7 | 961 |
| 6 | 5 | 906 | 21 | 6 | 970 |
| 7 | 11 | 928 | 22 | 9 | 905 |
| 8 | 5 | 948 | 23 | 9 | 962 |
| 9 | 7 | 994 | 24 | 8 | 900 |
| 10 | 8 | 960 | 25 | 11 | 998 |
| 11 | 7 | 982 | 26 | 5 | 935 |
| 12 | 6 | 921 | 27 | 6 | 970 |
| 13 | 7 | 938 | 28 | 6 | 967 |
| 14 | 5 | 1000 | 29 | 9 | 983 |
| 15 | 6 | 982 | 30 | 8 | 976 |

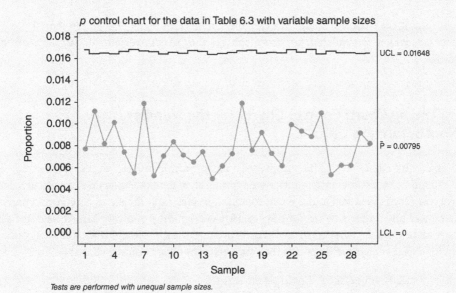

**Figure 6.2**  *p* chart for nonconforming APLs with variable sample sizes.

We can construct a *p* chart with variable sample sizes by using any of the statistical packages discussed in this book. To construct these charts using Minitab or R, we proceed as follows.

**Minitab**

1) Enter the nonconforming data in column C1 and the sample sizes in column C2.
2) From the menu bar select **Stat > Control Charts > Attribute Chart > p** to open the **p Chart** dialog box. In this dialog box, enter **C1** in the box under **Variables**, and in the box next to **Subgroup size**, enter **C2**. Then select any other desired option(s), and complete the required entries for each selected option. Click **OK**. The *p* chart shown in Figure 6.2 will appear in the **Session Window**.

**R**

As in Example 6.1, we can use the built-in R function "qcc()" to construct a *p* chart. The following R code can be used to generate the *p* chart for the data in Table 6.3.

```
library (qcc)
setwd("C:/Users/person/Desktop/SQC-Data folder")
Data = read.csv("Table6.3.csv")
qcc(Data[,1], sizes = Data[,2], type = "p")
```

The *p* control chart generated using R is exactly the same as shown in Figure 6.2.

Constructing the *p* control chart with variable sample sizes using JMP and Python is discussed in Chapter 10, which is available for download on the book's website: www.wiley.com/college/gupta/SQC.

## 6.5 The *np* Chart: Control Charts for the Number of Nonconforming Units

In an *np* control chart, we plot the number of nonconforming units in an inspected sample instead of the fraction of nonconforming units. Aside from a change in the scale on the vertical axis, the *np* control chart looks identical to the *p* control chart. In fact, both the *p* and *np* control charts can be implemented under the same circumstances. However, in the *p* control chart, the sample sizes can be constant or variable, whereas in the *np* control chart, the sample size is always kept constant. We summarize next some specific points that are pertinent for *np* control charts:

- The inspection sample sizes should be constant.
- The sample size should be large enough to include some nonconforming units.
- Record the sample size (sample size is constant) and the number of observed nonconforming units in each sample, and plot the number of nonconforming units on the control chart.

### 6.5.1 Control Limits for *np* Control Charts

Select $m$ samples, each of size $n$, from a given process under investigation, and then determine the number of nonconforming units in each sample. Let the number of nonconforming units found in each sample be denoted by $x_1, x_2, x_3, ..., x_m$, respectively. Then the control limits are determined as follows.

First, calculate $n\bar{p}$, the average number of nonconforming units per sample:

$$n\bar{p} = \frac{x_1 + x_2 + x_3 + ... + x_m}{m} \qquad (6.12)$$

Then the three-sigma control limits and CL for the $np$ control chart are given by

$$UCL = n\bar{p} + 3\sqrt{n\bar{p}(1-\bar{p})} \qquad (6.13)$$

$$CL = n\bar{p} \qquad (6.14)$$

$$LCL = np - 3\sqrt{n\bar{p}(1-\bar{p})} \qquad (6.15)$$

Note that in Eqs. (6.13) – (6.15), we use $\bar{p}$ as an estimator of $p$ when $p$ is not known. If, however, $p$ is known, then we use $p$ rather than $\bar{p}$.

To illustrate the construction of a $np$ control chart, we consider the data shown earlier in Table 6.2.

**Example 6.3**
Consider the data on APLs in Table 6.2 (Example 6.1). Construct an $np$ chart for these data, and verify whether the process is in statistical control.

From the control chart in Figure 6.3, we observe that all the points are well within the control limits. However, as in Figure 6.1, starting from point 7, nine successive points fall below the CL. This indicates that from day 7 through 15, the number of nonconforming was relatively low. The process conditions on these days should be investigated so that similar conditions can be extended for future use. Otherwise, since all the points of the current data fall within the control limits, the control limits can be extended for future use over the next 30 days after which the control chart should be re-evaluated.

We can construct an $np$ chart by using any of the statistical packages discussed in this book. To do so using Minitab or R, we proceed as follows.

**Minitab**
1) Enter the nonconforming data from Table 6.2 in column C1.
2) From the menu bar select **Stat > Control Charts > Attribute Chart > Np** to open the **Np Chart** dialog box. In this dialog box, enter **C1** in the box under **Variables**. And in the box next to **Subgroup size**, enter the sample size (e.g. **900** for this example). Then select any other desired option(s), and complete all the required entries for each selected option. Click **OK**. The $np$ chart shown in Figure 6.3 will appear in the **Session Window**.

**R**

As in Example 6.1, we can use the built-in R function "qcc()" to construct a *np* chart. The following R code can be used to generate the *np* chart for the data in Table 6.2.

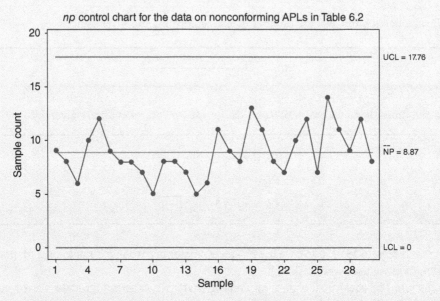

Figure 6.3 shows the *np* control chart for the data on nonconforming APLs in Table 6.2.

title on chart: *np* control chart for the data on nonconforming APLs in Table 6.2

UCL = 17.76

NP = 8.87

LCL = 0

Y-axis: Sample count; X-axis: Sample

**Figure 6.3** *np* control chart for the data on nonconforming APLs in Table 6.2.

```
library (qcc)
setwd("C:/Users/person/Desktop/SQC-Data folder")
Nonconforming = read.csv("Table6.2.csv")
qcc(Nonconforming, type = "np", size = 900)
```

The *np* control chart generated using R is exactly the same as shown in Figure 6.3.

Constructing an *np* control chart using JMP and Python is discussed in Chapter 10, which is available for download on the book's website: www.wiley.com/college/gupta/SQC.

## 6.6 The *c* Control Chart – Control Charts for Nonconformities per Sample

In many situations, we are interested in studying the number of nonconformities in a sample, which is also known as an *inspection unit*, rather than studying the fraction or total number of nonconforming units in a sample. This is particularly true when an inspected unit is nonconforming due to various types of nonconformities. For example, we may be interested in studying the quality of electric motors, which could be nonconforming due to defective bearings, a defective stator, a defective seal, or a defective terminal connection of the winding copper coil. As another example, suppose we are interested in studying the quality of printed circuit boards for laptops, which could be nonconforming due to mismatched or misaligned layers, solder flux corrosion, broken solder joints, incorrect controlled impedance, electrostatic discharge, and so on. In these cases, we may be more likely to study the nonconformities rather than the nonconforming units.

The control chart that is most commonly used to study nonconformities is the $c$ control chart. Comparing the $p$, $np$, and $c$ control charts, a $p$ control chart studies the fraction of nonconforming units, an $np$ control chart studies the total number of nonconforming units in each sample, and a $c$ control chart studies the total number of nonconformities in each sample (an inspection unit). The letter $c$ in $c$ control chart denotes the total number of nonconformities, which may be of one kind or of various kinds, in an inspection unit.

As an illustration, suppose that to develop trial control limits for a $c$ control chart for electric motors, we select samples of 100 motors, where each sample is considered one inspection unit or several inspection units, depending on how the inspection units are defined (the size of an inspection unit is purely a matter of convenience). The $c$ control chart is constructed with the samples sizes or number of inspection units in a sample being equal. It can be shown that under certain conditions or conditions of a Poisson process (see Gupta et al. (2020)), the number of nonconformities, $c$, is distributed according to a Poisson probability distribution with parameter $\lambda$, where $\lambda$ is the average number of nonconformities per inspection unit. The Poisson probability distribution is defined as

$$p(x) = \frac{e^{-\lambda}\lambda^x}{x!}, x = 0, 1, 2, 3, \ldots \tag{6.16}$$

where the mean and the variance of the Poisson distribution are equal to $\lambda$. For more details on this distribution, refer to Section 3.5. Now suppose that we select $m$ samples with each sample being one inspection unit, and let the number of nonconformities in these samples be equal to $c_1, c_2, c_3, \ldots, c_m$, respectively. Then the parameter $\lambda$, which is usually unknown, is estimated by

$$\hat{\lambda} = \bar{c} = \frac{c_1 + c_2 + c_3 + \ldots + c_m}{m} \tag{6.17}$$

Then the three-sigma control limits for the $c$ control chart are defined as follows:

$$\text{UCL} = \bar{c} + 3\sqrt{\bar{c}} \tag{6.18}$$

$$\text{CL} = \bar{c} \tag{6.19}$$

$$\text{LCL} = \bar{c} - 3\sqrt{\bar{c}} \tag{6.20}$$

Note that for small values of $\bar{c} (\leq 5)$, the Poisson distribution is asymmetric: the value of a type I error ($\alpha$) above the UCL and below the LCL usually are not the same. Thus, for small values of $\bar{c}$, it may be more prudent to use probability control limits rather than the three-sigma control limits. The probability control limits can be found by using Poisson distribution tables.

To illustrate the construction of $c$ control chart using three-sigma control limits, we consider the data in Table 6.4 of Example 6.4.

### Example 6.4  Paper Mill Data

A paper mill has detected that almost 90% of rejected paper rolls are due to nonconformities of two types: holes and wrinkles in the paper. The mill's team of quality control engineers decided to set up control charts to reduce or eliminate the number of these nonconformities. To do so, the team collected data by taking random sample of 5 rolls each day for 30 days and counting the number of nonconformities (holes and wrinkles) in each sample. The data are shown in Table 6.4. Set up a c control chart using these data, and check whether the process is in statistical control.

**Table 6.4**  Total number of nonconformities in samples consisting of five paper rolls each.

| Day | Total number of nonconformities | Day | Total number of nonconformities | Day | Total number of nonconformities |
|-----|--------------------------------|-----|--------------------------------|-----|--------------------------------|
| 1 | 8 | 11 | 7 | 21 | 9 |
| 2 | 6 | 12 | 6 | 22 | 6 |
| 3 | 7 | 13 | 6 | 23 | 8 |
| 4 | 7 | 14 | 8 | 24 | 7 |
| 5 | 8 | 15 | 6 | 25 | 6 |
| 6 | 7 | 16 | 6 | 26 | 9 |
| 7 | 8 | 17 | 8 | 27 | 9 |
| 8 | 7 | 18 | 9 | 28 | 7 |
| 9 | 6 | 19 | 8 | 29 | 7 |
| 10 | 9 | 20 | 9 | 30 | 8 |

### Solution

Using the data in Table 6.4, the estimate of the population parameter is given by

$$\hat{\lambda} = \bar{c} = \frac{\sum_{i=1}^{30} c_i}{30} = \frac{222}{30} = 7.4$$

Therefore, using Eqs. (6.18) – (6.20), the three-sigma control limits of the phase I c control chart are

$$\text{UCL} = 7.4 + 3\sqrt{7.4} = 15.56$$

$$\text{CL} = 7.4$$

$$\text{LCL} = 7.4 - 3\sqrt{7.4} = -0.76 = 0$$

Note that if the LCL turns out to be negative, as in this example, we set it to zero, since the number of nonconformities cannot be negative. The c control chart for the data in Table 6.4 is shown in Figure 6.4.

We can construct a c chart by using any of the statistical packages discussed in this book. To construct the c chart using Minitab or R, we proceed as follows.

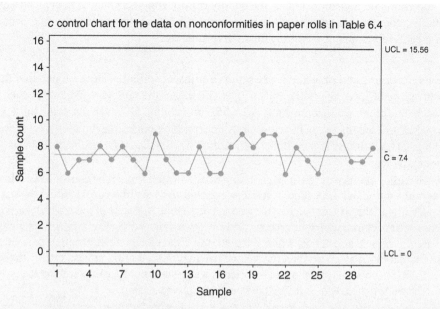

**Figure 6.4** *C* control chart of nonconformities for the data in Table 6.4.

**Minitab**

1) Enter the nonconformities data from Table 6.4 in column C1.
2) From the menu bar select **Stat > Control Charts > Attribute Chart > C** to open the **C Chart** dialog box. In this dialog box, enter **C1** in the box under **Variables**. Then select any other desired option(s), and complete all the required entries for each selected option. Click **OK**. The *c* chart shown in Figure 6.4 will appear in the **Session Window**.

**R**

As in earlier examples in this chapter, we can use the built-in R function "qcc()" to construct a *c* chart. The following R code can be used to generate the *c* chart for the data in Table 6.4:

```
library (qcc)
setwd("C:/Users/person/Desktop/SQC-Data folder")
Data = read.csv("Table6.4.csv")
qcc(Data[,2], type = "c")
```

The *c* control chart generated using R is exactly the same as shown in Figure 6.4. There may be minor differences in the control limits, which are due to rounding errors.

Constructing a *c* control chart using JMP and Python is discussed in Chapter 10, which is available for download on the book's website: www.wiley.com/college/gupta/SQC.

From the control chart shown in Figure 6.4, it is quite clear that the process is in statistical control. In other words, there are no special causes present in the process, and the only causes that may be affecting the process are common causes. Thus, to eliminate the nonconformities, the management must take action on the system, such as examining the quality of the wood chips used, changing old equipment, providing more training for the workers, etc. Also, to enhance the process further and eliminate nonconformities, the quality control engineers should apply techniques discussed in the analysis of the design of the experiment, which is beyond the scope of this book. The reader may refer to Gupta et al. (2020) for such techniques.

Notes:

- If economic factors and time allow, use samples or inspection units large enough that the LCL is positive. The LCL can be positive only if $\bar{c} > 9$. This means the sample size should be such that it can catch nine or more nonconformities with high probability. An advantage of having a positive LCL is that it allows us to see the conditions under which nonconformities are very low, thus providing an opportunity to perpetuate these conditions on-site and implement them elsewhere in the company.

- As noted earlier, the size of the inspection unit is usually determined by convenience. However, to determine the actual inspection unit size, we should also take into consideration the statistical characteristics of the process, such as the average run length, the state of the process (i.e. whether the process has deteriorated or improved), and any other factors that may require us to increase or decrease the sample size. Thus, while using control charts for nonconformities, particularly in phase I, situations may arise when the sample size may vary. But in $c$ control charts, the sample size is fixed. So when the sample size is variable, we use another type of control chart called a $u$ control chart. We discuss u control charts in the following section.

- If the samples consist of $n$ inspection units, the control limits for the $c$ control chart are given by

$$\text{UCL} = n\bar{c} + 3\sqrt{n\bar{c}} \tag{6.21}$$

$$\text{CL} = n\bar{c} \tag{6.22}$$

$$\text{LCL} = n\bar{c} - 3\sqrt{n\bar{c}} \tag{6.23}$$

## 6.7 The *u* Chart

A $u$ control chart is essentially the same as a $c$ control chart except that the $u$ control chart is always based on the number of nonconformities per inspection unit. For these charts, the number of units that each sample contains may vary, but the control limits are always determined based on one inspection unit. Thus, if $n$ is constant, we can use either a $c$ control chart or a $u$ control chart. In a $u$ control chart, the three-sigma control limits are given by

$$\text{UCL} = \bar{u} + 3\sqrt{\bar{u}/n} \tag{6.24}$$

$$\text{CL} = \bar{u} \tag{6.25}$$

$$\text{LCL} = \bar{u} - 3\sqrt{\bar{u}/n} \tag{6.26}$$

If the sample size varies, then we define $\bar{u}$ as

$$\bar{u} = \frac{c_1 + c_2 + \ldots + c_m}{n_1 + n_2 + \ldots + n_m} = \frac{\bar{c}}{\bar{n}} \tag{6.27}$$

where $m$ is the number of samples selected during the study period; $c_1, c_2, c_3, ..., c_m$ are the number of nonconformities in the $m$ samples; and $\bar{n}$ is the average sample size, which is given by

$$\bar{n} = \frac{n_1 + n_2 + ... + n_m}{m} \tag{6.28}$$

In this case, the CL is fixed, but the UCL and LCL are different for each sample point. The CL and three-sigma control limits for the $i$th sample are given by

$$UCL = \bar{u} + 3\sqrt{\bar{u}/n_i} \tag{6.29}$$

$$CL = \bar{u} \tag{6.30}$$

$$LCL = \bar{u} - 3\sqrt{\bar{u}/n_i} \tag{6.31}$$

Sometimes, if the samples do not vary too much, then the $n_i$'s in Eqs. (6.29) and (6.31) are replaced by $\bar{n}$, so that the control limits are

$$UCL = \bar{u} + 3\sqrt{\bar{u}/\bar{n}} \tag{6.32}$$

$$CL = \bar{u} \tag{6.33}$$

$$LCL = \bar{u} - 3\sqrt{\bar{u}/\bar{n}} \tag{6.34}$$

**Table 6.5** Number of nonconformities on printed boards for laptops per sample; each sample consists of five inspection units.

| Day | Number of nonconformities per sample | Day | Number of nonconformities per sample | Day | Number of nonconformities per sample |
|---|---|---|---|---|---|
| 1 | 48 | 11 | 40 | 21 | 47 |
| 2 | 49 | 12 | 44 | 22 | 33 |
| 3 | 38 | 13 | 43 | 23 | 37 |
| 4 | 49 | 14 | 35 | 24 | 33 |
| 5 | 43 | 15 | 31 | 25 | 34 |
| 6 | 37 | 16 | 42 | 26 | 49 |
| 7 | 45 | 17 | 34 | 27 | 50 |
| 8 | 48 | 18 | 30 | 28 | 49 |
| 9 | 39 | 19 | 49 | 29 | 35 |
| 10 | 46 | 20 | 44 | 30 | 39 |

To illustrate the construction of a $u$ chart, we consider the data in Table 6.5 of Example 6.5.

## Example 6.5   Printed Boards

A Six Sigma Green Belt team in the semiconductor industry found that the printed boards for laptops had nonconformities of several types, such as shorted traces, cold solder joints, and solder shorts. This number of nonconformities was considered unacceptable. Thus, to reduce the number of nonconformities in the printed boards, a team of engineers wants to set up a $u$ control chart. They collect data by selecting samples of 5 inspection units, where each inspection unit consists of 30 boards. The data, shown in Table 6.5, was collected over 30 days.

## Solution

Using the data in Table 6.5, we have

$$\bar{c} = 41.333$$

Therefore,

$$\bar{u} = \frac{\bar{c}}{5} = \frac{41.333}{5} = 8.2667$$

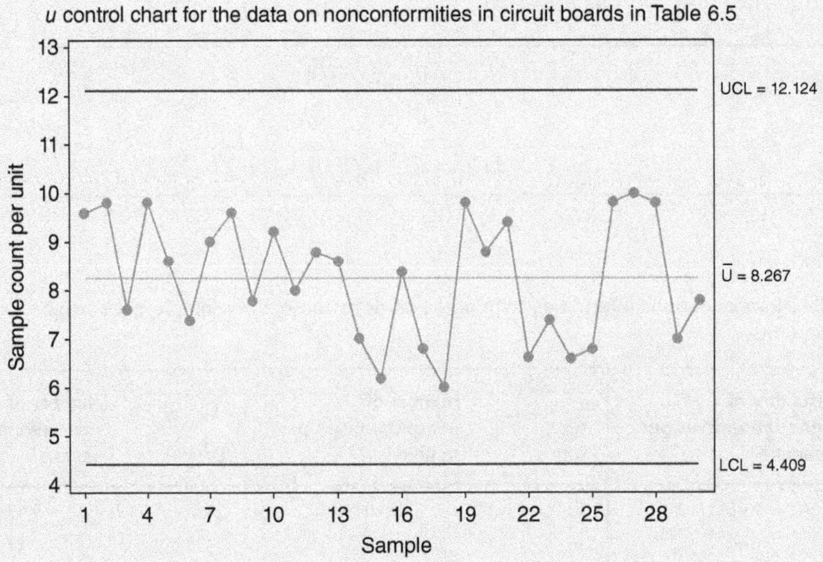

u control chart for the data on nonconformities in circuit boards in Table 6.5

**Figure 6.5**   $u$ control chart of nonconformities for the data in Table 6.5.

Hence, the control limits are given by

$$\text{UCL} = \bar{u} + 3\sqrt{\bar{u}/n} = 8.2667 + 3\sqrt{8.2667/5} = 12.124$$

$$\text{CL} = \bar{u} = 8.2667$$

$$\text{LCL} = \bar{u} - 3\sqrt{\bar{u}/n} = 8.2667 - 3\sqrt{8.2667/5} = 4.409$$

The $u$ chart for the data in Table 6.5 is as shown in Figure 6.5.

The $u$ chart shows that there are no assignable causes present in the process. In other words, only common causes are affecting the process. Therefore, management needs to take action on the system. Note that in this case, the $u$ control chart and the $c$ control chart are identical except for the scale on the vertical axis.

We can construct a $u$ chart using any of the statistical packages discussed in this book. To construct these charts using Minitab or R, we proceed as follows.

**Minitab**

1) Enter the nonconformities data from Table 6.5 in column C1.
2) From the menu bar select **Stat > Control Charts > Attribute Chart > U** to open the **U Chart** dialog box. In this dialog box, enter **C1** in the box under **Variables**, and in the box next to **Subgroup size**, enter the sample size. Then select any other desired option(s), and complete the required entries for each option you select. Click **OK**. The $u$ chart shown in Figure 6.5 will appear in the **Session Window**.

**R**

As in Example 6.4, we can use the built-in R function "qcc()" to construct a $u$ chart. The following R code can be used to generate the $u$ chart for the data in Table 6.5:

```
library (qcc)
setwd("C:/Users/person/Desktop/SQC-Data folder")
Data = read.csv("Table6.5.csv")
qcc(Data[,2], sizes = 5, type = "u")
```

The $u$ control chart generated using R is exactly the same as Figure 6.5, except that there may minor differences in the control limits due to rounding errors.

Constructing $u$ control charts using JMP and Python is discussed in Chapter 10, which is available for download on the book's website: www.wiley.com/college/gupta/SQC.

To illustrate the construction of a $u$ chart when the sample sizes vary, we consider the following example.

**Example 6.6**

Suppose that in Example 6.5, for some administrative reasons, it is not possible to examine five inspection units every day. In other words, the sample size varies, and the data obtained are as shown in Table 6.6. Construct and interpret a $u$ chart for the data in Table 6.6.

**Solution**

Using the data in Table 6.6, we have

$$\bar{u} = \frac{c_1 + c_2 + \dots + c_m}{n_1 + n_2 + \dots + n_m}$$

$$= \frac{33 + 40 + 38 + \dots + 40}{3 + 5 + 3 + \dots + 3} = \frac{1162}{126} = 9.22$$

The control limits are calculated for each individual sample. Thus, for example, the control limits for sample 1 are given by

$$UCL = \bar{u} + 3\sqrt{\bar{u}/n_1} = 9.22 + 3\sqrt{9.22/3} = 14.48$$

$$CL = \bar{u} = 9.22$$

$$LCL = \bar{u} - 3\sqrt{\bar{u}/n_1} = 9.22 - 3\sqrt{9.22/3} = 3.96$$

The control limits for the other samples are calculated in the same manner. The CL, however, remains the same. The *u* chart for the data in Table 6.6 is shown in Figure 6.6.

We can construct a *u* chart with variable sample sizes using any of the statistical packages discussed in this book. To construct a *u* chart using Minitab or R, we proceed as follows.

**Minitab**

1) Enter the nonconformities data from Table 6.6 in column C1 and the sample sizes in column C2.
2) From the menu bar select **Stat > Control Charts > Attribute Chart > U** to open the U Chart dialog box. In this dialog box, enter **C1** in the box under **Variables**, and in the box next to **Subgroup size**, type **C2**. Then select any other desired option(s), and complete all the required entries for each selected option. Click **OK**. The *u* chart shown in Figure 6.6 will appear in the **Session Window**.

**R**

As discussed in Example 6.4, we can use the built-in R function "qcc()" to construct a *u* chart. The following R code can be used to generate the *u* chart for the data in Table 6.6.

**Table 6.6**  Number of nonconformities on printed boards with varying sample sizes.

| Day | Sample size | Number of nonconformities per sample | Sample size | Day | Number of nonconformities per sample |
|-----|-------------|--------------------------------------|-------------|-----|--------------------------------------|
| 1 | 3 | 33 | 5 | 16 | 40 |
| 2 | 5 | 40 | 4 | 17 | 40 |
| 3 | 3 | 38 | 5 | 18 | 37 |
| 4 | 5 | 43 | 4 | 19 | 40 |
| 5 | 5 | 45 | 5 | 20 | 37 |
| 6 | 5 | 35 | 4 | 21 | 39 |
| 7 | 5 | 35 | 5 | 22 | 48 |
| 8 | 3 | 41 | 5 | 23 | 39 |
| 9 | 3 | 40 | 5 | 24 | 36 |
| 10 | 5 | 30 | 3 | 25 | 39 |
| 11 | 5 | 36 | 3 | 26 | 36 |
| 12 | 5 | 40 | 5 | 27 | 43 |
| 13 | 3 | 33 | 3 | 28 | 36 |
| 14 | 4 | 36 | 4 | 29 | 49 |
| 15 | 4 | 38 | 3 | 30 | 40 |

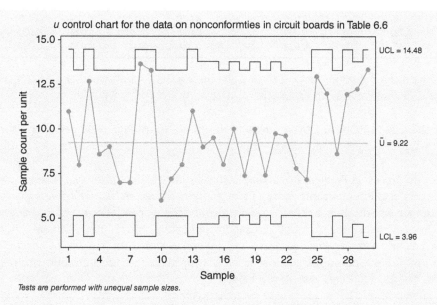

**Figure 6.6** *u* control chart for the data on nonconformities in circuit boards in Table 6.6.

```
library (qcc)
setwd("C:/Users/person/Desktop/SQC-Data folder")
Data = read.csv("Table6.6.csv")
qcc(Data[,3], sizes = Data[,2], type = "u")
```

The *u* control chart generated using R is exactly the same as shown in Figure 6.6, except there may be minor differences in the control limits due to rounding errors.

Constructing the control chart using JMP and Python is discussed in Chapter 10, which is available for download on the book's website: www.wiley.com/college/gupta/SQC.

The *u* control chart shows that all the sample points are within the control limits. However, several points fall beyond the warning control limits (see Chapter 5). Moreover, sample points 8 and 9 fall very close to the UCL. All these observations about the process indicate that there may be special causes present in the process. In other words, the process may be on the verge of becoming unstable. Therefore, precautions should be taken to prevent the process from becoming unstable (but care should be taken not to "over-control" the process).

## Review Practice Problems

1 Explain the difference between a *p* control chart and an *np* control chart.

2 A quality manager asked his team to implement *p* control charts for a process that was recently introduced. The team collected samples of size $n = 1000$ parts hourly over a period of 30 hours and determined that the fractions of nonconformance are as follows:

| 0.021 | 0.021 | 0.028 | 0.029 | 0.026 | 0.029 | 0.027 | 0.022 | 0.027 | 0.023 |
| 0.021 | 0.022 | 0.024 | 0.027 | 0.030 | 0.030 | 0.020 | 0.022 | 0.024 | 0.022 |
| 0.029 | 0.030 | 0.030 | 0.024 | 0.027 | 0.024 | 0.029 | 0.024 | 0.025 | 0.027 |

Construct a $p$ chart for the process and plot these points. Assess whether the process is under statistical control. Read the data column-wise.

**3** Use the data in Problem 2 to construct an $np$ control chart. Plot all 30 points on your control chart, and make your conclusions. Are your conclusions different from those in Problem 2?

**4** Since the process in Problem 2 was completely new, the quality control team was not aware that the parts could be nonconforming due to many nonconformities, which was unacceptable. Thus, to reduce the number of nonconformities, the quality control team had to change its strategy from implementing $p$ control charts to implementing $c$ charts. To implement this, they collected hourly samples over the next 30 hours and recorded the number of nonconformities (instead of nonconforming parts per inspection unit). Each inspection unit consisted of 100 parts. The data collected is as shown below.

| 57 | 43 | 58 | 50 | 52 | 44 | 55 | 53 | 56 | 58 |
| 59 | 44 | 58 | 59 | 59 | 40 | 57 | 53 | 59 | 50 |
| 57 | 57 | 47 | 57 | 55 | 44 | 53 | 55 | 45 | 54 |

(a) Determine the UCL and LCL for a $c$ chart.
(b) Construct a $c$ chart for the process, and plot all the data in it. Judge whether the process is under statistical control.

**5** In Problem 4, the control team now faced another difficulty: the process of determining all the nonconformities was very time-consuming, and sometimes they could not examine all 100 parts in one hour. Consequently, the sample sizes had to be varied. They then decided to define an inspection unit of 20 parts and collect 3 to 5 inspection units depending on the time available. Prepare an appropriate control chart for the following data.

| Hour | Subgroup size | Nonconformities per subgroup | Hour | Subgroup size | Nonconformities per subgroup |
|---|---|---|---|---|---|
| 1 | 3 | 25 | 16 | 4 | 36 |
| 2 | 3 | 22 | 17 | 5 | 44 |
| 3 | 5 | 41 | 18 | 4 | 32 |
| 4 | 4 | 34 | 19 | 4 | 36 |
| 5 | 5 | 38 | 20 | 5 | 42 |
| 6 | 5 | 43 | 21 | 5 | 46 |
| 7 | 4 | 30 | 22 | 5 | 42 |
| 8 | 4 | 3 | 23 | 4 | 34 |
| 9 | 5 | 42 | 24 | 4 | 30 |
| 10 | 5 | 45 | 25 | 5 | 39 |
| 11 | 5 | 40 | 26 | 3 | 33 |
| 12 | 4 | 33 | 27 | 3 | 32 |
| 13 | 3 | 27 | 28 | 5 | 40 |
| 14 | 5 | 38 | 29 | 5 | 36 |
| 15 | 4 | 36 | 30 | 4 | 36 |

**6** The following data give the number of nonconforming catalytic converters in a sample of 100. Construct a fraction nonconforming control chart by determining the trial control limits. If any points fall beyond the control limits, assume that special causes are found, and determine the revised control limits by ignoring the points beyond the control limits.

| Sample # | Nonconforming catalytic converters | Sample # | Nonconforming catalytic converters |
|---|---|---|---|
| 1 | 8 | 16 | 7 |
| 2 | 6 | 17 | 3 |
| 3 | 10 | 18 | 6 |
| 4 | 8 | 19 | 4 |
| 5 | 5 | 20 | 3 |
| 6 | 12 | 21 | 6 |
| 7 | 16 | 22 | 3 |
| 8 | 10 | 23 | 9 |
| 9 | 8 | 24 | 6 |
| 10 | 7 | 25 | 11 |
| 11 | 9 | 26 | 12 |
| 12 | 7 | 27 | 5 |
| 13 | 6 | 28 | 8 |
| 14 | 4 | 29 | 7 |
| 15 | 2 | 30 | 4 |

**7** Use the data in Problem 6 to set up a nonconforming $np$ control chart, and examine whether the process is under control. Compare this control chart with the one in Problem 6 having trial control limits, and comment on what you observe.

**8** The following data show the number of nonconformities per 100 square yards (one unit) of woolen cloth. Select and set up an appropriate control chart that you would use to determine whether the process is in statistical control. Give your conclusion.

| Sample # | Nonconformities | Sample # | Nonconformities | Sample # | Nonconformities |
|---|---|---|---|---|---|
| 1 | 3 | 9 | 15 | 17 | 3 |
| 2 | 7 | 10 | 14 | 18 | 5 |
| 3 | 5 | 11 | 6 | 19 | 11 |
| 4 | 10 | 12 | 9 | 20 | 13 |
| 5 | 2 | 13 | 1 | 21 | 16 |
| 6 | 1 | 14 | 17 | 22 | 9 |
| 7 | 8 | 15 | 14 | 23 | 10 |
| 8 | 7 | 16 | 9 | 24 | 8 |

**9** Suppose that in Problem 8, it was not convenient to take samples of 100 square yards (one unit). The quality control engineer decided to take samples of varied sizes. The new data obtained is shown next. The sample sizes (in units of 100 square yards) are given in parentheses in the sample number columns.

| Sample # | Nonconformities | Sample # | Nonconformities | Sample # | Nonconformities |
|---|---|---|---|---|---|
| 1 (1.10) | 4 | 9 (1.15) | 17 | 17 (1.15) | 13 |
| 2 (1.25) | 9 | 10 (1.25) | 12 | 18 (1.10) | 7 |
| 3 (1.00) | 6 | 11 (1.26) | 16 | 19 (1.09) | 12 |
| 4 (1.05) | 14 | 12 (1.28) | 7 | 20 (1.20) | 15 |
| 5 (1.15) | 12 | 13 (1.20) | 7 | 21 (1.26) | 12 |
| 6 (1.20) | 3 | 14 (1.10) | 16 | 22 (1.28) | 7 |
| 7 (1.24) | 11 | 15 (1.08) | 17 | 23 (1.15) | 12 |
| 8 (1.28) | 13 | 16 (1.12) | 20 | 24 (1.20) | 18 |

What are the trial control limits and the CL for a control chart appropriate for monitoring the process under investigation? Assuming that assignable causes, if any, have been detected and corrected, set up a revised control chart for monitoring future production by controlling the number of nonconformities.

**10** Use the data in Problem 9 to set up a control chart using average group size $\bar{n} = \sum n_i/m$ (assuming group sizes do not vary much). What conclusion can you draw by using the average sample size? Compare your conclusion with the one you made in Problem 9.

**11** A fraction nonconforming control chart has the following three-sigma control limits and CL:

$$\text{UCL} = 0.105, \text{CL} = 0.05, \text{LCL} = 0$$

(a) Using a normal approximation, determine the probability of type I errors.
(b) Using a normal approximation, determine the probability of type II errors, assuming that the true process fraction nonconforming is .065.
(c) Redo (a) and (b) using the Poisson approximation to the binomial distribution.

**12** In Problem 11, find the ARL when the process is in statistical control, and then find the ARL when the process nonconforming fraction is (a) .065; (b) .070; (c) .075.

**13** The following data give the number of nonconforming thermostat seals in 24 samples of size 200.
(a) Using a statistical package, set up a $p$ control chart for these data.

| Sample number | Nonconforming | Sample number | Nonconforming |
|---|---|---|---|
| 1 | 7 | 13 | 5 |
| 2 | 7 | 14 | 9 |
| 3 | 7 | 15 | 9 |
| 4 | 5 | 16 | 7 |
| 5 | 8 | 17 | 10 |
| 6 | 8 | 18 | 6 |
| 7 | 11 | 19 | 7 |
| 8 | 10 | 20 | 6 |
| 9 | 11 | 21 | 11 |
| 10 | 5 | 22 | 11 |
| 11 | 8 | 23 | 7 |
| 12 | 11 | 24 | 9 |

(b) Ignoring any points that fall beyond the control limits, and assuming that assignable causes are found for them, determine the revised control limits for future use.

**14** In Problem 13, suppose that we take 20 additional samples of size 200 each and observe the number of nonconforming thermostat seals in each sample as being

| | | | | | | | | | |
|---|---|---|---|---|---|---|---|---|---|
| 8 | 12 | 8 | 9 | 11 | 3 | 7 | 7 | 9 | 8 |
| 11 | 9 | 8 | 10 | 7 | 9 | 6 | 9 | 4 | 7 |

Plot the new points on the revised control chart. Is the process in statistical control?

**15** A team of engineers in a manufacturing company that produces detonation sensors, which are used in car emission control systems, wants to minimize the number of nonconforming sensors produced by the company. They collect 20 size-150 samples of sensors and find the number of nonconforming sensors in each sample:

| Sample no. | 1 | 2 | 3 | 4 | 5 | 6 | 7 | 8 | 9 | 10 |
|---|---|---|---|---|---|---|---|---|---|---|
| Nonconforming sensors | 8 | 6 | 9 | 7 | 7 | 5 | 9 | 8 | 4 | 8 |

| Sample no. | 11 | 12 | 13 | 14 | 15 | 16 | 17 | 18 | 19 | 20 |
|---|---|---|---|---|---|---|---|---|---|---|
| Nonconforming Sensors | 10 | 9 | 7 | 6 | 7 | 8 | 5 | 6 | 8 | 9 |

(a) Set up a $p$ control chart for these data.
(b) Ignoring any points that fall beyond the control limits, assuming that assignable causes are found, determine the revised control limits for future use.

**16** Suppose in Problem 15 that due to manufacturing difficulties, a team of engineers cannot collect any more samples of equal size. The subsequent samples they collected produced the following data.

| Sample no. | 1 | 2 | 3 | 4 | 5 | 6 | 7 | 8 | 9 | 10 |
|---|---|---|---|---|---|---|---|---|---|---|
| Sample size | 170 | 150 | 175 | 140 | 164 | 140 | 172 | 150 | 157 | 163 |
| Nonconforming | 9 | 6 | 8 | 7 | 7 | 5 | 10 | 8 | 4 | 9 |

| Sample no. | 11 | 12 | 13 | 14 | 15 | 16 | 17 | 18 | 19 | 20 |
|---|---|---|---|---|---|---|---|---|---|---|
| Sample size | 180 | 165 | 155 | 162 | 174 | 145 | 140 | 158 | 160 | 165 |
| Nonconforming | 11 | 8 | 7 | 8 | 9 | 8 | 5 | 7 | 6 | 8 |

(a) Set up a $p$ control chart for these data.
(b) Ignoring any points that fall beyond the control limits, and assuming that assignable causes are found, determine the revised control limits for future use.

**17** Apply the Western Electric rules to the control charts prepared in Problem 15. Identify if any of these control charts show a pattern that would imply the process is out of control.

**18** Set up a probability control chart for a binomial proportion $p$ with $n = 80$, $p = p_0 = 0.4$, and using the level of significance $\alpha = 0.05$.

**19** A manufacturing company produces fan belts for a car in batches of size 1000. The following data provide the number of nonconforming belts in the last 20 batches.

| Sample number | 1 | 2 | 3 | 4 | 5 | 6 | 7 | 8 | 9 | 10 |
|---|---|---|---|---|---|---|---|---|---|---|
| Nonconforming fan belts | 19 | 23 | 24 | 20 | 17 | 17 | 19 | 22 | 21 | 17 |
| Sample number | 11 | 12 | 13 | 14 | 15 | 16 | 17 | 18 | 19 | 20 |
| Nonconforming fan belts | 17 | 17 | 22 | 18 | 16 | 18 | 18 | 22 | 18 | 24 |

(a) Compute the trial limits needed to set up an appropriate control chart with three-sigma control limits for the number of nonconforming belts.

(b) Ignoring any points that fall beyond the control limits, and assuming that assignable causes are found, determine the revised CL and control limits for future use.

**20** Set up an *np* control chart for the data in Problem 15. Compare your conclusion with that found in Problem 15.

**21** A manufacturing company produces cylinder heads for car engines. The company asks its team of quality control engineers to set up a control chart to improve the cylinder head quality. The team collects 25 samples of 60 cylinder heads each and finds a total of 50 nonconforming cylinder heads in all of the samples.

(a) Find an estimate of the process average fraction of nonconforming cylinder heads, assuming that the process is in statistical control.

(b) Compute the trial control limits for an *np* control chart.

(c) If there is no change in the average fraction of nonconforming cylinder heads, find the probability that 20 subsequent samples of size 60 each contain a total of 48 nonconforming cylinder heads.

**22** A startup company making computer chips decides to go for 100% inspection until it becomes clear that the chips are meeting the specifications. A team of engineers finds a total of 450 nonconforming chips during the first 25 days of production. During this period, the total number of chips manufactured is 23,775. Set up a *p* control chart with trial control limits based on the average number of chips produced per day.

**23** In Problem 22, the engineer finds that two points are beyond the UCL. After an investigation, she determines an assignable cause that helps to explain why the points fall beyond the control limits. She also determines that on these two days, a total of 2155 chips were manufactured, of which 80 chips are nonconforming. Assume that the assignable causes have been removed. Set up the new control limits for the *p* control chart by ignoring the production of these two days.

**24** Consider a three-sigma $\overline{X}$ control chart. Draw an OC curve for a process that has upward shifts of 0.1, 0.3, 0.6, 0.8, 1.0, 1.5, 1.8, 2.0, and 2.5 standard deviations. Assume that the sample size is 4.

**25** A manufacturer of circuit boards finds that the boards have too many nonconformities. A quality control engineer at the company decides to select random samples of five circuit boards each and count the total number of nonconformities in each sample. The data obtained are shown next.

| Sample number | 1 | 2 | 3 | 4 | 5 | 6 | 7 | 8 | 9 | 10 |
|---|---|---|---|---|---|---|---|---|---|---|
| Nonconformities | 15 | 18 | 21 | 16 | 22 | 14 | 19 | 20 | 20 | 15 |

| Sample number | 11 | 12 | 13 | 14 | 15 | 16 | 17 | 18 | 19 | 20 |
|---|---|---|---|---|---|---|---|---|---|---|
| Nonconformities | 13 | 23 | 21 | 21 | 25 | 19 | 12 | 16 | 21 | 19 |

Determine $\bar{c}$, compute three-sigma trial control limits for a $c$ chart, and plot all the points. Ignoring any points that fall beyond the control limits, and assuming that assignable causes are found, determine the revised control limits for future use.

**26** Suppose that in Problem 25, the process average had a one-sigma upward shift. Find the probability of not detecting the shift in the first subsequent sample or the first two subsequent samples.

**27** Suppose that in Problem 25, due to administrative difficulties, it is not always possible to inspect five circuit boards, so the sample size varies. The new data obtained are as follows:

| Subgroup size | 5 | 4 | 3 | 4 | 4 | 3 | 5 | 6 | 4 | 3 |
|---|---|---|---|---|---|---|---|---|---|---|
| Nonconformities | 15 | 14 | 11 | 12 | 22 | 10 | 19 | 21 | 18 | 11 |

| Subgroup size | 4 | 5 | 3 | 4 | 5 | 4 | 6 | 4 | 5 | 5 |
|---|---|---|---|---|---|---|---|---|---|---|
| Nonconformities | 13 | 21 | 11 | 13 | 21 | 17 | 24 | 12 | 17 | 18 |

Compute three-sigma trial control limits, and plot the $u$ control chart. Ignoring any points that fall beyond the control limits, and assuming that assignable causes are found, determine the revised control limits for future use.

**28** Determine 0.95 and 0.05 probability control limits for a $c$ control chart when the process mean is equal to 20 nonconformities.

**29** Determine three-sigma control limits for a $u$ control chart for the average number of nonconformities per inspection unit, given that the CL is 10 and the subgroup size is equal to 8 inspection units.

**30** In Problem 29, find the new control limits if we want 0.975 and 0.025 probability control limits instead of three-sigma control limits.

# 7

# Phase II Quality Control Charts for Detecting Small Shifts

## 7.1 Introduction

In Chapters 5 and 6, we studied *Shewhart control charts,* also known as phase I control charts, for variables and attributes. Phase I control charts are used to detect large shifts, which usually occur either when the process is not very mature or when the implementation of SPC is in the initial stages. During these early stages, a process may be highly influenced by special causes, and consequently, large shifts in the process are likely to occur. By the time the phase I implementation of SPC is over, most of the special causes are eliminated or reduced, and as a result, large shifts become rare. That is, after the phase I implementation of SPC is complete and the phase II implementation of SPC commences, the process is usually under control, and only small shifts typically occur. In this chapter, we study phase II control charts, which are used in a matured process and when the SPC implementation of phase I is over. The phase II control charts are usually employed when we want to detect shifts that are smaller than one-and-half times the process standard deviation ($1.5\,\sigma$). In phase II, we study the next generation of control charts, namely, cumulative sum (CUSUM), moving average (MA), and exponentially weighted moving average (EWMA) control charts.

Shewhart control charts are known to be not very effective in detecting small shifts or shifts that are less than $1.5\sigma$. When the shifts are small, a remedial approach is to increase the sensitivity of the Shewhart control charts by using the following criteria from the *Western Electric Handbook* rules (1956):

- Two out of three consecutive points fall outside the two-sigma warning limits.
- Four out of five consecutive points fall at a distance of one sigma from the center line (CL).
- Eight consecutive points fall above or below the CL.

If any point falls outside the three-sigma control limits, or if any one of these conditions occurs, then the process is considered to be out of control. Subsequently, appropriate action is taken, such as increasing the sampling frequency, increasing the sample size, or both. However, note that even though these actions do increase sensitivity, as Champ and Woodall (1987) pointed out, using these rules, the in-control average run length (ARL) is only 91.75. Thus, under normality, the false alarm rate is little more than four times the false alarm rate of a Shewhart chart when *Western Electric Handbook* rules are not used. Moreover, the simplicity of Shewhart charts, which is an important advantage of these charts, is lost. Keeping all this in mind, it is fair to say that when the shifts are small, Shewhart control charts with these modifications are not a viable alternative to CUSUM

*Statistical Quality Control: Using Minitab, R, JMP, and Python*, First Edition. Bhisham C. Gupta.
© 2021 John Wiley & Sons, Inc. Published 2021 by John Wiley & Sons, Inc.
Companion website: www.wiley.com/go/college/gupta/SQC

control charts. Also, the CUSUM control chart has another major advantage over the Shewhart control charts: in the CUSUM control chart, the $i$th point uses information that is contained in the $i$th sample and all the samples collected before that point. The Shewhart control chart, on the other hand, uses only the information contained in the $i$th sample. So, the Shewhart chart ignores all the information provided by the samples collected before the $i$th sample. We note that this feature of the CUSUM control chart makes the plotted points dependent. Therefore, it is difficult to interpret any pattern other than when the point exceeds the *decision interval* for an upward or downward shift. Other control charts that are very effective in detecting the small shifts are the MA control charts and EWMA control charts. In this chapter, we discuss all three control charts (CUSUM, MA, and EWMA), which are very desirable alternatives to the Shewhart control chart for controlling processes in phase II.

Relatively speaking, the MA control chart is not as effective as CUSUM and EWMA charts. CUSUM and EWMA control charts can be used to monitor the process mean as well as the process variance, fraction nonconforming, and nonconformities. However, our primary focus of discussion for these control charts will be the process mean. A good reference for a detailed study of CUSUM control charts is Hawkins and Olwell (1998).

## 7.2 Basic Concepts of CUSUM Control Charts

### 7.2.1 CUSUM Control Charts vs. Shewhart $\overline{X}$ and R Control Charts

Basic assumptions of a CUSUM control chart are:

- The observations are independently and identically normally distributed with mean $\mu$ and standard deviation $\sigma$.
- The mean, $\mu$, and standard deviation, $\sigma$, are known.

Note that in practice, the parameters $\mu$ and $\sigma$ are rarely known and must be estimated. It is very important to remember that while estimating these parameters, precision is essential. Even a small error, particularly in estimating $\mu$, will have a cumulative effect, which may result in false alarms. One way to avoid such problems is to take very large samples when estimating these parameters.

Before we begin our formal discussion of designing a CUSUM control chart, we first use a small set of data to show how a CUSUM chart is more effective than a Shewhart chart in detecting a small shift in the process mean. Note that the CUSUM chart we discuss in this section has no decision interval and is thus not a formal CUSUM control chart.

**Example 7.1 Medical Equipment Data**
Consider a manufacturing process for medical equipment parts. We are interested in studying a quality characteristic of one of the parts manufactured by the process. Let the quality characteristic when the process is under control be normally distributed with mean 15 and standard deviation 2. The data shown in Table 7.1 give the first 10 random samples of size 4, which are taken when the process is stable and producing parts with mean value 15 and standard deviation 2. The last 10 random samples, again of size 4, are taken from the process after its mean experienced an upward shift of one standard deviation, resulting in a new process with mean 17 and standard deviation 2. Construct the Shewhart $\overline{X} - R$ control chart and the CUSUM control chart for the data in Table 7.1. Then comment on the outcomes of the $\overline{X} - R$ and CUSUM control charts.

**Table 7.1** Data from a manufacturing process of medical equipment parts before and after its mean has experienced a shift of one sigma (sample size four).

| Sample ($i$) | Sample | $\bar{X}_i$ | $\bar{Z}_i = \dfrac{\bar{X}_i - 15}{\sigma/\sqrt{n}}$ | $S_i = \bar{Z}_i + S_{i-1}$ |
|---|---|---|---|---|
| 1 | 15.35 15.05 13.92 10.70 | 13.7550 | −1.2450 | −1.2450 |
| 2 | 12.50 14.37 12.03 14.35 | 13.3125 | −1.6875 | −2.9325 |
| 3 | 17.99 13.61 13.35 12.77 | 14.4300 | −0.5700 | −3.5025 |
| 4 | 17.83 14.56 16.14 18.50 | 16.7575 | 1.7575 | −1,7450 |
| 5 | 13.70 16.26 14.71 14.07 | 14.6850 | −0.3150 | −2.0600 |
| 6 | 15.70 12.90 17.05 15.62 | 15.3175 | 0.3175 | −1.7425 |
| 7 | 17.80 14.80 15.15 16.01 | 15.9400 | 0.9400 | −0.8025 |
| 8 | 15.28 12.15 19.81 14.68 | 15.4800 | 0.4800 | −0.3225 |
| 9 | 11.87 17.37 13.91 13.43 | 14.1450 | −0.8550 | −1.1775 |
| 10 | 13.96 16.16 14.74 15.56 | 15.1050 | 0.1050 | −1.0725 |
| 11 | 18.96 13.75 17.48 20.85 | 17.7600 | 2.7600 | 1.6875 |
| 12 | 14.20 18.18 19.17 15.77 | 16.8300 | 1.8300 | 3.5175 |
| 13 | 16.18 19.33 14.78 21.46 | 17.9375 | 2.9375 | 6.4550 |
| 14 | 15.42 15.74 19.66 16.70 | 16.8800 | 1.8800 | 8.3350 |
| 15 | 17.76 16.47 15.36 17.84 | 16.8575 | 1.8575 | 10.1925 |
| 16 | 16.01 16.48 16.86 14.99 | 16.0850 | 1.0850 | 11.2775 |
| 17 | 18.12 18.82 16.42 16.96 | 17.5800 | 2.5800 | 13.8575 |
| 18 | 16.37 17.05 20.24 17.70 | 17.8400 | 2.8400 | 16.6975 |
| 19 | 16.84 18.89 10.97 17.67 | 16.0925 | 1.0925 | 17.7900 |
| 20 | 13.49 16.65 17.88 19.67 | 16.9225 | 1.9225 | 19.7125 |

## Solution

As described in the introductory section, the basic principle of a CUSUM chart is to detect small shifts in the process mean at time $i$. This is done by calculating the accumulated sum of deviations ($S_i$) of the sample means $\bar{X}_1, \bar{X}_2, ..., \bar{X}_i$ from the target value ($\mu_0$) measured in standard deviation units. That is,

$$S_i = \sum_{j=1}^{i} \frac{(\bar{X}_j - \mu_0)}{\sigma/\sqrt{n}} = \frac{(\bar{X}_i - \mu_0)}{\sigma/\sqrt{n}} + S_{i-1} = \bar{Z}_i + S_{i-1}, \ i = 1, 2, \cdots, 20 \qquad (7.1)$$

where $\bar{Z}_1, \bar{Z}_2, ..., \bar{Z}_i$ are the standardized values of $\bar{X}_1, \bar{X}_2, ..., \bar{X}_i$, respectively. In the present example, the target value is 15, and the accumulated sum of deviations of the sample means from the target value at time $i$ measured in standard deviation units is given in column 5 of Table 7.1. From Eq. (7.1), we can easily see that if the samples come from a stable process with a mean of, say, $\mu_1$, then the points ($i, S_i$) when plotted on a chart will either

- Scatter randomly around the zero line,
- Show clearly an upward trend, or
- Show clearly a downward trend,

depending on whether $\mu_1 = \mu_0$, $\mu_1 > \mu_0$, or $\mu_1 < \mu_0$.

The $\overline{X}$–R and CUSUM control charts are shown in Figures 7.1 and 7.2, respectively. The $\overline{X}$–R control chart shows that the process is clearly stable and exhibits no abnormality. In other words, it is not detecting the fact that the samples starting from the 11th sample came from the process after its mean had experienced an upward shift of one standard deviation, which is obviously a small shift since it is less than $1.5\sigma$. However, the CUSUM control chart clearly shows an upward trend, which, in fact, started at the 11th sample.

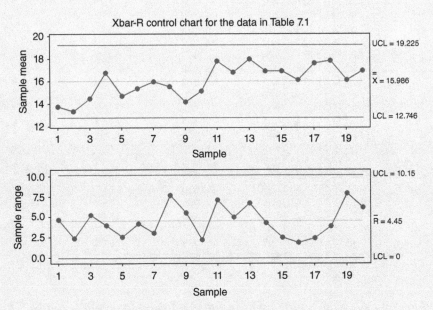

**Figure 7.1** $\overline{X}-R$ control chart for the data in Table 7.1.

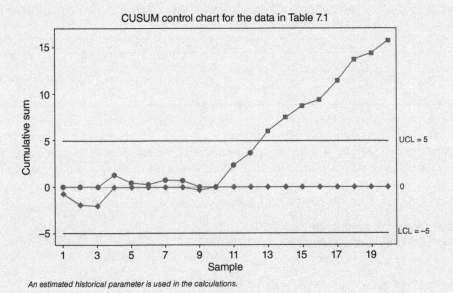

*An estimated historical parameter is used in the calculations.*

**Figure 7.2** CUSUM chart for the data in Table 7.1.

As noted by Lucas (1982), CUSUM control charts give tighter process control than classical control charts such as Shewhart charts. With the tighter control provided by CUSUM control charts, more emphasis is placed on keeping the process on target rather than allowing it to drift within the limits. Moreover, the sensitivity of the CUSUM control chart would not be seriously affected if we took samples of size $n = 1$ instead of $n > 1$. CUSUM control charts, in fact, more frequently use $n = 1$ or individual values. In CUSUM charts for individual values, the quantities $\overline{X}_i$, $\overline{Z}_i$, and $\sigma/\sqrt{n}$ in Table 7.1 are replaced by $X_i$, $Z_i$, and $\sigma$, respectively. We develop the CUSUM control chart in the next section.

## 7.3  Designing a CUSUM Control Chart

In section 7.2, we saw that CUSUM control charts are quite effective for detecting small shifts. Unlike the $\overline{X}$–$R$ Shewhart control charts, CUSUM charts can be designed to detect *one-sided* or *two-sided shifts*. These charts are defined by two parameters: $k$ and $h$, or the *reference value* and the *decision interval,* respectively, and are defined in section 7.3.1.

Two-sided process control using a CUSUM control chart is achieved by concurrently using two one-sided CUSUM control charts. In both one-sided CUSUM control charts, we can use the same or different reference values, depending on whether the upward and downward shifts are equally important, or if one is more important than the other. Furthermore, in designing a CUSUM control chart, the average run length (ARL) usually plays an important role. The ARL of a two-sided control chart is obtained by combining the ARLs of two one-sided control charts, where the ARL of one-sided control charts for upward shifts and downward shifts are not necessarily the same. Again, note that the values of the ARLs depend solely on the sensitivity of the shifts. Sometimes the size of the shift on one side may be more serious than the other. For example, consider the tensile strength of a copper wire as the quality characteristic. Obviously, in this case, any upward shift in mean tensile strength may not be as serious as a downward shift. It can be seen from Van Dobben de Bruyn (1968) that

$$\frac{1}{ARL(two-sided)} = \frac{1}{ARL^+} + \frac{1}{ARL^-} \tag{7.2}$$

where $ARL^+$ and $ARL^-$ are the ARLs of two one-sided control charts when the shift is upward and downward, respectively. In this section, we focus our attention on a two-sided control chart.

CUSUM control charts can be implemented by using either a *numerical procedure* or a *graphical procedure*. The numerical procedure consists of constructing a table known as a *tabular CUSUM,* whereas the graphical procedure consists of preparing a graph known as a *V-mask*. Practitioners usually prefer to use a numerical procedure. For visual observation, however, the tabular CUSUM can be plotted in a graph. The tabular CUSUM in Table 7.3 is plotted in Figure 7.3. In this book, we discuss only the construction of a tabular CUSUM. The use of a V-mask CUSUM chart is usually not preferable because some of its characteristics are not desirable. Also, in a V-mask control chart, we cannot use the fast-initial response (FIR) feature (see Section 7.3.2) of CUSUM, a very desirable characteristic of the chart. A thorough discussion of the V-mask CUSUM control chart is beyond the scope of this book.

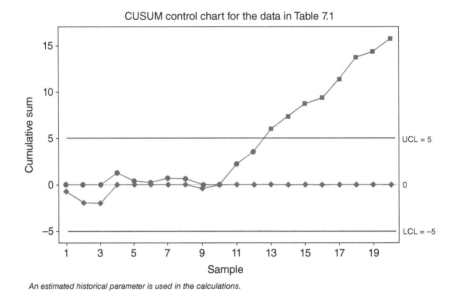

An estimated historical parameter is used in the calculations.

**Figure 7.3** Minitab printout of a two-sided CUSUM control chart for the data in Table 7.1.

### 7.3.1 Two-Sided CUSUM Control Charts Using the Numerical Procedure

As mentioned earlier, tabular CUSUM control charts are defined by two parameters $k$ and $h$, respectively referred to as the *reference value* and *decision interval*. The parameter $k$ is defined as

$$k = \frac{|\mu_1 - \mu_0|}{2\sigma} \tag{7.3}$$

where $\mu_0$ is the target value or process mean when the process is under control, $\mu_1$ is the process mean after it has experienced a shift, and $\sigma$ is the process standard deviation. Note that if we are using the standardized values $Z_i$, $\sigma$ is replaced by 1. In Example 7.1, it can easily be verified that $k = 0.5$, since the process had experienced a shift of one standard deviation, $\sigma$. As a rule of thumb, the parameter $h$ is usually selected to be equal to 4 or 5 (if the observations are not standardized, then $h$ is equal to 4 or 5 times the process standard deviation, $\sigma$). In practice, however, $h$ should be chosen so that the value of the ARL is neither too small nor too large. This is due to the fact that small ARLs may result in too many false alarms, whereas large ARLs may cause the process to work even when it has already experienced a shift and is producing nonconforming products. Preferably, $h$ should be such that when $\mu_1 = \mu_0$, when the process has experienced no shift, the value of ARL, which we denote by $ARL_0$, is approximately equal to 370. The value of 370 for $ARL_0$ is intentionally chosen to match the $ARL_0$ of the Shewhart control chart under normal distribution with three-sigma control limits. Hawkins (1993) gives a table of $k$ and $h$ values associated with two-sided CUSUM control chart $ARL_0$ values equal to 370. We provide some of these values below in Table 7.2a. Note that $h = 5$ with $k = 0.5$ will produce $ARL_0$ value slightly greater than 370. In Table 7.2b, values of ARL are given for certain shifts in the process mean for CUSUM and Shewhart control charts with $k = 0.5$ and $h = 4$ or 5.

**Table 7.2a**  Values of *h* for given value of *k* when ARL$_0$ = 370 for two-sided tabular CUSUM.

| *k* | 0.25 | 0.50 | 0.75 | 1.00 | 1.25 | 1.50 |
|---|---|---|---|---|---|---|
| *h* | 8.01 | 4.77 | 3.34 | 2.52 | 1.99 | 1.61 |

**Table 7.2b**  ARL values for certain mean shifts for CUSUM and Shewhart control charts with k = 0.5 and h = 4 or 5.

| Shift in mean (multiple of $\sigma$) | h = 4 | h = 5 | Shewhart $\overline{X}$ control chart |
|---|---|---|---|
| 0.00 | 168 | 465 | 371.00 |
| 0.25 | 74.2 | 139 | 281.14 |
| 0.50 | 26.6 | 38.0 | 155.22 |
| 0.75 | 13.3 | 17.0 | 81.22 |
| 1.00 | 8.38 | 10.4 | 44.00 |
| 1.50 | 4.75 | 5.75 | 14.97 |
| 2.00 | 4.34 | 4.01 | 6.30 |
| 2.50 | 2.62 | 3.11 | 3.24 |
| 3.00 | 2.19 | 2.57 | 2.01 |
| 4.00 | 1.71 | 2.01 | 1.19 |

This table shows that for shifts of 1.5 standard deviations or smaller, CUSUM control charts are much more efficient than Shewhart $\overline{X}$ charts. However, as the mean shift becomes larger than 1.5 standard deviations, this advantage is lost. Hawkins and Olwell (1998, pp. 48–49) give very extensive tables of one-sided CUSUM control charts for ARL values as a function of *k* and *h*, as well as for *h* as a function of *k* and ARL. Hawkins and Olwell (1998) also provided some computer programs that can be used to generate values of ARL and *h* that are not included in these tables. Having defined the parameters *k* and *h*, we are now ready to define those statistics needed to implement two one-sided CUSUM control charts using a numerical procedure. These statistics are defined as follows:

$$S_i^+ = \text{Max} \left[ 0, z_i - k + S_{i-1}^+ \right] \tag{7.4}$$

$$S_i^- = \text{Min} \left[ 0, z_i + k + S_{i-1}^- \right] \tag{7.5}$$

where the initial values of $S_0^+$ and $S_0^-$ are zero. The statistics $S_i^+$ and $S_i^-$ are used to implement one-sided CUSUM control charts when the shifts are upward and downward, respectively. For a two-sided CUSUM control chart, $S_i^+$ and $S_i^-$ are used simultaneously. We illustrate the implementation of the numerical procedure for the CUSUM control chart by using the data in Table 7.1 of Example 7.1.

### Example 7.2

In Table 7.3, we reproduce columns corresponding to Sample ($i$), $\overline{X}_i$ and $\overline{Z}_i$, from Table 7.1 and append two new columns: one for $S_i^+$ and another for $S_i^-$, where $S_i^+$ and $S_i^-$ are as defined in Eqs. 7.4 and 7.5, respectively, with k = 0.5 and h = 5.

### Solution

From column 3 of Table 7.3, we have

$$S_1^+ = \text{Max}\left[0, \overline{Z}_1 - k + S_0^+\right]$$

$$= \text{Max}\left[0, -1.2450 - 0.50 + 0\right] = 0.00$$

$$S_1^- = \text{Min}\left[0, \overline{Z}_1 + k + S_0^-\right]$$

$$= \text{Min}\left[0, -1.2450 + 0.50 + 0\right] = -0.7450$$

$$S_2^+ = \text{Max}\left[0, \overline{Z}_2 - k + S_1^+\right]$$

$$= \text{Max}\left[0, -1.6875 - 0.50 + 0\right] = 0.00$$

$$S_2^- = \text{Min}\left[0, \overline{Z}_2 + k + S_1^-\right]$$

$$= \text{Min}\left[0, -1.6875 + 0.50 - 0.7450\right] = -1.9325$$

**Table 7.3** Tabular CUSUM control chart for the data given in Table 7.1.

| Sample ($i$) | $\overline{X}_i$ | $\overline{Z}_i = \dfrac{\overline{X}_i - 20}{\sigma/\sqrt{n}}$ | $S_i^+ = Max\left(0, Z_i - k + S_{i-1}^+\right)$ | $S_i^- = Min\left(0, Z_i + k + S_{i-1}^-\right)$ |
|---|---|---|---|---|
| 1 | 13.7550 | −1.2450 | 0.0000 | −0.7450 |
| 2 | 13.3125 | −1.6875 | 0.0000 | −1.9325 |
| 3 | 14.4300 | −0.5700 | 0.0000 | −2.0025 |
| 4 | 16.7575 | 1.7575 | 1.2575 | 0.0000 |
| 5 | 14.6850 | −0.3150 | 0.4425 | 0.0000 |
| 6 | 15.3175 | 0.3175 | 0.2600 | 0.0000 |
| 7 | 15.9400 | 0.9400 | 0.7000 | 0.0000 |
| 8 | 15.4800 | 0.4800 | 0.6800 | 0.0000 |
| 9 | 14.1450 | −0.8550 | 0.0000 | −0.3550 |
| 10 | 15.1050 | 0.1050 | 0.0000 | 0.0000 |
| 11 | 17.7600 | 2.7600 | 2.2600 | 0.0000 |
| 12 | 16.8300 | 1.8300 | 3.5900 | 0.0000 |
| 13 | 17.9375 | 2.9375 | 6.0275 | 0.0000 |
| 14 | 16.8800 | 1.8800 | 7.4075 | 0.0000 |
| 15 | 16.8575 | 1.8575 | 8.7650 | 0.0000 |
| 16 | 16.0850 | 1.0850 | 9.3500 | 0.0000 |
| 17 | 17.5800 | 2.5800 | 11.4300 | 0.0000 |
| 18 | 17.8400 | 2.8400 | 13.7700 | 0.0000 |
| 19 | 16.0925 | 1.0925 | 14.3625 | 0.0000 |
| 20 | 16.9225 | 1.9225 | 15.7850 | 0.0000 |

Similarly, we can calculate $S_i^+$ and $S_i^-$ for $i = 3, 4, \cdots, 20$. All of these values are listed in columns 4 and 5 of Table 7.3.

Table 7.3 gives a complete summary of a two-sided CUSUM control chart for the data in Example 7.1. From column 4, we see that the value of $S_i^+$ at $i = 13$ is 6.0275, which is clearly greater than the decision interval $h = 5$. We conclude that at this point, the process is out of control due to an upward mean shift. Also, note that in column 5, no value of $S_i^-$ has gone below the decision interval $h = -5$, which implies that no downward mean shift has occurred. Moreover, a clear trend of nonzero values of $S_i^+$ occurred at sample 11, which implies that the process mean-shifted at some point between samples 10 and 11. These conclusions are also confirmed by the CUSUM control chart, shown in Figure 7.3.

Once we conclude that the process is out of control, it is critical to immediately estimate the new process mean, or the shifted mean, $\mu_1$. Otherwise, without recalculating such an estimate, it is difficult to make any appropriate changes or adjustments in the process to bring it back to its target value, $\mu_0$. This estimate can be found simply by determining the average of the observations taken during shifted periods and when the process was out of control. In this example, since the sample size $n$ is greater than 1, we have

$$\hat{\mu}_1 = \overline{\overline{X}} = \frac{1}{m}\sum_{i=1}^{m} \overline{X}_i \simeq (17.76 + 16.83 + 17.94)/3 = 17.51$$

where $m$ is the number of periods between the time when the process mean shift occurred and when it was concluded that the process was out of control. In this example, $m = 3$ since the process mean shift occurred before sample 11 was taken, and we concluded that the process was out of control after sample 13 was taken.

## Example 7.3

Reconsider the data presented in Example 7.1. Here we take random samples of size one (the fourth sample point from each sample in Example 7.1) to see whether our conclusions change. This data is given in Table 7.4. As in Example 7.1, we note that the first 10 observations were taken when the process was stable, manufacturing medical equipment parts with mean value 15 and standard deviation 2. The last 10 observations were taken from the same process after its mean experienced an upward shift of one standard deviation, resulting in a new process with mean 17 and standard deviation 2.

## Solution

In this example, we construct both the $\overline{X}$ Shewhart control chart for individual values and the CUSUM control chart. The Minitab printouts of the $\overline{X}$ and CUSUM control charts are shown in Figures 7.4 and 7.5. The $\overline{X}$ control chart shows no abnormalities in the process, or it has not detected the one-standard-deviation shift in the process mean. The CUSUM control chart, on the other hand, as we can see from Figure 7.5, has detected this shift. We first discuss the construction of the CUSUM control chart for individual values.

Following the same steps as shown in Example 7.2, and using the data in Table 7.4, the entries in other columns of Table 7.4 are obtained as shown here:

**Table 7.4** Data from a manufacturing process of medical equipment before and after its mean experienced a shift of one process standard deviation (sample size 1).

| Sample # | $X_i$ | $Z_i = (X_i - \mu)/\sigma$ | $S_i^+ = Max(0, Z_i - k + S_{i-1}^+)$ | $S_i^- = Min(0, Z_i + k + S_{i-1}^-)$ |
|---|---|---|---|---|
| 1 | 10.70 | -2.150 | 0.000 | -1.650 |
| 2 | 14.35 | -0.325 | 0.000 | -1.475 |
| 3 | 12.77 | -1.115 | 0.000 | -2.090 |
| 4 | 18.50 | 1.750 | 1.250 | 0.000 |
| 5 | 14.07 | -0.465 | 0.285 | 0.000 |
| 6 | 15.62 | 0.310 | 0.095 | 0.000 |
| 7 | 16.01 | 0.505 | 0.100 | 0.000 |
| 8 | 14.68 | -0.160 | 0.000 | 0.000 |
| 9 | 13.43 | -0.785 | 0.000 | -0.285 |
| 10 | 15.56 | 0.280 | 0.000 | 0.000 |
| 11 | 20.85 | 2.925 | 2.425 | 0.000 |
| 12 | 15.77 | 0.385 | 2.310 | 0.000 |
| 13 | 21.46 | 3.230 | 5.040 | 0.000 |
| 14 | 16.70 | 0.850 | 5.390 | 0.000 |
| 15 | 17.84 | 1.420 | 6.310 | 0.000 |
| 16 | 14.99 | -0.005 | 5.805 | 0.000 |
| 17 | 16.96 | 0.980 | 6.285 | 0.000 |
| 18 | 17.70 | 1.350 | 7.135 | 0.000 |
| 19 | 17.67 | 1.335 | 7.970 | 0.000 |
| 20 | 19.67 | 2.335 | 9.805 | 0.000 |

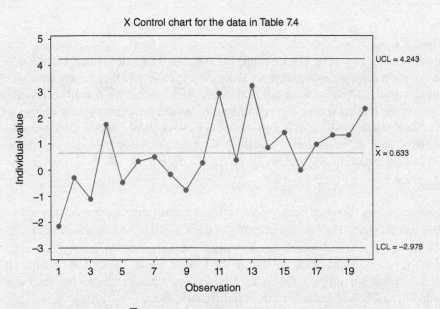

**Figure 7.4** $\bar{X}$ control chart for individual values in Table 7.4.

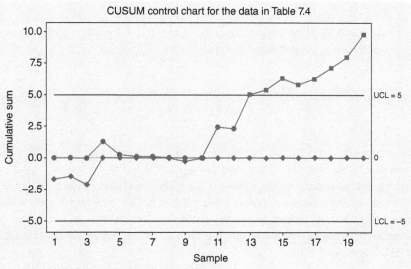

An estimated historical parameter is used in the calculations.

**Figure 7.5** CUSUM control chart for individual values in Table 7.4.

$$S_1^+ = \text{Max}\left[0, Z_1 - k + S_0^+\right]$$
$$= \text{Max}\left[0, -2.150 - 0.50 + 0\right] = 0.00$$
$$S_1^- = \text{Min}\left[0, Z_1 + k - S_0^-\right]$$
$$= \text{Min}\left[0, -2.150 + 0.50 + 0\right] = -1.650$$
$$S_2^+ = \text{Max}\left[0, Z_2 - k + S_1^+\right]$$
$$= \text{Max}\left[0, -0.325 - 0.50 + 0\right] = 0.00$$
$$S_2^- = \text{Min}\left[0, Z_2 + k + S_1^-\right]$$
$$= \text{Min}\left[0, -0.325 + 0.50 - 1.650\right] = -1.475$$

We can calculate the other values of $S_i^+$ and $S_i^-$ for $i = 3, 4, \cdots, 20$ in the same manner. These values are listed in columns 4 and 5 of Table 7.4, which gives us a complete summary of a two-sided CUSUM control chart for the data in this example. From column 4 of Table 7.4, we see that the value of $S_i^+$ at point 13 is 5.040, which is greater than the decision interval of $h = 5$. Therefore, we conclude that at point 13, the process is out of control due to an upward process mean shift. Also, note that in column 5, no value of $S_i^-$ has gone below the decision interval of $h = -5$, which implies that no downward process mean shift has occurred. Moreover, a clear upward trend of nonzero values of $S_i^+$ occurred at sample 11, which implies that the process mean-shifted at some point between the times when samples 10 and 11 were taken.

These conclusions are also confirmed by the CUSUM control chart shown in Figure 7.5.

Since we have concluded that the process is out of control, it is prudent to find an estimate of the new process mean, $\mu_1$, which is given by

$$\hat{\mu}_1 = \overline{X} = \frac{\sum\limits_{i=11}^{13} X_i}{m} = (20.85 + 15.77 + 21.46)/3 = 19.36$$

where $m$ is the number of periods between the time when the process mean shift occurred and when it was concluded that the process was out of control.

Here, for the sake of comparison, we present the $X$ Shewhart control chart and CUSUM control chart. These control charts are prepared using Minitab. The Shewhart $X$ control chart given in Figure 7.4 shows that the process is under statistical control. In other words, it does not detect that the process mean has experienced a one-standard-deviation shift, whereas the CUSUM control chart did detect this shift.

Summarizing the results of Examples 7.1, 7.2, and 7.3, we have the following:

- The regular $\overline{X}$–$R$ control chart in Example 7.1 and the $X$ control chart for individual values in Example 7.3 did not detect the process mean shift of one standard deviation.
- In Example 7.2, the CUSUM control chart indicated that the process was out of control at period $i = 13$, just at the third sample taken after the actual shift had occurred. In total, we selected 12 parts before we knew that the process was out of control.
- In Example 7.3, the CUSUM control chart again indicated that the process was out of control at period $i = 13$. In other words, it indicated when the third sample was taken after the actual shift had occurred. In total, we selected only three parts before we knew that the process was out of control.

Note that in Examples 7.2 and 7.3, if the parts were selected at the same frequency – say, every half hour – then it would be quicker to determine that the process is out of control with samples of size one rather than samples of size four. This result may be true in general. Finally, to avoid any confusion about using different values of $\sigma$, and for the sake of simplicity, we recommend using standardized values ($z_i$) rather than actual values ($x_i$). Note that when we use standardized values ($z_i$), the target value is zero, and the standard deviation is equal to one.

### 7.3.2 The Fast Initial Response (FIR) Feature for CUSUM Control Charts

Lucas and Crosier (1982) introduced the FIR feature, which improves the sensitivity either at process startup or immediately after any corrective control action has been taken to improve the process. In standard CUSUM control charts, we use the initial value $S_0 = 0$, but in CUSUM with the FIR feature, the starting value of $S_0$ is set at some value greater than zero. Lucas and Crosier pointed out that, following a possibly ineffective control action, control of the process is less certain, and therefore extra sensitivity is desired. The FIR feature in CUSUM control charts does exactly this. Furthermore, Lucas and Crosier recommended a starting value of $h/2$ when setting up CUSUM. For an illustration of FIR, we rework Example 7.3.

## Example 7.4

Reconsider the data in Table 7.4 of Example 7.3, where the observations represent quality characteristics of a medical equipment part manufactured by a given process. These data are reproduced in Table 7.5. Construct the tabular CUSUM control chart for these data using the FIR feature.

## Solution

Table 7.5 gives the complete tabular CUSUM control chart using FIR for the data in Example 7.3. The graphical form of the new CUSUM control chart is shown in Figure 7.6.

In Example 7.3, the CUSUM chart parameters were $k = 0.5$, $h = 5$, and the starting CUSUM value was $S_0^+ = S_0^- = 0$. In this example, we used $S_0^+ = h/2 = 2.5$, $S_0^- = -h/2 = -2.5$. Note that by using FIR, the CUSUM for a downward shift in column 5 showed that, at period $i = 3$, it came close to the lower decision interval. The CUSUM for the upward shift in column 4 was not affected by the starting value of 2.5. It is important to note, however, that using the FIR feature helps to detect an early shift when the process has experienced a change at the very start.

To see the effect of FIR when the process is out of control, we construct a CUSUM chart for the previous process after it has experienced an upward shift of one sigma. For comparative purposes, we use the observations starting from the 11th observation in Table 7.5, as they were taken after the process had experienced an upward shift of one sigma. These observations and the corresponding CUSUM with $S_0^+ = h/2 = 2.5$, $S_0^- = -h/2 = -2.5$ are shown in Table 7.6.

**Table 7.5** Tabular CUSUM control chart using FIR feature for data in Table 7.4.

| Sample # | $X_i$ | $Z_i = (X_i - \mu)/\sigma$ | $S_i^+ = Max(0, Z_i - k + S_{i-1}^+)$ | $S_i^- = Min(0, Z_i + k + S_{i-1}^-)$ |
|---|---|---|---|---|
| 1 | 10.70 | -2.150 | 0.000 | -4.150 |
| 2 | 14.35 | -0.325 | 0.000 | -3.975 |
| 3 | 12.77 | -1.115 | 0.000 | -4.590 |
| 4 | 18.50 | 1.750 | 1.250 | -2.340 |
| 5 | 14.07 | -0.465 | 0.285 | -2.305 |
| 6 | 15.62 | 0.310 | 0.095 | -1.495 |
| 7 | 16.01 | 0.505 | 0.100 | -0.490 |
| 8 | 14.68 | -0.160 | 0.000 | -0.150 |
| 9 | 13.43 | -0.785 | 0.000 | -0.435 |
| 10 | 15.56 | 0.280 | 0.000 | 0.000 |
| 11 | 20.85 | 2.925 | 2.425 | 0.000 |
| 12 | 15.77 | 0.385 | 2.310 | 0.000 |
| 13 | 21.46 | 3.230 | 5.040 | 0.000 |
| 14 | 16.70 | 0.850 | 5.390 | 0.000 |
| 15 | 17.84 | 1.420 | 6.310 | 0.000 |
| 16 | 14.99 | -0.005 | 5.805 | 0.000 |
| 17 | 16.96 | 0.980 | 6.285 | 0.000 |
| 18 | 17.70 | 1.350 | 7.135 | 0.000 |
| 19 | 17.67 | 1.335 | 7.970 | 0.000 |
| 20 | 19.67 | 2.335 | 9.805 | 0.000 |

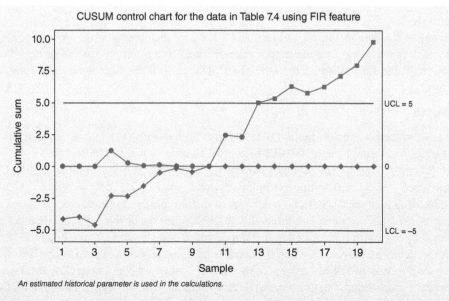

**Figure 7.6**  Two-sided CUSUM control chart using the FIR feature for the data in Table 7.5.

**Example 7.5**

Reproduce observations 11–20 from Table 7.5, shown in Table 7.6, and construct the tabular CUSUM control chart for these data using the FIR feature. Compare the effects of the FIR feature in this example and Example 7.4.

**Solution**

Table 7.6 gives the complete tabular CUSUM control chart for the data in Table 7.5 after it experienced an upward shift of one sigma. Clearly, the FIR feature has a noticeable effect on the CUSUM chart in the sense that the first two sample points came very close to the upper decision interval and the third sample point falls beyond it. However, since the shift was upward, the CUSUM chart for the downward shift had no effect. To summarize these results, we can say that the FIR feature for the CUSUM control chart is very valuable whenever there is a change in the process; otherwise, it may not have a significant effect.

We can construct a CUSUM chart using any one of the statistical packages discussed in this book. Thus, to construct a CUSUM chart for the data in Table 7.1 (Example 7.1) using Minitab or R, we proceed as follows.

**Minitab**

1) Enter all the data in Table 7.1 in columns C1−C4: that is, use one row for each subgroup.
2) From the menu bar, select **Stat > Control Charts > Time Weighted Charts > CUSUM**.

**Table 7.6** Tabular CUSUM control chart using the FIR feature for data in Table 7.4, after the process had experienced an upward shift of one sigma.

| Sample # | $X_i$ | $Z_i = (X_i - \mu)/\sigma$ | $S_i^+ = Max(0, Z_i - k + S_{i-1}^+)$ | $S_i^- = Min(0, Z_i + k + S_{i-1}^-)$ |
|---|---|---|---|---|
| 1 | 20.85 | 2.925 | 4.925 | 0.000 |
| 2 | 15.77 | 0.385 | 4.810 | 0.000 |
| 3 | 21.46 | 3.230 | 7.540 | 0.000 |
| 4 | 16.70 | 0.850 | 7.890 | 0.000 |
| 5 | 17.84 | 1.420 | 8.810 | 0.000 |
| 6 | 14.99 | −0.005 | 8.305 | 0.000 |
| 7 | 16.96 | 0.980 | 8.785 | 0.000 |
| 8 | 17.70 | 1.350 | 9.635 | 0.000 |
| 9 | 17.67 | 1.335 | 10.470 | 0.000 |
| 10 | 19.67 | 2.335 | 12.305 | 0.000 |

3) In the **CUSUM Chart** dialog box, select the option **Observations for a subgroup are in one row of columns**, and enter **C1-C4** in the box next to it.
4) Enter the sample size (**4**) in the box next to **Sample size**.
5) Enter the target value (**15**) in the box next to **Target.**
6) Select any option(s) available in this dialog box. For example, if we select option *CUSUM Chart Options*, then in the new dialog box that appears, we have various new options: **Parameters, Estimate, Plan/Type,** and **Storage**.
   a) **Parameter**: Enter the specified value in the box next to **Standard deviation**.
   b) **Estimate**: Use this option to estimate the standard deviation. Under this option, we select one of the possible options of estimating the standard deviation: method of moving range, R bar, S bar, or pooled standard deviation and others.
   c) **Plan/Type**: Under this option, we can choose the type of CUSUM chart that we would like: **Tabular CUSUM, V-mask CUSUM,** and then select the sub-option (if applicable) select fast initial response **(FIR) CUSUM**, choose the appropriate values of the parameters $h$ and $k$.
   d) **Storage**: In this dialog box, select any appropriate storage options: for example, sample means, sample standard deviations, etc. Then click **OK**. The desired CUSUM control chart shown in Figure 7.7 will appear in the **Session Window**.

**R**

The "qcc" package for constructing the CUSUM chart contains the built-in function "cusum()". We can use it as follows:

```
library (qcc)
# Read the data from a local directory by using the following working directory in R.
 setwd("C:/Users/person/Desktop/SQC-Data folder")
 parts = read.csv("Table7.1.csv")
 cusum(parts, center =15, std.dev = 2)
# If standard deviation is not given then we can use the command std.dev ="RMSDF", where
RMSDF stands for 'root-mean-squared estimate'.
```

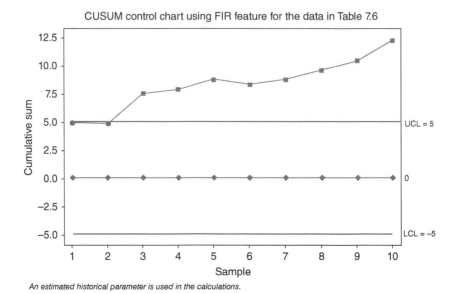

**Figure 7.7** Two-sided CUSUM control chart using the FIR feature for the data in Table 7.6.

The construction of CUSUM control charts using JMP and Python is discussed in Chapter 10, which is available for download on the book's website: www.wiley.com/college/gupta/SQC.

### 7.3.3   One-Sided CUSUM Control Charts

Sometimes we are interested in detecting a shift in one direction only: that is, either in the upward or downward direction instead of both directions simultaneously. This is particularly important for situations in which a shift in one direction is more serious than the other. For example, consider the case where we are interested in measuring the compressive strength of concrete blocks. In this case, any upward shift may be valuable, but any downward shift may have serious consequences. To construct the corresponding CUSUM control chart in such a case, we need to calculate only $S^+$ or $S^-$, depending on whether we are seeking to detect an upward shift or downward shift. Furthermore, to construct a one-sided CUSUM chart, we must note that the values of $k$ and $h$ are not the same as for a two-sided CUSUM control chart with the same ARL value. If we keep the same values of $k$ and $h$ for one-sided CUSUM as for two-sided CUSUM, then the ARL value will be much larger. For example, using Siegmund's (1985) formula, we can see that the ARL value for a stable process with $k = 0.5$ and $h = 5$, is equal to 938.2.

### 7.3.4   Combined Shewhart-CUSUM Control Charts

As Hawkins and Olwell (1998) put it,

> *CUSUMs are excellent diagnostics for detecting and diagnosing step changes in process parameters. However, these are not the only changes that can occur. Transient special causes are also an important reality and an important source of quality problems. They cannot be ignored, and relying solely on CUSUM charts for SPC is shortsighted. Just as Shewhart charts are not particularly effective in detecting less-than-massive persistent changes in process parameter, the CUSUM charts are not particularly effective in detecting massive transient changes in process*

*parameters. In fact, the CUSUMs are meant to detect persistent but small changes in process parameters. Proper SPC requires the use of both types of controls: CUSUMs for persistent but small changes and Shewhart charts for large transient problems.*

Lucas (1982) introduced a combined Shewhart-CUSUM quality control scheme. The combined Shewhart-CUSUM control chart gives an out-of-control signal if the most recent sample is either outside the Shewhart control limits or beyond the CUSUM decision interval value. Lucas recommends using the standardized values $z_i$ for the combined chart, which is

$$z_i = \frac{(x_i - \mu_0)}{\sigma}$$

where $\mu_0$ is the target mean value and $\sigma$ is the process standard deviation. Note that if rational subgroups of size n > 1 are used, then $z_i, x_i$, and $\sigma$ are replaced by $\bar{z}_i, \bar{x}_i$ and $\sigma/\sqrt{n}$, respectively.

In the combined Shewhart-CUSUM control chart, an out-of-control signal at time $i$ occurs when either $S_i^+$ or $S_i^-$ exceeds the decision interval value $h$, or the $z_i$ or $\bar{z}_i$ value falls outside the Shewhart control limits.

It can be seen from Lucas (1982) that $ARL_0$ for the combined Shewhart-CUSUM chart with $h = 5$, $k = 0.5$, and Shewhart control limits $\pm 3$ is 223.9 without the FIR feature and 206.5 with the FIR feature ($S_o = 2.5$). Clearly, both of these values are much smaller than 370, the $ARL_0$ for a Shewhart control chart. This could cause the process to have more frequent false alarms. For example, if samples are taken every hour, the Shewhart control chart would cause a false alarm every 15.4 days, the combined Shewhart-CUSUM control chart without the FIR feature would cause a false alarm every 9.3 days, and the combined Shewhart-CUSUM control chart with the FIR feature every 8.5 days. To avoid this scenario, Lucas recommends using Shewhart control limits $\pm 3.5$, which results in $ARL_0$ being approximately 392.7 without the FIR feature and 359.7 with the FIR feature and $S_o = 2.5$. These values of $ARL_0$ are certainly more comparable to 370. Thus the combined Shewhart-CUSUM chart can be implemented just by using the basic CUSUM chart so that an out-of-control signal occurs whenever either the CUSUM value exceeds the decision interval value of $h = 5$ or the absolute value of $z_i$ or $\bar{z}_i$ becomes greater than 3.5.

### 7.3.5   CUSUM Control Charts for Controlling Process Variability

No process is immune to changes in variability, and this is important to note when CUSUM control charts are implemented. Any unnoticed changes in process variability can adversely affect the conclusion made about the process mean using CUSUM charts. Hence, when using CUSUM charts to control the process mean, it is also important to control process variability.

Consider a process with quality characteristic $X_i$ distributed as normal with mean $\mu$ and standard deviation $\sigma$. Then the standardized random variable

$$Z_i = \frac{X_i - \mu}{\sigma}$$

is distributed as standard normal. Hawkins (1981) showed that the random variable $\sqrt{|Z_i|}$ is approximately distributed as normal with mean 0.822 and standard deviation 0.349: that is, the random variable

$$V_i = \frac{\sqrt{|Z_i|} - 0.822}{0.349}$$

is distributed as standard normal. Furthermore, when the standards deviation of $X_i$ increases, the mean of $V_i$ increases. However, any change in the mean of $V_i$ can be detected by designing a two-sided CUSUM chart as follows:

$$V_i^+ = \text{Max}\left[0, v_i - k + V_{i-1}^+\right] \tag{7.6}$$

$$V_i^- = \text{Min}\left[0, v_i + k + V_{i-1}^-\right] \tag{7.7}$$

where $V_0^+ = V_0^- = 0$, and $k$ and $h$ are the respective reference value and the decision interval. As Hawkins and Olwell (1998) point out, if desired, we can use the FIR feature in the same manner as in the basic CUSUM chart: that is, by taking $V_0^+ = V_0^- = h/2$. The interpretation of this CUSUM chart is very similar to the CUSUM control chart in Section 7.3.1. Thus, if either of the CUSUM values $V_i^+$ or $V_i^-$ exceeds the decision interval, an out-of-control signal for the variance occurs. For more details and examples of CUSUM control charts for process variability, readers are referred to Hawkins (1993).

## 7.4  Moving Average (MA) Control Charts

Consider the process of manufacturing medical equipment parts, discussed earlier in this chapter. Let $x_1, x_2, x_3, ..., x_n$ be the observation of the quality characteristics under investigation. Then the *moving average $M_i$* of span $m$ at time $i$ is defined as

$$M_i = \frac{x_1 + x_2 + ... + x_i}{i}, i = 1, 2, ..., m-1 \tag{7.8}$$

$$M_i = \frac{x_{i-m+1} + x_{i-m+2} + ... + x_i}{m}, i = m, m+1, ..., n \tag{7.9}$$

**Example 7.6**
Suppose that during the manufacturing process of medical equipment parts, we collect five observations on the quality characteristic under investigation. Find the moving averages $M_i$, $i = 1, 2, ..., 5$ of span 3.

**Solution**

Since $n = 5$ and $m = 3$, we have

$$M_1 = x_1$$

$$M_2 = \frac{x_1 + x_2}{2}$$

$$M_3 = \frac{x_1 + x_2 + x_3}{3}$$

$$M_4 = \frac{x_2 + x_3 + x_4}{3}$$

$$M_5 = \frac{x_3 + x_4 + x_5}{3}$$

The variance of the moving average $M_i$ is given by the following:

For $i < m$,

$$V(M_i) = V\left(\frac{x_1 + x_2 + \dots + x_i}{i}\right)$$

$$= \frac{1}{i^2} \sum_{j=1}^{i} V(x_j) = \frac{\sigma^2}{i}, \quad j > 0 \tag{7.10}$$

For $i \geq m$,

$$V(M_i) = V\left(\frac{x_{i-m+1} + x_{i-m+2} + \dots x_i}{m}\right)$$

$$= \frac{1}{m^2} \sum_{j=i-m+1}^{i} V(x_j) = \frac{\sigma^2}{m} \tag{7.11}$$

where $\sigma^2$ is the process variance. Thus the variance of the moving averages in Example 7.6 is

$$V(M_1) = \sigma^2$$

$$V(M_2) = \frac{\sigma^2}{2}$$

$$V(M_3) = \frac{\sigma^2}{3}$$

$$V(M_4) = \frac{\sigma^2}{3}$$

$$V(M_5) = \frac{\sigma^2}{3}$$

Now, consider a quality characteristic of a process that is normally distributed with a target mean value $\mu_0$ and variance $\sigma^2$. The probability control limits for $M_i$ with type I error $\alpha$ are, for $i < m$,

$$\text{UCL} = \mu_0 + z_{\alpha/2} \frac{\sigma}{\sqrt{i}} \tag{7.12}$$

$$\text{LCL} = \mu_0 - z_{\alpha/2} \frac{\sigma}{\sqrt{i}} \tag{7.13}$$

and for $i \geq m$,

$$\text{UCL} = \mu_0 + z_{\alpha/2} \frac{\sigma}{\sqrt{m}} \tag{7.14}$$

$$\text{LCL} = \mu_0 - z_{\alpha/2} \frac{\sigma}{\sqrt{m}} \tag{7.15}$$

In order to have three-sigma control limits, replace $z_{\alpha/2}$ in Eqs. (7.12–7.15) with 3. The process is considered out-of-control as soon as $M_i$ falls outside the control limits. It is important, however, to note that since the plotted points $(i, M_i)$ are not independent, no reasonable interpretation can be made about the patterns in the MA graph. For detecting smaller shifts, $m$ should be large. Note that a large value of $m$ would result in a longer time required to detect bigger shifts. For an illustration, we consider the data in Table 7.4 of Example 7.3.

Note: In the earlier discussion, we considered the MA chart for individual values. Sometimes the process may allow us to take samples of size $n > 1$. In such situations, the individual values in Table 7.7 are replaced by the sample means. Furthermore, the probability control limits in Eqs. (7.12–7.15) are defined as follows for $i < m$

$$\text{UCL} = \mu_0 + z_{\alpha/2}\frac{\sigma}{\sqrt{in}} \tag{7.16}$$

$$\text{LCL} = \mu_0 - z_{\alpha/2}\frac{\sigma}{\sqrt{in}} \tag{7.17}$$

and for $i \geq m$

$$\text{UCL} = \mu_0 + z_{\alpha/2}\frac{\sigma}{\sqrt{mn}} \tag{7.18}$$

$$\text{LCL} = \mu_0 - z_{\alpha/2}\frac{\sigma}{\sqrt{mn}} \tag{7.19}$$

**Example 7.7**
Consider the data in Table 7.4 of Example 7.3 (reproduced in Table 7.7), from a process that is manufacturing medical equipment parts. Design a moving average (MA) control chart for these data with span $m = 4$, and interpret the results.

**Solution**

The $M_i$ values are determined as shown here and are then entered in Table 7.7:

$$M_1 = X_1 = 10.70$$

**Table 7.7** Moving average chart ($M_i$) for data in Table 7.4 with $\mu = 15$, $\sigma = 2$.

| Sample # | 1 | 2 | 3 | 4 | 5 | 6 | 7 | 8 | 9 | 10 |
|---|---|---|---|---|---|---|---|---|---|---|
| $X_i$ | 10.70 | 14.35 | 12.77 | 18.50 | 14.07 | 15.62 | 16.01 | 14.68 | 13.43 | 15.56 |
| $M_i$ | 10.70 | 12.52 | 12.61 | 14.08 | 14.92 | 15.24 | 16.05 | 15.10 | 14.94 | 14.92 |
| Sample # | 11 | 12 | 13 | 14 | 15 | 16 | 17 | 18 | 19 | 20 |
| $X_i$ | 20.85 | 15.77 | 21.46 | 16.70 | 17.84 | 14.99 | 16.96 | 17.70 | 17.67 | 19.67 |
| $M_i$ | 16.13 | 16.40 | 18.41 | 18.70 | 17.94 | 17.75 | 16.62 | 16.87 | 16.83 | 18.00 |

$$M_2 = (X_1 + X_2)/2 = (10.70 + 14.35)/2 = 12.52$$

$$M_3 = (X_1 + X_2 + X_3)/3 = (10.70 + 14.35 + 12.77)/3 = 12.61$$

$$M_4 = (X_1 + X_2 + X_3 + X_4)/4 = (10.70 + 14.35 + 12.77 + 18.50)/4 = 14.08$$

$$M_5 = (X_2 + X_3 + X_4 + X_5)/4 = (14.35 + 12.77 + 18.50 + 14.07)/4 = 14.92$$

$$\cdots$$

$$M_{20} = (X_{17} + X_{18} + X_{19} + X_{20})/4$$
$$= (16.96 + 17.70 + 17.67 + 19.67)/4 = 18.00$$

Now the points $(i, M_i)$ from Table 7.7 are plotted in the MA control chart. The MA control chart for the data in Table 7.7 is prepared using Minitab and is as shown in Figure 7.8. The 13th point falls outside the control limits, so the process is out of control at time $i = 13$. Note that the control limits for $i < m$ are wider and vary quite a bit, but from $i = m$ onward, the control limits become narrower and fixed. The CL, however, is fixed for all $i$. The three-sigma control limits in Figure 7.8 are determined using Eqs. (7.12–7.15), as shown next. The CUSUM control chart also shows that the process is out of control at time $i = 13$. For the time period $i = 1, 2, \cdots,$ 20, the control limits are calculated as shown.

$i = 1$:

$$UCL = 15 + 3\frac{2}{\sqrt{1}} = 21$$

$$LCL = 15 - 3\frac{2}{\sqrt{1}} = 9$$

$i = 2$:

$$UCL = 15 + 3\frac{2}{\sqrt{2}} = 19.24$$

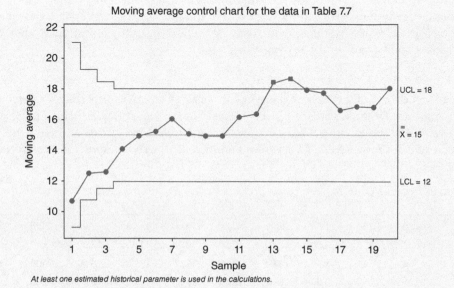

Moving average control chart for the data in Table 7.7

*At least one estimated historical parameter is used in the calculations.*

**Figure 7.8** Minitab printout MA control chart for the data in Table 7.7.

$$LCL = 15 - 3\frac{2}{\sqrt{2}} = 10.76$$

$$\cdots$$

$i = 20$:

$$UCL = 15 + 3\frac{2}{\sqrt{4}} = 18$$

$$LCL = 15 - 3\frac{2}{\sqrt{4}} = 12.00$$

Values of all LCLs and UCLs are listed in Table 7.8.

Note that both the LCL and UCL for the MA control chart starting from sample point $4 (= m)$ onward are fixed.

We can construct an **MA** control chart using any of the statistical packages discussed in this book. To do so using Minitab or R, we proceed as follows.

**Minitab**

1) Enter all the data from Table 7.4 in column C1.
2) From the menu bar, select **Stat > Control Charts > Time Weighted Charts > Moving Average**. The **Moving Average Chart** dialog box appears on the screen. In this dialog box, choose the option **All observations for a chart are in one column**, and enter **C1** in the next box. If the sample size is $n > 1$, then enter the data in columns C1–C$n$ and choose the option **Observations for a subgroup are in one row of columns**.
3) Enter **1** in the box next to **Subgroup size** or the appropriate value of $n$ if the sample size is greater than 1.
4) Enter the specified value of the span in the box next to **Length of MA** (**4** in this example).
5) Select any desirable options, such as **MA Options**. A dialog containing various sub-options opens: **Parameters**, **Estimate**, **S Limits**, and **Storage**. Use any appropriate sub-options, complete all required entries, and then click **OK.** The MA control chart shown in Figure 7.8 will appear in the **Session Window**.

**R**

Currently, there are no packages available in R to construct an MA control chart. A customized program written for the MA control chart is not given here for lack of space, but it is provided in Chapter 10. The construction of MA control charts using JMP and Python is also discussed in Chapter 10, which is available for download on the book's website: www.wiley.com/college/gupta/SQC.

**Table 7.8** Control limits for each sample point.

| LCL | 9.00 | 10.76 | 11.54 | 12.00 | 12.00 | 12.00 | 12.00. | 12.00 | 12.00 | 12.00 |
|-----|------|-------|-------|-------|-------|-------|--------|-------|-------|-------|
| LCL | 12.00 | 12.00 | 12.00 | 12.00 | 12.00 | 12.00 | 12.00 | 12.00 | 12.00 | 12.00 |
| UCL | 21.00 | 19.24 | 18.46 | 18.00 | 18.00 | 18.00 | 18.00 | 18.00 | 18.00 | 18.00 |
| UCL | 18.00 | 18.00 | 18.00 | 18.00 | 18.00 | 18.00 | 18.00 | 18.00 | 18.00 | 18.00 |

## 7.5 Exponentially Weighted Moving Average (EWMA) Control Charts

A EWMA control chart is very useful for processes when only one observation per period is available. Like a CUSUM control chart, it has the capability of detecting small process mean shifts. The EWMA control chart was introduced by Roberts (1959), who called it a *geometric moving average control chart*. Lately, however, it has come to be known as an *exponentially weighted moving average control chart*. The EWMA control chart uses information from all observations made up to and including time $i$, as is done with CUSUM control charts. In EWMA charts, we plot the values of the statistics (*not a standard normal variate*) $z_i$, defined as

$$z_i = \lambda x_i + (1 - \lambda) z_{i-1} \tag{7.20}$$

where $0 < \lambda \le 1$ is a constant, the initial value $z_0$ is set at the process target mean value, $\mu_0$ (if $\mu_0$ is not given, we set $z_0$ at $\bar{x}$), and $\lambda$ is referred to as the *weight*. The statistics $z_i$ and $z_{i-1}$ in Eq. (7.20) are the exponentially weighted moving averages at time $i$ and $i - 1$, respectively, and $x_i$ is the value of the observation at time $i$. Furthermore, note that $z_i$ is the weighted average of all the previous observations. This can easily be shown using an iterative method, as follows:

$$z_i = \lambda x_i + (1 - \lambda) z_{i-1}$$
$$= \lambda x_i + (1 - \lambda)[\lambda x_{i-1} + (1 - \lambda) z_{i-2}]$$
$$= \lambda x_i + \lambda(1 - \lambda) x_{i-1} + (1 - \lambda)^2 z_{i-2}$$
$$\cdots$$

$$z_i = \lambda \sum_{j=0}^{i-1} (1 - \lambda)^j x_{i-j} + (1 - \lambda)^i z_0 \tag{7.21}$$

At any time $i$, the most recent observation, $x_i$, is assigned the weight $\lambda$; while at any earlier time $(i - j), j = 1, 2, ..., i - 1$, the observation $x_{i-j}$ is assigned the weight $\lambda(1 - \lambda)^j$, which decreases geometrically as $j$ increases. Thus, for example, if $\lambda = 0.4$, then the current observation is assigned a weight of 0.4 and the preceding observations are assigned a weight of 0.24, 0.144, 0.0864, 0.0518, 0.0311, and so on. Clearly, we see that the older observations carry less weight. This makes sense, particularly when the process mean is changing slowly but continuously. Moreover, it is also interesting to note that since EWMA uses the weighted *average* of all past and present observations, it is not particularly sensitive to the normality assumption.

If we assume the observations $x_i$ are independent, with process mean $\mu$ and variance $\sigma^2$, then the mean of the statistic $z_i$ is given by

$$E(z_i) = \lambda \sum_{j=0}^{i-1} (1 - \lambda)^j \mu + (1 - \lambda)^i E(z_0)$$

$$= \lambda \left[ 1 + (1 - \lambda) + (1 - \lambda)^2 + ... + (1 - \lambda)^{i-1} \right] \mu + (1 - \lambda)^i E(z_0)$$

The quantity inside the brackets is a geometric series with a common ratio equal to $(1 - \lambda)$; and since the sum of a finite geometric series is $[1 + r + r^2 + ... + r^{n-1} = (1 - r^n)/(1 - r)]$, the quantity in brackets is equal to $\dfrac{1 - (1 - \lambda)^i}{\lambda}$.

Thus we have

$$E(z_i) = \mu + (1-\lambda)^i(\mu_0 - \mu) \qquad (7.22)$$

Note that if $z_0$ is set at $\bar{x}$, then Eq. (7.22) reduces to

$$E(z_i) = \mu \qquad (7.23)$$

since $E(\bar{x}) = \mu$. It is customary to set the target value $\mu_o$ at $\mu$. In any case, in all applications, we set the expected value of $z_i$ at $\mu$, as defined in Eq. (7.23). If $\mu$ is not known, or if the target value is not given, then $\mu$ is replaced with $\bar{x}$. The variance of the statistic $z_i$ is given by

$$V(z_i) = \left(\lambda^2 + \lambda^2(1-\lambda)^2 + \lambda^2(1-\lambda)^4 + \ldots + \lambda^2(1-\lambda)^{2i}\right)\sigma^2$$

$$= \lambda^2\left[1 + (1-\lambda)^2 + (1-\lambda)^4 + \ldots + (1-\lambda)^{2i}\right]\sigma^2$$

Again, the quantity inside the brackets is a geometric series with the common ratio equal to $(1-\lambda)^2$. Thus we have

$$V(z_i) = \lambda^2 \frac{\left[1-(1-\lambda)^{2i}\right]}{\left[1-(1-\lambda)^2\right]}\sigma^2$$

$$= \lambda^2 \frac{\left[1-(1-\lambda)^{2i}\right]}{\left[2\lambda-\lambda^2\right]}\sigma^2$$

$$= \left(\frac{\lambda}{2-\lambda}\right)\left[1-(1-\lambda)^{2i}\right]\sigma^2 \qquad (7.24)$$

From Eqs. (5.23) and (5.24), it follows that the CL and the control limits for EWMA control charts are given by

$$UCL = \mu_0 + 3\sigma\sqrt{\left(\frac{\lambda}{2-\lambda}\right)\left[1-(1-\lambda)^{2i}\right]} \qquad (7.25)$$

$$CL = \mu_0 \qquad (7.26)$$

$$LCL = \mu_0 - 3\sigma\sqrt{\left(\frac{\lambda}{2-\lambda}\right)\left[1-(1-\lambda)^{2i}\right]} \qquad (7.27)$$

Note that the control limits in Eqs. (7.25) and (7.27) are variable. However, when $i$ becomes very large, the quantity $[1 - (1 - \lambda)^{2i}]$ approaches unity; consequently, the EWMA control limits become constant and are given by

$$UCL = \mu_0 + 3\sigma\sqrt{\left(\frac{\lambda}{2 - \lambda}\right)} \tag{7.28}$$

$$CL = \mu_0 \tag{7.29}$$

$$LCL = \mu_0 - 3\sigma\sqrt{\left(\frac{\lambda}{2 - \lambda}\right)} \tag{7.30}$$

In both cases, an appropriate value of $\lambda$ needs to be determined, as we will discuss shortly. The control limits of the general form of the EWMA control chart are given by

$$UCL = \mu_0 + L\sigma\sqrt{\left(\frac{\lambda}{2 - \lambda}\right)\left[1 - (1 - \lambda)^{2i}\right]} \tag{7.31}$$

$$CL = \mu_0 \tag{7.32}$$

$$LCL = \mu_0 - L\sigma\sqrt{\left(\frac{\lambda}{2 - \lambda}\right)\left[1 - (1 - \lambda)^{2i}\right]} \tag{7.33}$$

and when $i$ is very large, the earlier control limits reduce to

$$UCL = \mu_0 + L\sigma\sqrt{\left(\frac{\lambda}{2 - \lambda}\right)} \tag{7.34}$$

$$CL = \mu_0 \tag{7.35}$$

$$LCL = \mu_0 - L\sigma\sqrt{\left(\frac{\lambda}{2 - \lambda}\right)} \tag{7.36}$$

**Table 7.9**   A selection of EWMA charts with $ARL_0 \cong 500$.

| $\delta$ | $\lambda = 0.05$ $L = 2.615$ | $\lambda = 0.10$ $L = 2.814$ | $\lambda = 0.20$ $L = 2.962$ | $\lambda = 0.25$ $L = 2.998$ | $\lambda = 0.40$ $L = 3.054$ |
|---|---|---|---|---|---|
| 0.25 | 84.1 | 106.0 | 150.0 | 170.0 | 224.0 |
| 0.50 | 28.8 | 31.3 | 41.8 | 48.8 | 71.2 |
| 0.75 | 16.4 | 15.9 | 18.2 | 20.1 | 28.4 |
| 1.00 | 11.4 | 10.3 | 10.5 | 11.1 | 14.3 |
| 1.50 | 7.1 | 6.1 | 5.5 | 5.5 | 5.9 |
| 2.00 | 5.2 | 4.4 | 3.7 | 3.6 | 3.5 |
| 2.50 | 4.2 | 3.4 | 2.9 | 2.7 | 2.5 |
| 300 | 3.5 | 2.9 | 2.4 | 2.3 | 2.0 |
| 4.00 | 2.7 | 2.2 | 1.9 | 1.7 | 1.4 |

Several authors, including Crowder (1987, 1989) and Lucas and Saccucci (1990), have studied the problem of determining the ARL for different values of the parameters $\lambda$ and $L$. Quesenberry (1997) has discussed some of their results in detail. Lucas and Saccucci (1990) provided tables of ARL as a function of $\lambda$, $L$, and $\delta$, where $\delta$ is a shift in the process mean measured in units of the process standard deviation. Some of their results are given in Table 7.9.

**Example 7.8**
Consider the data in Table 7.4 of Example 7.3 for a manufacturing process of medical equipment parts. Design an EWMA control chart for these data with $\lambda = 0.20$, $L = 2.962$, $\mu_0 = 15$, $\sigma = 2$, and interpret the results.

**Solution**

The data from Table 7.4 is reproduced in Table 7.10. The $z_i$ values are determined as shown below and then entered in Table 7.10. Thus, using Eq. (7.20), we have

$$z_1 = \lambda x_1 + (1 - \lambda)z_0$$
$$= 0.20 \times 10.70 + (1 - 0.20) \times 15 = 14.14$$
$$z_2 = \lambda x_2 + (1 - \lambda)z_1$$
$$= 0.20 \times 14.35 + (1 - 0.20) \times 14.14 = 14.18$$
$$z_3 = \lambda x_3 + (1 - \lambda)z_2$$
$$= 0.20 \times 12.77 + (1 - 0.20) \times 14.18 = 13.90$$

$$\cdots$$

$$z_{20} = \lambda x_{20} + (1 - \lambda)z_{19}$$
$$= 0.20 \times 19.67 + (1 - 0.20) \times 17.11 = 17.72$$

**Table 7.10** Exponentially weighted moving average control chart ($z_i$) for data in Table 7.4 with $\lambda = 0.20$, $L = 2.962$.

| Sample # | 1 | 2 | 3 | 4 | 5 | 6 | 7 | 8 | 9 | 10 |
|---|---|---|---|---|---|---|---|---|---|---|
| $x_i$ | 10.70 | 14.35 | 12.77 | 18.50 | 14.07 | 15.62 | 16.01 | 14.68 | 13.43 | 15.56 |
| $z_i$ | 14.14 | 14.18 | 13.90 | 14.82 | 14.67 | 14.86 | 15.09 | 15.01 | 14.69 | 14.87 |

| Sample # | 11 | 12 | 13 | 14 | 15 | 16 | 17 | 18 | 19 | 20 |
|---|---|---|---|---|---|---|---|---|---|---|
| $x_i$ | 20.85 | 15.77 | 21.46 | 16.70 | 17.84 | 14.99 | 16.96 | 17.70 | 17.67 | 19.67 |
| $z_i$ | 16.06 | 16.00 | 17.10 | 17.02 | 17.18 | 16.74 | 16.79 | 16.97 | 17.11 | 17.72 |

The control limits in Figure 7.9 with $\lambda = 0.20$, $L = 2.962$, $\mu_0 = 15$, $\sigma = 2$ are determined using Eqs. (7.31–7.36) as shown.

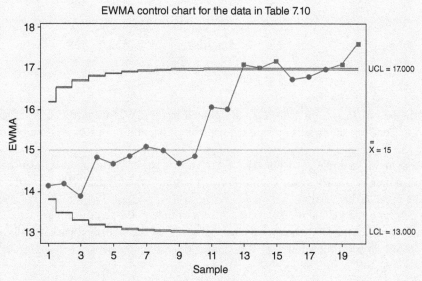

At least one estimated historical parameter is used in the calculations.

**Figure 7.9** Minitab printout of the EWMA control chart for the data in Table 7.10.

$i = 1$:

$$\text{UCL} = 15 + 2.962 \times 2\sqrt{\left(\frac{0.2}{2-0.2}\right)\left[1 - (1 - 0.2)^2\right]}$$

$$= 16.20$$

$$\text{LCL} = 15 - 2.962 \times 2\sqrt{\left(\frac{0.2}{2-0.2}\right)\left[1 - (1 - 0.2)^2\right]}$$

$$= 13.80$$

$i = 2$:

$$\text{UCL} = 15 + 2.962 \times 2\sqrt{\left(\frac{0.2}{2-0.2}\right)\left[1-(1-0.2)^4\right]}$$

$$= 16.54$$

$$\text{LCL} = 15 - 2.962 \times 2\sqrt{\left(\frac{0.2}{2-0.2}\right)\left[1-(1-0.2)^4\right]}$$

$$= 13.46$$

...

$i = 20$ (since $i$ is fairly large):

$$\text{UCL} = 15 + 2.962 \times 2\sqrt{\left(\frac{0.2}{2-0.2}\right)}$$

$$t = 17.00$$

$$\text{LCL} = 15 - 2.962 \times 2\sqrt{\left(\frac{0.2}{2-0.2}\right)}$$

$$= 13.00$$

Now the points $(i, z_i)$ from Table 7.10 are plotted in the EWMA control chart, as shown in Figure 7.9. The 13th point falls outside the control limits, and the process is out of control at time $i= 13$.

Note that in this example, the CUSUM and MA control charts also show that the process is out of control at time $i = 13$. The control limits for each sample point for the EWMA control chart are shown in Table 7.11.

Note that both the LCL and UCL, starting from sample point 12 onward, are fixed.

We can construct the EWMA control chart using any of the statistical packages discussed in this book. We do so using Minitab or R by taking the following steps.

**Minitab**

1) Enter all the data ($z_i$) from Table 7.10 in column C1 of the Worksheet window.
2) From the menu bar, click **Stat > Control Charts > Time Weighted Charts > EWMA**. The **EWMA Charts** dialog box appears. In this dialog box, choose the option **All observations for a chart are in one column**.

**Table 7.11** Control limits for each sample point.

| LCL | 13.80 | 13.46 | 13.28 | 13.18 | 13.11 | 13.07 | 13.04 | 13.03 | 13.02 | 13.01 |
|-----|-------|-------|-------|-------|-------|-------|-------|-------|-------|-------|
| LCL | 13.00 | 13.00 | 13.00 | 13.00 | 13.00 | 13.00 | 13.00 | 13.00 | 13.00 | 13.00 |
| UCL | 16.20 | 16.54 | 16.72 | 16.82 | 16.89 | 16.93 | 16.96 | 16.97 | 16.98 | 16.99 |
| UCL | 16.99 | 17.00 | 17.00 | 17.00 | 17.00 | 17.00 | 17.00 | 17.00 | 17.00 | 17.00 |

3) Enter **C1** in the next box. If the sample size is n > 1, then enter the data in columns C1–C*n* and, in step 2, choose the option **Observations for a subgroup are in one row of columns**. Then enter **1** or the appropriate value of $n$ in the box next to **Subgroup size**.

4) Enter the specified value of $\lambda$ in the box next to **Weight of EWMA**.

5) Select any desirable options. For example, select **EWMA Options**. Another dialog box opens, containing various sub-options: **Parameters**, **Estimate**, **S Limits**, and **Storage**. Use any appropriate sub-options, and complete all the required entries. Then, in each of the two dialog boxes, click **OK**. The EWMA control chart shown in Figure 7.9 appears in the **Session Window**.

## R

The "qcc" package contains the function "EWMA()" for constructing EWMA charts. We can use it as follows:

```
library (qcc)
# Read the data from a local directory by using the following working directory in R
setwd("C:/Users/person/Desktop/SQC-Data folder")
parts = read.csv("Table7.4.csv")
ewma(parts, lambda =0.2, center =15, std.dev=2)
```

The construction of MA control charts using JMP and Python is discussed in Chapter 10, which is available for download on the book's website: www.wiley.com/college/gupta/SQC.

## Review Practice Problems

Note: Sometimes when $\sigma = 1$ we use the term standardized reference value k and decision interval h instead of using simply reference value k and decision interval h because k and h are normally defined in terms of sigma.

**1**   Briefly discuss the difference between the $\overline{X}$ control chart and the CUSUM control chart.

**2**   When would you recommend the use of a CUSUM control chart instead of an $\overline{X}$ control chart to someone in a quality control department?

**3**   What values are plotted in a CUSUM control chart?

**4**   What is the smallest sample size that you would recommend for use in a CUSUM control chart?

**5**   Consider a data set of 10 samples of the size recommended in Problem 4. Determine the points that are plotted in this CUSUM control chart.

**6**   The following data are measurements of compressive strengths (in units of 100 psi) of concrete slabs used in a housing project. Twenty random samples of size 4 each were taken, and the data obtained is shown.

| $x_1$ | $x_2$ | $x_3$ | $x_4$ |
|---|---|---|---|
| 34 | 36 | 38 | 38 |
| 36 | 40 | 36 | 40 |
| 34 | 37 | 43 | 38 |
| 34 | 45 | 42 | 46 |
| 44 | 40 | 35 | 34 |
| 35 | 43 | 41 | 38 |
| 34 | 44 | 40 | 36 |
| 36 | 41 | 45 | 41 |
| 46 | 43 | 34 | 36 |
| 41 | 34 | 42 | 36 |
| 43 | 34 | 35 | 40 |
| 41 | 39 | 38 | 36 |
| 45 | 35 | 46 | 45 |
| 41 | 39 | 40 | 38 |
| 34 | 35 | 46 | 44 |
| 34 | 38 | 42 | 39 |
| 35 | 43 | 34 | 38 |
| 42 | 37 | 39 | 37 |
| 35 | 44 | 34 | 38 |
| 42 | 34 | 42 | 40 |

The target compressive strength is 40; and from past experience, we know the process standard deviation is $\sigma = 2$. Construct a two-sided tabular CUSUM control chart for these data using decision interval and reference values $h = 5$, and $k = 0.5$, respectively.

7   Repeat Problem 6 using a standardized reference value $k = 0.25$ and a decision interval $h = 8.01$. Compare your conclusions with those obtained in Problem 6, and comment on the outcomes of the two charts.

8   Suppose in Problem 6 that samples of size one were taken, and the measurements are those listed under $x_1$. Set up a tabular CUSUM control chart using a standardized reference value $k = 0.5$ and a decision interval $h = 4.77$.

9   Use the data in Problem 8 to set up a tabular CUSUM control chart using a standardized reference value $k = 0.5$ and a decision interval $h = 5$.

10   In Problem 6, set up a CUSUM control chart using the FIR feature. Examine whether this feature helped to detect any shift in the process mean more quickly.

11   Use the data in Problem 6 to construct a CUSUM control chart for controlling process variability. State your conclusions.

12   The following data give the length (in cm) of a radiator mounting bolt for 20 samples of 4 cars each. The target length of the bolt is 16 cm, and from past experience, we know the process standard deviation is $\sigma = 0.6$. Determine the standardized mean $\bar{Z}_i$ for each sample, and then set up a tabular CUSUM control chart to detect a shift of one sigma. Use the standardized reference value $k = 0.5$ and a decision interval $h = 5$.

| $x_1$ | $x_2$ | $x_3$ | $x_4$ |
|---|---|---|---|
| 17.0 | 16.4 | 16.5 | 15.8 |
| 16.2 | 15.1 | 16.3 | 15.2 |
| 15.1 | 15.7 | 16.3 | 15.0 |
| 16.3 | 16.6 | 17.0 | 16.9 |
| 15.5 | 16.7 | 15.1 | 16.3 |
| 16.0 | 16.8 | 15.2 | 16.5 |
| 16.0 | 15.9 | 15.7 | 16.0 |
| 16.4 | 15.5 | 16.6 | 15.7 |
| 16.5 | 15.1 | 15.2 | 15.6 |
| 15.2 | 16.0 | 15.3 | 16.6 |
| 16.7 | 16.6 | 15.6 | 16.6 |
| 16.2 | 15.9 | 17.0 | 15.5 |
| 16.5 | 16.9 | 15.3 | 15.0 |
| 16.8 | 16.3 | 15.2 | 16.1 |
| 15.4 | 16.9 | 16.0 | 15.1 |
| 16.2 | 15.5 | 15.3 | 16.8 |
| 16.1 | 16.6 | 16.6 | 16.8 |
| 16.1 | 16.6 | 15.4 | 15.2 |
| 16.3 | 16.5 | 15.9 | 15.6 |
| 16.2 | 15.5 | 15.3 | 16.2 |

**13**  Construct a tabular CUSUM control chart for the data in Problem 6, using the FIR feature by setting the starting CUSUM values of $S_0^+$ and $S_0^-$ at $h/2$ and $-h/2$, respectively. Use the standardized reference value $k = 0.5$ and decision interval $h = 5$.

**14**  Use the data in Problem 12 to construct a CUSUM control chart for controlling process variability. State your conclusions.

**15**  Repeat Problem 12 using a standardized reference value of $k = 0.25$ and a decision interval of $h = 8.01$. Compare your conclusions with those obtained in Problem 12, and comment on the outcomes of the two charts.

**16**  Consider the following 25 observations from a process under investigation (read the observations in each row from left to right). Use these data to construct a moving average (MA) control chart with span $m = 4$, and then interpret the results.

| 15 | 11 | 8 | 15 | 6 |
|---|---|---|---|---|
| 14 | 16 | 11 | 14 | 7 |
| 13 | 6 | 9 | 5 | 10 |
| 15 | 15 | 9 | 15 | 7 |
| 15 | 10 | 12 | 12 | 16 |

**17**  Reconsider the data in Problem 16. Set up moving average control charts for spans $m = 3$ and 5. Compare the results of using these different span values, and comment on how the different values of $m$ affect the final conclusions.

**18** The following (coded) data give the diameter of the gear ring for the landing gear of a plane (read the observations in each row from left to right). Use these data to design a MA control chart with span $m = 3$, and then interpret the results.

| 35.0 | 36.9 | 36.6 | 35.9 | 35.9 | 36.5 | 35.5 | 36.6 | 35.5 | 36.8 |
|------|------|------|------|------|------|------|------|------|------|
| 35.9 | 35.7 | 35.8 | 35.5 | 36.3 | 36.1 | 35.0 | 36.1 | 35.1 | 35.9 |

**19** Reconsider the data in Problem 18. Set up a MA control chart with span $m = 4$. How has the result changed by changing the span size from 3 to 4?

**20** Repeat Problem 19 using the span size 5. Compare the results of this problem with those of Problems 18 and 19, and make a general statement about the effect of span size on MA control charts.

**21** Consider the data in Problem 16. Construct an EWMA control chart for these data with $\lambda = 0.20$, $L = 2.962$, $\mu_0 = 12$, and $\sigma = 3.8$, and then interpret the results.

**22** Repeat Problem 21, now letting the value of $L$ vary and holding the value of $\lambda$ fixed; specifically, use $L = 3.0$, 3.5, and 4.0, and $\lambda = 0.25$. Discuss how the different values of $L$ affect the final outcomes of the three control charts.

**23** Set up EWMA control charts for the data in Problem 16 using $L = 3$ and $\lambda = 0.20$, 0.25, and 0.3. Comment on how the different values of $\lambda$ affect the form of the three control charts.

**24** Reconsider the data on gear rings in Problem 18. Design an EWMA control chart for these data with $\lambda = 0.20$, $L = 2.9$, $\mu_0 = 36$, and $\sigma = 0.6$, and then interpret the results.

**25** Reconsider the data in Problem 18. Use these data to construct an EWMA control chart using $\lambda = 0.1$, $L = 2.8$.

**26** Reconsider the data in Problem 18. Use these data to construct EWMA control charts using $L = 2.8$; $\lambda = 0.2$, 0.25, 0.4. Discuss the effect of the $\lambda$ values on the control limits of EWMA control charts.

**27** Reconsider the data in Problem 18. Use these data to construct an EWMA control chart using $\lambda = 0.05$, $L = 2.6$.

**28** A manufacturing process is set up to produce voltage regulators with an output voltage distributed as normal with mean 12.0 volts and a standard deviation of 0.2 volts. The following data give the first 10 random samples of size 4, which were taken when the process was stable and producing voltage regulators with an output voltage of mean value 12.0 volts and standard deviation 0.2 volts. The last 10 random samples, again of size 4, were taken from that process after its mean experienced an upward shift of one standard deviation, resulting in the production of voltage regulators with an output voltage distributed as normal with mean 12.2 volts and standard deviation 0.2 volts. Construct a Shewhart $\overline{X} - R$ control chart and a tabular CUSUM control chart for these data using a standardized decision interval $h = 5$ and standardized reference value $k = 0.5$. Comment on the outcomes of the $\overline{X} - R$ and CUSUM charts.

| Sample # | Observations | | | |
|----------|---------|---------|---------|---------|
| 1 | 12.2271 | 11.9393 | 12.2361 | 12.0965 |
| 2 | 12.2039 | 11.9401 | 12.0969 | 11.7367 |
| 3 | 11.9173 | 12.3727 | 11.8975 | 12.2109 |
| 4 | 12.2746 | 12.1076 | 11.4397 | 11.8610 |
| 5 | 11.7907 | 11.9185 | 11.8867 | 11.6639 |
| 6 | 12.2992 | 11.6647 | 11.8495 | 11.9401 |
| 7 | 11.9241 | 11.7489 | 12.1205 | 12.1652 |
| 8 | 12.2346 | 11.7144 | 11.7734 | 12.2993 |
| 9 | 11.9334 | 12.4345 | 11.9171 | 12.0369 |
| 10 | 11.7504 | 11.7916 | 11.8530 | 12.0098 |
| 11 | 12.1004 | 12.5736 | 12.1211 | 12.5222 |
| 12 | 12.5091 | 12.3449 | 12.2997 | 12.2390 |
| 13 | 12.2813 | 11.9147 | 12.2503 | 12.2183 |
| 14 | 12.1541 | 11.8863 | 12.1174 | 12.5731 |
| 15 | 12.3297 | 12.1774 | 12.2993 | 12.1301 |
| 16 | 12.3095 | 12.1205 | 12.3604 | 12.5297 |
| 17 | 11.9462 | 11.8434 | 12.2388 | 12.4401 |
| 18 | 12.6445 | 12.0190 | 12.4178 | 12.1172 |
| 19 | 12.2876 | 12.0814 | 12.0820 | 12.1975 |
| 20 | 12.1029 | 12.4406 | 12.1419 | 12.3893 |

**29** Referring to Problem 28, construct a tabular CUSUM control chart for samples of size one by selecting the first observation from each sample so that the voltage data obtained is as follows. Use $h = 5$ and $k = 0.5$. Compare the results you obtain in this problem with those obtained in Problem 28.

| | | | | | | | | | |
|---|---|---|---|---|---|---|---|---|---|
| 12.2271 | 12.2039 | 11.9173 | 12.2746 | 11.7907 | 12.2992 | 11.9241 | 12.2346 | 11.9334 | 11.7504 |
| 12.1004 | 12.5091 | 12.2813 | 12.1541 | 12.3297 | 12.3095 | 11.9462 | 12.6445 | 12.2876 | 12.1029 |

**30** Use the data from Problem 29 to set up a moving average control chart with m = 5, and interpret your results.

**31** Construct an EWMA control chart for the data in Problem 29 with $\lambda = 0.4$, $L = 2.58$, $\mu_0 = 12$, and $\sigma = 0.2$, and interpret your results.

**32** Consider a manufacturing process in the semiconductor industry. A quality characteristic of this process is normally distributed with mean 25 and standard deviation 3. Twenty randomly selected observations on this quality characteristic are given here. Construct an EWMA control chart for these data with $\lambda = 0.4$, $L = 3.054$, $\mu_0 = 25$, and $\sigma = 3$, and interpret your results.

| | | | | | | | | | |
|---|---|---|---|---|---|---|---|---|---|
| 17.0161 | 26.1820 | 24.8380 | 25.5103 | 24.2961 | 24.4898 | 22.5242 | 19.8843 | 24.0591 | 29.1806 |
| 26.9380 | 24.4623 | 24.3371 | 26.5040 | 25.2357 | 22.9902 | 19.8725 | 26.4289 | 24.6567 | 24.8073 |

**33** Use the data from Problem 32 to set up a MA control chart with m = 3 and control limits of probability 99%. Use the estimated value of the process mean as the target value. Interpret your results.

**34** Set up a tabular CUSUM control chart for the data in Problem 32. Use the standardized reference value $k = 0.5$ and the decision interval $h = 4.77$. Assume that the target value of the quality characteristic is 24 and that past experience indicates that the process standard deviation is 2.70.

**35** Redo Problem 34 using the FIR feature with starting values $S_0^+ = h/2$ and $S_0^- = -h/2$. Comment on the results.

**36** Use the data from Problem 32 to set up a MA control chart with m = 5 and control limits of probability 99%. Use the estimated value of the process mean as the target value. Interpret your results.

**37** The following data give the diameter in mm of 20 randomly selected ball bearings from a production line. The target value of the diameter is 46 mm, and the process standard deviation is known to be 8.5 mm. Set up a tabular CUSUM control chart for this process using the standardized parameter values $h = 5$ and $k = 0.5$.

| 45 | 31 | 49 | 44 | 52 | 53 | 51 | 60 | 51 | 54 |
|----|----|----|----|----|----|----|----|----|----|
| 34 | 53 | 52 | 44 | 39 | 55 | 52 | 53 | 32 | 31 |

**38** Redo Problem 37 using the FIR feature with starting values equal to $h/2$. Comment on the results.

**39** Use the data in Problem 37 to set up a tabular CUSUM control chart for the process using the standardized parameter values $h = 8.01$ and $k = 0.25$. Compare the results you obtain in this problem with those obtained in Problem 37.

**40** Reconsider the data in Problem 37. Set up MA control charts with $m = 3, 4$, and 5. Compare the results obtained by using these different span values, and comment on how the different values of $m$ affect the final conclusions. Use $\alpha = 0.01$.

**41** Set up EWMA control charts for the data in Problem 37 using $L = 3$ and $\lambda = 0.20, 0.25$, and 0.3. Comment on how the different values of $\lambda$ affect the final outcomes of the three control charts you constructed.

**42** Repeat Problem 41, now letting the values of $L$ vary and the value of $\lambda$ remain fixed; specifically, use $L = 2.5, 3.0, 3.5$, and $\lambda = 0.25$. Discuss how the different values of $L$ affect the final results of the three control charts.

**43** Discuss how the results in Problem 38 will change if the starting values for the FIR feature are set using $h/4$.

**44** Discuss how the results in Problem 43 will change if the starting values for the FIR feature are now set using $3h/4$.

**45** The following data show the tread depth (in mm) of 20 tires selected randomly from a large lot manufactured using a new process:

| | | | | | | | | | |
|---|---|---|---|---|---|---|---|---|---|
| 6.28 | 7.06 | 6.50 | 6.76 | 6.82 | 6.92 | 6.86 | 7.15 | 6.57 | 6.48 |
| 6.64 | 6.94 | 6.49 | 7.14 | 7.16. | 7.10 | 7.08 | 6.48 | 6.40 | 6.54 |

a) Estimate the process mean and process standard deviation.
b) Set up a tabular CUSUM control chart for this process using the standardized parameter values $h = 8.01$ and $k = 0.25$, and the estimated values of the process mean and process standard deviation in (a).

**46** Set up a tabular CUSUM control chart for the data in Problem 45 using the parameter values $h = 5$ and $k = 0.5$, and the FIR feature with starting values $S_0^+ = h/2$ and $S_0^- = -h/2$.

**47** Set up EWMA control charts for the data in Problem 45 using $L = 3$ and $\lambda = 0.15, 0.20$, and $0.25$. Comment on how the different values of $\lambda$ affect the final outcomes of the three control charts you constructed.

**48** Reconsider Problem 47, now letting the value of $L$ vary and keeping the value of $\lambda$ fixed; that is, use $L = 2.5, 3.0$, and $3.5$, and $\lambda = 0.25$. Discuss how the different values of $L$ affect the final outcomes of the three control charts.

**49** Reconsider the data in Problem 45. Set up MA control charts with spans $m = 2$ and $4$. Compare the results obtained by using these different span values, and comment on how the different values of $m$ affect the final conclusions. Use $\alpha = 0.05$.

**50** Redo Problem 45 using standardized parameter values $h = 4.77$ and $k = 0.5$. Comment on the conclusions reached with these values of $h$ and $k$.

**51** Pull-strength tests are applied to 24 soldered leads in electronic apparatus. The test records the force required in pounds to break the bond. The data obtained in this experiment are given here.

| | | | | | | | | | | | |
|---|---|---|---|---|---|---|---|---|---|---|---|
| 21.5 | 20.6 | 18.7 | 22.3 | 24.1 | 20.6 | 19.8 | 18.7 | 24.2 | 22.3 | 19.5 | 20.6 |
| 24.6 | 23.5 | 22.5 | 23.5 | 22.7 | 21.5 | 20.5 | 23.6 | 22.5 | 23.5 | 21.7 | 19.9 |

a) Estimate the process mean and standard deviation.
b) The target value of the pull-strength test is 22 pounds. Set up two one-sided CUSUM control charts for these data to detect a $1\sigma$ shift in the process mean. Use parameter values so that $ARL_0$ is approximately 370. Interpret your results.

**52** Reconsider the data in Problem 51. Set up MA control charts with $m = 3$ and $5$. Compare the results obtained by using the different span values, and comment on how the different values of $m$ affect the final conclusions. Use $\alpha = 0.05$.

**53** Set up EWMA control charts for the data in Problem 51: one using $L = 2.962$, $\lambda = 0.20$, and another using $L = 3.054$, $\lambda = 0.40$ (The two sets of parameters result in $ARL_0 \cong 500$). Comment on how the different values of $L$ and $\lambda$ affect the final outcomes of the control charts.

# 8

# Process and Measurement System Capability Analysis

## 8.1  Introduction

In Chapter 1, we defined a *process* as a series of actions or operations performed in producing manufactured or non-manufactured products. A *process* may also be defined as a combination of the workforce, equipment, raw material, methods, and environment that work together to produce output. For a better understanding, refer to the flow chart diagram shown in Figure 1.1 in Chapter 1; it shows where each component fits in a process.

The quality of the final product depends on both how the process is designed and how it is executed. However, as we noted in earlier chapters, no matter how perfectly a process is designed or how well it is executed, no two items produced by the process are identical. The difference between the two items is called *variation*. The goal of the process, then, is to reduce this variation in a very systematic way.

In Chapters 5–7, we studied various types of control charts, including Shewhart, cumulative sum (CUSUM), moving average (MA), and exponentially weighted moving average (EWMA) control charts, and we noted that they can be used to help us bring a process under statistical control. Once the process is under statistical control, the variation in the process is reduced significantly. The next step is to check whether the process can produce products that meet the specifications set by the customer. If a process is producing products that can meet these specifications, then we can assume that process is capable. To check whether a process is capable, in the first part of this chapter we study *process capability indices*. Based on these indices, we can make an objective decision about whether the process under investigation is capable. It is important to remind ourselves that the outcomes of the process capability indices are valid only if they are applied to a process under statistical control. *If a process is not under statistical control, the application of the process capability indices is not appropriate.* Process capability indices, just like quality control charts, are excellent tools for reducing the variation in a process.

In the second part of this chapter, we discuss another statistical quality control (SQC) technique called *pre-control*. Pre-control, if it is used at all, is used *after* a process is determined to be capable and operating in a state of statistical control. In the years following the development and introduction of pre-control, numerous research papers have been written both in support of it and to criticize it. For more information about the development and implementation of pre-control, see the references in the Bibliography.

Pre-control is used as a mechanism to reduce the amount of sampling and inspection required to validate that a process is producing products and services consistent with customer's expectations, as defined by specifications. In many, if not most, applications, statistical process control (SPC) and process capability analysis are used to validate process performance by monitoring statistically

*Statistical Quality Control: Using Minitab, R, JMP, and Python*, First Edition. Bhisham C. Gupta.
© 2021 John Wiley & Sons, Inc. Published 2021 by John Wiley & Sons, Inc.
Companion website: www.wiley.com/go/college/gupta/SQC

significant signals that indicate the presence or absence of variation due to assignable causes. SPC, however, does require regular sampling of rational subgroups (typically, sample sizes of five or more), inspection, and measurements. Thus an inspection can become expensive, particularly when production volume is high. Pre-control follows the use of SPC and process capability analysis, and it requires comparably less sampling and inspections and measurements. Therefore, pre-control is intended to indicate process performance and is less expensive than SPC once a process has achieved long-term statistical control. That said, pre-control is not intended to be a substitute for SPC (see Ledolter and Swersey (1997)).

In the third and last part of this chapter, we study *measurement system analysis* (MSA). The measurement system plays an important role in any process improvement activities. Generally, whenever we measure items or units, we will encounter two types of variability: some variability is inherent in the units being measured, and other variability comes from the measurement system that is being used. We noted in earlier chapters that the inherent variability in units is caused by variations due to common causes (assuming that all assignable causes are either completely or mostly eliminated), and we also discussed how variations due to common causes can be addressed. In this chapter, we discuss variability due to the measurement system. Such variability may consist of several components: an instrument or gauge that is used to measure the units, an operator who uses the instrument, the conditions under which units are being measured, and other factors that may influence the measurement system. As noted earlier, the measurement system constitutes an integral part of any process quality control, so it makes sense to determine how much variation due to the measurement system contributes to overall variability.

The purpose of MSA is to sort out the various components of variability in the measurement system and determine what portion of total variation from the measurement system is due to the instrument (i.e. equipment or gauge). From this, we can evaluate whether the measurement system is capable of producing the intended results.

## 8.2 Development of Process Capability Indices

Once a process is under statistical control and is thus predictable, the next concern of a manufacturer is to determine whether the process is capable of delivering the desired product. One way to address such a concern is to quantify its capability. A *process capability index* (PCI) is a unitless measure that quantifies process capability: specifically, the relationship between the output and the designed tolerance or specifications. This measure has become an essential tool in process capability analysis. Moreover, since the PCI is a unitless measure of capability, it can be communicated between manufacturer and supplier. More and more manufacturers and suppliers are using PCI as an important part of their contracts to ensure quality.

A process capability analysis is simply the comparison of the distribution of process output with the product tolerance. As noted by Kotz and Lovelace (1998), the results of process capability analysis can be very valuable in many ways. Deleryd (1996) developed a list of the 13 most common ways of using the results of process capability analysis:

1) As a basis in the improvement process.
2) As an alarm clock.
3) As specifications for investments. Giving specifications for levels of PCIs expected to be reached by new machines facilitates the purchasing process.

4) As a certificate for customers. Along with the delivery, the supplier can attach the results from the process capability studies conducted when the actual products were produced.
5) As a basis for new construction. By knowing the capability of the production processes, the designer knows how to set reasonable specifications to make the product manufacturable.
6) For control of maintenance efforts. By continuously conducting process capability studies, it is possible to see if some machines are gradually deteriorating.
7) As specifications for introducing new products.
8) To assess the reasonableness of customer demands.
9) To motivate co-workers.
10) To decide on priorities in the improvement process.
11) As a basis for inspection activities.
12) As a receipt for improvement.
13) To formulate quality improvement programs.

To implement a process capability analysis, we need to consider the following:

- The target value specification, which is usually defined by the customer.
- The specification limits, which should be defined either by the customer or the technical staff, and should be agreed on by the manufacturer. Furthermore, the specification limits should be such that they allow manufacturing variability without jeopardizing the proper function of the product.
- An analysis of the process that allows the manufacturer to determine if the product can meet the customer's specifications.

Once production starts, the manufacturer conducts capability studies to compare the measures of the quality characteristic of the manufactured product with the specification limits. This is the point where PCIs are used.

The first-generation PCIs were established in Japan in the 1970s and were as follows:

$C_p$ – Inherent capability of a process
$k$ – Position of the process in relation to the target value
$C_{pk}$ – Position of the $3\sigma$ process in relation to the target value
$C_{pl}$ – Position of the $3\sigma$ process in relation to the lower specification limit
$C_{pu}$ – Position of the $3\sigma$ process in relation to the upper specification limit

In this chapter, we discuss these and various other PCIs that are frequently used in process capability analysis. However, as the topic is very dense, we cannot address every aspect of these indices. For further reading, the reader is directed to Kotz and Lovelace (1998).

Throughout the study of these indices, we are going to assume that the process producing the desired quality characteristic is under statistical control and thus is predictable.

## 8.3 Various Process Capability Indices

### 8.3.1 Process Capability Index: $C_p$

Let X be the process quality characteristic that is being evaluated. And let USL and LSL be the *upper specification limit* and *lower specification limit*, respectively. The performance of the process with respect to these limits is defined as follows.

The percentage of nonconforming units produced by the process at the upper end = P (X > USL). The percentage of nonconforming produced by the process at the lower end = P (X < LSL). Thus, the total percentage of nonconforming units produced by the process is defined as

$$P \ (X < LSL \ or \ X > USL) = 1 - P \ (LSL < X < USL)$$

This probability gives us the performance of a process with respect to the specification limits. Now we look into the performance of the process with respect to the *natural tolerance limits*: that is, the *upper natural tolerance limit* (UNTL) and the *lower natural tolerance limit* (LNTL), where UNTL and LNTL are defined as $3\sigma$ above and below the process mean, respectively.

The performance of a process with respect to the natural tolerance limits is the percentage of the product produced by the process with its quality characteristic falling within the interval ($\mu - 3\sigma$, $\mu + 3\sigma$), where $\mu$ *and* $\sigma$ are, respectively, the mean and standard deviation of the process quality characteristic. Assuming that the process quality characteristic is normally distributed and the process is under statistical control, the percentage of the product produced by the process with its quality characteristic falling within the interval ($\mu - 3\sigma$, $\mu + 3\sigma$) is approximately 99.74%. As noted earlier, a PCI is nothing more than a comparison between what a process is expected to produce and what it is actually producing. Thus, we now define a PCI $C_p$, one of the first five indices established in Japan and proposed by Juran et al. (1974), is as follows:

$$C_p = \frac{USL - LSL}{UNTL - LNTL} = \frac{USL - LSL}{(\mu + 3\sigma) - (\mu - 3\sigma)} = \frac{USL - LSL}{6\sigma} \qquad (8.1)$$

Note that the numerator in Eq. (8.1) is the desired range of the process quality characteristic, whereas the denominator is the actual range of the process quality characteristic. From this definition, we see that a process can produce products of desired quality and the process is capable only if the range in the numerator is at least as large as that in the denominator. In other words, the process is capable only if $C_p \geq 1$. Larger values of $C_p$ are indicative of a capable process. For a $6\sigma$ process, $C_p = 2$. A predictable process, which is normally distributed with $C_p = 1$ and mean located at the center of the specification limits, is expected to produce approximately 0.27% nonconforming units. Montgomery (2013) has given a comprehensive list of PCI ($C_p$) values and associated process nonconforming unit proportions for both one- and two-sided specifications.

Since $C_p$ is very easy to calculate, it is widely used in industry. However, its major drawback is the fact that it does not take into consideration the position of the process mean. A process could be incapable even if the value of $C_p$ is large (>1). In other words, a process could produce 100% defectives, for example, if the process mean falls outside the specification limits and is far from the target value. Furthermore, note that the value of $C_p$ will become even larger if the value of the process standard deviation, $\sigma$, decreases while the process mean moves away from the target value.

Note that the numerator in Eq. (8.1) is always known, but the denominator is usually unknown. This is because in almost all practical applications, the process standard deviation, $\sigma$, is unknown. Thus, to calculate $C_p$, we replace $\sigma$ in Eq. (8.1) with its estimator, $\hat{\sigma}$. From Chapter 5, we know that $\sigma$ can be estimated either by the sample standard deviation $S$ or by $\overline{R}/d_2$. However, note that the estimate $\overline{R}/d_2$ is usually used only when the process is under statistical control and the sample size is less than 10. Thus, an estimated value of $C_p$ is given by

$$\hat{C}_p = \frac{USL - LSL}{6\hat{\sigma}} \qquad (8.2)$$

We illustrate the computation of $\hat{C}_p$ with the following example.

### Example 8.1   Car Engine Data

Table 8.1 gives the summary statistics for $\overline{X}$ and $R$ for 25 samples of size $n = 5$ collected from a process producing tie rods for a particular type of car engine. The measurement data are the lengths of the tie rods, and the measurement scale is in mm.

The target value and the specification limits for the length of the rods are 272 and 272 ± 8, respectively. Calculate the value of $\hat{C}_p$, assuming that the tie rod lengths are normally distributed. Find the percentage of nonconforming tie rods produced by the process.

### Solution

To find the value of $\hat{C}_p$ and the percentage of nonconforming tie rods produced by the process, we first need to estimate the process mean, $\mu$, and the process standard deviation, $\sigma$. These estimates may be found by using $\overline{\overline{X}}$ and $\overline{R}/d_2$, respectively. Thus, we get

$$\hat{\mu} = \overline{\overline{X}} = \frac{1}{m}\sum_{i=1}^{m}\overline{X}_i$$

$$= \frac{1}{25}(274 + 265 + 269 + \cdots + 273) = 272$$

$$\hat{\sigma} = \frac{\overline{R}}{d_2} = \frac{1}{d_2}\left(\frac{1}{m}\sum_{i=1}^{m}R_i\right)$$

$$= \frac{1}{2.326}\left(\frac{1}{25}(6 + 8 + \cdots + 5 + 6)\right) = \frac{6}{2.326} = 2.58$$

**Table 8.1**   Measurement data for tie rods.

| Sample number | $\overline{X}$ | $R$ | Sample number | $\overline{X}$ | $R$ |
|---|---|---|---|---|---|
| 1 | 274 | 6 | 14 | 268 | 8 |
| 2 | 265 | 8 | 15 | 271 | 6 |
| 3 | 269 | 6 | 16 | 275 | 5 |
| 4 | 273 | 5 | 17 | 274 | 7 |
| 5 | 270 | 8 | 18 | 272 | 4 |
| 6 | 275 | 7 | 19 | 270 | 6 |
| 7 | 271 | 5 | 20 | 274 | 7 |
| 8 | 275 | 4 | 21 | 273 | 5 |
| 9 | 272 | 6 | 22 | 270 | 4 |
| 10 | 273 | 8 | 23 | 274 | 6 |
| 11 | 269 | 6 | 24 | 273 | 5 |
| 12 | 273 | 5 | 25 | 273 | 6 |
| 13 | 274 | 7 | | | |

where the value of $d_2$ for sample size 5 is found from Table A.2 in the Appendix. Substituting the values of USL, LSL, and $\hat{\sigma}$ in Eq. (8.2), we get

$$\hat{C}_p = \frac{280 - 264}{6(2.58)} = \frac{16}{15.48} = 1.03$$

which indicates that the process is capable. To find the percentage of nonconforming tie rods produced by the process, we proceed as follows:

$$\text{Percentage of nonconforming units} = (1 - P\,(264 \leq X \leq 280)) \times 100\%$$
$$= (1 - P\,(-3.1 \leq Z \leq 3.1)) \times 100\%$$
$$= (1 - 0.9980) \times 100\% = 0.2\%$$

Thus, in this example, the percentage of nonconforming tie rods is 0.2%.

This result may be expected when we consider the value of $\hat{C}_p(= 1.03)$. However, as noted earlier, this may not always be the case since $C_p$ does not take into consideration where the process mean is located. To better explain this, we use the following example.

**Example 8.2**

Suppose that the process in Example 8.1 had a setback, and as a result the process mean had an upward shift. Now suppose that after the process experienced this shift, we took another set of 25 random samples of size $n = 5$, and these samples produced $\overline{\overline{X}} = 276$ and $\overline{R} = 6$. Clearly, in this example, the value of $\overline{\overline{X}}$ has changed from 272 to 276 while $\overline{R}$ has remained the same. Recalculate the value of $\hat{C}_p$, find the percentage of nonconforming rods, and comment on your findings. Assume that the quality characteristic of the process is still normally distributed.

**Solution**

From the given values of $\overline{\overline{X}}$ and $\overline{R}$, we obtain $\hat{\mu} = 276$ and $\hat{\sigma} = 2.58$. Since the process standard deviation did not change, the value of $\hat{C}_p$ remained the same: that is, $\hat{C}_p = 1.03$. However, the percentage of nonconforming tie rods produced by the process will be different. Thus, we have

$$\text{Percentage of nonconforming units} = (1 - P\,(264 \leq X \leq 280)) \times 100\%$$
$$= \left(1 - P\left(\frac{264 - 276}{2.58} \leq \frac{X - 276}{2.58} \leq \frac{280 - 276}{2.58}\right)\right) \times 100\%$$
$$= (1 - P\,(-4.65 \leq Z \leq 1.55)) \times 100\%$$
$$\cong 6.06\%$$

Thus, even though the value of $\hat{C}_p$ did not change after the process mean experienced a shift, the process is producing nonconforming units at a rate of approximately 30 times that in the previous example. In other words, $\hat{C}_p$ did not measure the effect of the upward shift that had made the process unable to produce products within the specification limits. Of course, the same would be true if the process had experienced a downward shift. This major drawback of $C_p$ makes it less reliable than many other PCIs available in the literature. We will study several of them here. However, before we study other PCIs, let us see another alternative but equivalent interpretation of $C_p$ by finding the percentage of the specification band used

$$\text{Percentage of specification band used} = \frac{1}{C_p} \times 100$$

A smaller percentage of the specification band used indicates a better process. Again, for reasons discussed earlier, this interpretation can sometimes be misleading.

Two other PCIs first used in Japan are $C_{pl}$ and $C_{pu}$. These indices are related to the LSL and USL, respectively, and are defined as follows:

$$C_{pl} = \frac{\mu - \text{LSL}}{3\sigma} \tag{8.3}$$

$$C_{pu} = \frac{\text{USL} - \mu}{3\sigma} \tag{8.4}$$

The estimates of $C_{pl}$ and $C_{pu}$ are given by

$$\hat{C}_{pl} = \frac{\overline{\overline{X}} - \text{LSL}}{3\hat{\sigma}} \tag{8.5}$$

$$\hat{C}_{pu} = \frac{\text{USL} - \overline{\overline{X}}}{3\hat{\sigma}} \tag{8.6}$$

To illustrate the computation of $\hat{C}_{pl}$ and $\hat{C}_{pu}$, we use the information given in Example 8.1.

**Example 8.3**

Using the information given in Example 8.1, compute $\hat{C}_{pl}$ and $\hat{C}_{pu}$.

**Solution**

$$\hat{C}_{pl} = \frac{\overline{\overline{X}} - \text{LSL}}{3\hat{\sigma}} = \frac{272 - 264}{3(2.58)} = 1.03$$

$$\hat{C}_{pu} = \frac{\text{USL} - \overline{\overline{X}}}{3\hat{\sigma}} = \frac{280 - 272}{3(2.58)} = 1.03$$

Note that both $\hat{C}_{pl}$ and $\hat{C}_{pu}$ are equal to $\hat{C}_p$, which will always be the case when the process mean is located at the center of the specification limits. Moreover, when both $\hat{C}_{pl}$ and $\hat{C}_{pu}$ are equal, the percentage of nonconforming units below the LSL and above the USL is the same.

**Example 8.4**

Using the information given in Example 8.2, compute $\hat{C}_{pl}$ and $\hat{C}_{pu}$.

**Solution**

$$\hat{C}_{pl} = \frac{\overline{\overline{X}} - \text{LSL}}{3\hat{\sigma}} = \frac{276 - 264}{3(2.58)} = 1.55$$

$$\hat{C}_{pu} = \frac{\text{USL} - \overline{\overline{X}}}{3\hat{\sigma}} = \frac{280 - 276}{3(2.58)} = 0.52$$

In this case, the value of $\hat{C}_{pl}$ is much larger than 1, whereas the value of $\hat{C}_{pu}$ is much smaller than 1. These observations indicate that most of the nonconforming units produced by the process are falling above the USL. Clearly, our observation is valid since the process had suffered an upward shift, which means the process mean had moved closer to the USL. Finally, we may note that both $C_{pl}$ and $C_{pu}$ are sensitive to where the process mean is located.

### 8.3.2 Process Capability Index: $C_{pk}$

To overcome the centering problem in $C_p$ that was discussed in the preceding subsection, another PCI was introduced: $C_{pk}$. This is another PCI of the first five indices used in Japan and is defined as

$$C_{pk} = \min \left( C_{pl}, C_{pu} \right) \tag{8.7}$$

For instance, in Example 8.4, the value of $C_{pk}$ is equal to 0.52, which means the process is not capable.

Furthermore, note that $C_{pk}$ is related to $C_p$ as follows:

$$C_{pk} = (1-k)C_p \tag{8.8}$$

where

$$k = \frac{|((USL + LSL)/2) - \mu|}{(USL - LSL)/2} \tag{8.9}$$

It can easily be seen that $0 \le k \le 1$, so that $C_{pk}$ is always less than or equal to $C_p$. Also, note that when the process mean, $\mu$, coincides with the center of the specification limits, then $k = 0$, and therefore $C_{pk}$ equals $C_p$.

Note that $C_{pk}$ takes care of the centering problem only if the process standard deviation remains the same. For example, consider a normally distributed process with standard deviation $\sigma = 1$, LSL = 20, and USL = 32. Now suppose that the process mean, $\mu$, changes, but the process standard deviation, $\sigma$, remains the same. Table 8.2 shows how the values of $C_{pk}$ and $C_p$ are affected.

Table 8.2 shows that the value of $C_p$ remains the same, but the value of $C_{pk}$ changes as the mean changes. It is also clear from the table that as the mean changes and moves away from the *center* (26) of the specification limits, the value of $C_{pk}$ changes. As soon as the mean moves within $3\sigma$ of one of the specifications limits, the process tends to become incapable, and the value of $C_{pk}$ becomes less than one. For example, from the earlier table, we note that when the mean is equal to 30 it is only two standard deviations away from the USL, and $C_{pk} = 0.67$. It can easily be seen that the process at that point is producing approximately 2.25%

**Table 8.2** Values of $C_{pk}$ and $C_p$ as the process mean changes.

| Process mean, $\mu$ | $C_{pk}$ | $C_p$ | Process mean, $\mu$ | $C_{pk}$ | $C_p$ |
|---|---|---|---|---|---|
| 26 | 2 | 2 | 29 | 1.00 | 2 |
| 27 | 1.63 | 2 | 30 | 0.67 | 2 |
| 28 | 1.33 | 2 | 31 | 0.33 | 2 |

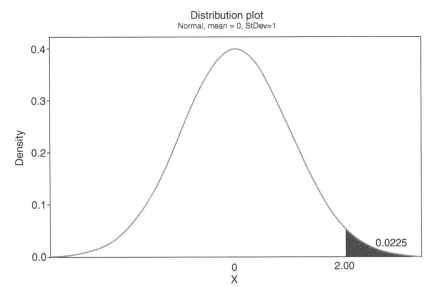

**Figure 8.1** Normal distribution graph showing the percentage of nonconforming units.

**Table 8.3** Different processes with the same value of $C_{pk}$.

| Process | LSL | USL | Center | $\mu$ | $\sigma$ | $C_{pk}$ |
|---------|-----|-----|--------|-------|----------|----------|
| 1 | 12 | 36 | 24 | 24 | 4 | 1.00 |
| 2 | 12 | 36 | 24 | 26 | 3.33 | 1.00 |
| 3 | 12 | 36 | 24 | 28 | 2.67 | 1.00 |
| 4 | 12 | 36 | 24 | 30 | 2.00 | 1.00 |

nonconforming units (see the normal distribution graph in Figure 8.1). The shaded area in the graph represents the percentage of nonconforming units.

Note that if the process standard deviation also changes as the mean changes, then the value of $C_{pk}$ may not change even when the process mean moves away from the center. It can easily be seen that this will always be the case provided the distance of the process mean from the nearest specification limit in units of $\sigma$ remains the same. For example, in Table 8.3, we have four processes with the same specification limits but different process means ($\mu$) and different process standard deviations ($\sigma$), yet the value of $C_{pk}$ in each case remains the same. This is because in each case, the distance between the process mean and the nearest specification limit (in this case, USL) is three times the process standard deviation.

Thus we can say that in some ways, $C_{pk}$ is also not an adequate measure of centering. Assuming that the process characteristic is normally distributed, Table 8.4 gives the parts per million (PPM) of nonconforming units for different values of $C_{pk}$.

From Table 8.4, we can see that each of the processes in Table 8.3 will produce 1350 nonconforming PPM. This is only possible if the process standard deviation is shrinking while the process mean is shifting, and the *natural tolerance limits* remain within the specification limits. For each case in Table 8.3, the process mean is off from the USL by $3\sigma$. Finally, note that by definition, the process

**Table 8.4** PPM of nonconforming units for different values of $C_{pk}$.

| $C_{pk}$ | 1.00 | 1.33 | 1.67 | 2.00 |
|---|---|---|---|---|
| PPM | 1350 | 30 | 1 | .001 |

with standard definition $\sigma = 2$ will be of six-sigma quality if the process mean is at the center point. In the earlier case, however, the process is producing 1350 nonconforming PPM because the process mean is off-center by $3\sigma$, a shift larger than $1.5\sigma$, which would be tolerable only in a Six Sigma quality process.

Boyles (1991) gave the upper and lower bounds on the percentage of conforming units associated with values of $C_{pk}$ as follows:

$$100 \times \left( (2P(Z \leq 3C_{pk}) - 1) \right)\% \leq (\% \text{ of conforming units}) \leq 100 \times P(Z \leq 3C_{pk})\% \qquad (8.10)$$

Thus, for example, the bounds on the percentage of conforming units associated with $C_{pk} = 1$ are

$$100 \times \left( (2P(Z \leq 3) - 1) \right) \leq (\% \text{ of conforming units}) \leq 100 \times P(Z \leq 3)$$

or

$$99.73 \leq (\% \text{ of conforming units}) \leq 99.865$$

Using this inequality, we can easily determine that the bounds on the percentage of nonconforming units associated with $C_{pk} = 1$ are

$$0.135 \leq (\% \text{ of nonconforming units}) \leq 0.27$$

or

$$1350 \leq (\text{nonconforming PPM}) \leq 2700$$

Several authors, including Pearn et al. (1992), Kotz and Lovelace (1998), and Montgomery (2013), have given $100(1 - \alpha)\%$ confidence intervals for $C_p$ and $C_{pk}$. That is,

$$\hat{C}_p \sqrt{\frac{\chi^2_{1-\alpha/2,n-1}}{n-1}} \leq C_p \leq \hat{C}_p \sqrt{\frac{\chi^2_{\alpha/2,n-1}}{n-1}} \qquad (8.11)$$

$$\hat{C}_{pk} \left[ 1 - Z_{\alpha/2} \sqrt{\frac{1}{9n\hat{C}^2_{pk}} + \frac{1}{2(n-1)}} \right] \leq C_{pk} \leq \hat{C}_{pk} \left[ 1 + Z_{\alpha/2} \sqrt{\frac{1}{9n\hat{C}^2_{pk}} + \frac{1}{2(n-1)}} \right] \qquad (8.12)$$

If the quantity $(9n)$ is significantly large so that the first factor under the square root in Eq. (8.12) is very small, then it reduces to the more simplified form given in Eq. (8.13). Note that the quantities $\chi^2_{1-\alpha/2,n-1}$ and $\chi^2_{\alpha/2,n-1}$ are the percentage points of the Chi-square distributions, and they can be found by using any one of the statistical packages discussed in this book.

$$\hat{C}_{pk}\left[1 - Z_{\alpha/2}\sqrt{\frac{1}{2(n-1)}}\right] \leq C_{pk} \leq \hat{C}_{pk}\left[1 + Z_{\alpha/2}\sqrt{\frac{1}{2(n-1)}}\right] \qquad (8.13)$$

### 8.3.3  Process Capability Index: $C_{pm}$

The PCI $C_{pk}$ was introduced because $C_p$ did not take into consideration the position of the process mean relative to the target value, which usually coincides with the center of the specification limits. However, from our discussion in Section 8.3.2, it is clear that even $C_{pk}$ does not adequately serve the intended purpose when the process standard deviation is also changing. To meet this challenge, a new capability index was introduced: $C_{pm}$.

Taguchi (1985, 1986) applied the loss function to quality improvement, where the characteristics of Taguchi's loss function are as follows:

- Quality is best when on target.
- Quality decreases as products deviate.
- Customer dissatisfaction grows with deviation.
- Dissatisfaction can be expressed as a dollar loss to customers and society.
- Dollar loss can often be approximated by a simple quadric function.

Taguchi (1986), focusing on the reduction of deviation from the target value, defined the capability index $C_p$ as follows:

$$C_p = \frac{USL - LSL}{6\tau} \qquad (8.14)$$

where

$$\tau^2 = E(X - T)^2 = \sigma^2 + (\mu - T)^2 \qquad (8.15)$$

Chan et al. (1988) independently introduced Taguchi's index, $C_p$, and called it $C_{pm}$. They noted that $C_{pm}$ reacts to changes in the process in much the same manner as $C_{pk}$. However, the changes in $C_{pm}$ are more drastic than in $C_{pk}$. This can be seen by comparing the values of $C_{pk}$ and $C_{pm}$ presented in Table 8.5.

Consider, for example, USL = 22, LSL = 10, T = 16, and $\sigma = 1$, and let $\mu$ vary while $\sigma$ remains the same. Then the changes in the values of $C_{pk}$ and $C_{pm}$ as $\mu$ deviates from the target value are as shown in Table 8.5.

**Table 8.5**  The values of $C_{pk}$ and $C_{pm}$ as $\mu$ deviates from the target.

| $\mu$ | 16 | 17 | 18 | 19 | 20 |
|---|---|---|---|---|---|
| $C_{pk}$ | 2.00 | 1.67 | 1.33 | 1.00 | 0.67 |
| $C_{pm}$ | 2.00 | 1.41 | 0.89 | 0.63 | 0.49 |

Since, in practice, $\mu$ and $\sigma$ are not known, to measure the capability of a process necessitates the estimation of $\mu$ and $\sigma$. Thus, using the estimates of $\mu$ and $\sigma$, the estimator for $C_{pm}$ that is most commonly used is

$$\hat{C}_{pm} = \frac{USL - LSL}{6\sqrt{s^2 + (\bar{x} - T)^2}} \tag{8.16}$$

where

$$s^2 = \frac{1}{n-1}\sum_{i=1}^{n}(x_i - \bar{x})^2 \tag{8.17}$$

Taguchi (1985) proposed a slightly different estimator for $C_{pm}$:

$$\hat{C}_{pm} = \frac{d}{3\sqrt{\left[fs^2 + n(\bar{x} - T)^2\right]/(f + 1)}} \tag{8.18}$$

or

$$\hat{C}_{pm} = \frac{USL - LSL}{6\sqrt{\left[(s_1^2 + (\bar{x} - T)^2\right]}} \tag{8.18a}$$

where

$$s_1^2 = \frac{1}{n}\sum_{i=1}^{n}(x_i - \bar{x})^2 \tag{8.19}$$

Kotz and Lovelace (1998) preferred this estimator over the one given in Eq. (8.16) due to some statistical properties.

**Example 8.5**
Consider a stable process that has LSL = 24 and USL = 44 and a target value at $T = 34$. Suppose a random sample of size 25 from this process produced a sample mean of $\bar{x} = 36$ and sample standard deviation of $s = 2.5$. Determine point estimates for $C_p$, $C_{pk}$, and $C_{pm}$. Assume that the quality characteristic of the process is normally distributed.

**Solution**

Using the information given in this example and the formulas presented earlier, we obtain the following point estimates:

$$\hat{C}_p = \frac{USL - LSL}{6s} = \frac{44 - 24}{6 \times 2.5} = 1.33$$

$$\hat{C}_{pk} = Min\left(\frac{\overline{x} - LSL}{3s}, \frac{USL - \overline{x}}{3s}\right) = Min(1.6, 1.06) = 1.06$$

$$\hat{C}_{pm} = \frac{USL - LSL}{6\sqrt{s^2 + (\overline{x} - T)^2}} = \frac{44 - 24}{6\sqrt{(2.5)^2 + (36 - 34)^2}} = 1.04$$

Using Taguchi's definition, we have

$$\hat{C}_{pm} = \frac{d}{3\sqrt{[fs^2 + n(\overline{x} - T)^2]/(f + 1)}}$$

$$= \frac{(44 - 24)/2}{3\sqrt{[24(2.5)^2 + 25(36 - 34)^2]/25}} = 1.054$$

### 8.3.4 Process Capability Index: $C_{pmk}$

This capability index was introduced as a further improvement over $C_{pm}$ by Pearn et al. (1992). The index $C_{pmk}$ is usually known as a third-generation capability index and is defined as

$$C_{pmk} = Min\left(\frac{\mu - LSL}{3\tau}, \frac{USL - \mu}{3\tau}\right) \tag{8.20}$$

where $\tau$ is as defined in Eq. (8.15). Thus $C_{pmk}$ can also be written as

$$C_{pmk} = Min\left(\frac{\mu - LSL}{3\sqrt{\sigma^2 + (\mu - T)^2}}, \frac{USL - \mu}{3\sqrt{\sigma^2 + (\mu - T)^2}}\right) \tag{8.21}$$

The relationship between $C_{pmk}$ and $C_{pm}$ is similar to that between $C_{pk}$ and $C_p$. In other words,

$$C_{pmk} = (1 - k)C_{pm} \tag{8.22}$$

where $k$ is as defined in Eq. (8.9).

A point estimator for $C_{pmk}$ is given by

$$\hat{C}_{pmk} = \frac{(USL - LSL)/2 + |\overline{x} - T|}{3\hat{\tau}} \tag{8.23}$$

where

$$\hat{\tau} = \sqrt{s^2 + (\overline{x} - T)^2} \tag{8.24}$$

**Example 8.6**

Find an estimator $\hat{C}_{pmk}$ of $C_{pmk}$ for the process in Example 8.5.

**Solution**

Substituting the values of LSL, USL, $T$, $s$, and $\bar{x}$ in Eq. (8.24), we get

$$\hat{C}_{pmk} = \frac{(44-24)/2 + |36-34|}{3\sqrt{2.5^2 + (36-34)^2}}$$

$$= \frac{12}{3\sqrt{10.25}} = 1.25$$

Chen and Hsu (1995) showed that the previous estimator is asymptotically normal. Wallgren (1996) emphasizes that $C_{pmk}$ is more sensitive to the deviation of the process mean from the target value than $C_{pk}$ or $C_{pm}$. Later in this chapter, using numerical data, we will study this characteristic of $C_{pmk}$, $C_{pm}$, and $C_{pk}$, as well as another capability index introduced by Gupta (2005). Pearn and Kotz (1994) and Vannman (1995) have ranked the four indices in terms of their sensitivity with respect to the differences between the process mean and the target value as follows: (1) $C_{pmk}$, (2) $C_{pm}$, (3) $C_{pk}$, (4) $C_p$.

### 8.3.5 Process Capability Index: $C_{pnst}$

The capability index $C_{pnst}$ was recently introduced by the author, Gupta (2005). Before we formally define this capability index, we will give some background and rationale.

Deming (1985) points out that "There is no process, no capability and no meaningful specifications, except in statistical control. When a process has been brought into a statistical control, it has a definable capability, expressible as the economic level of quality of the process." Thus, *natural tolerance limits* (NTLs) should play an important role in defining the PCI. Some indices, such as the ones we have discussed so far, have successfully solved the problem of process variation ($\sigma$) and process centering (i.e. the difference between the process mean, $\mu$, and the target value, $T$). In particular, these are second- and third-generation indices: $C_{pk}$, $C_{pm}$, and $C_{pmk}$, which are defined as we saw earlier in terms of process variation ($\sigma$), process centering ($\mu - T$), and the distance between the process mean and the specification limits, i.e., $\mu - LSL$ and $USL - \mu$. However, none of these indices take into consideration the distance between the NTLs and the specification limits. A process becomes barely capable, or even incapable, as the NTLs coincide with or cross the *specification limits* (SLs). A process should be deemed incapable before the process mean coincides or crosses the specification limits. Thus, once the position of the NTLs is determined by the position of the mean, $\mu$, the distance between the NTLs and SLs is more important than the distance between $\mu$ and the SLs.

The capability index $C_{pnst}$ is defined as

$$C_{pnst} = \min\left(\frac{LNTL - LSL}{3\tau}, \frac{USL - UNTL}{3\tau}\right) \tag{8.25}$$

where LNTL and UNTL are the lower and the upper natural tolerance limits and $\tau$ is as defined in Eq. (8.15). Additionally, using the customary definitions of LNTL and UNTL, that is,

$$\text{LNTL} = \mu - 3\sigma, \text{UNTL} = \mu + 3\sigma$$

$C_{pnst}$ can be expressed as

$$C_{pnst} = \min \left( \frac{\mu - 3\sigma - LSL}{3\sqrt{\sigma^2 + (\mu - T)^2}}, \frac{USL - \mu - 3\sigma}{3\sqrt{\sigma^2 + (\mu - T)^2}} \right)$$

$$= \min \left( \frac{\frac{\mu - LSL}{3\sigma} - 1}{\sqrt{1 + \left(\frac{\mu - T}{\sigma}\right)^2}}, \frac{\frac{USL - \mu}{3\sigma} - 1}{\sqrt{1 + \left(\frac{\mu - T}{\sigma}\right)^2}} \right)$$

$$C_{pnst} = \frac{1}{\sqrt{1 + \left(\frac{\mu - T}{\sigma}\right)^2}} \min \left( \frac{\mu - LSL}{3\sigma} - 1, \frac{USL - \mu}{3\sigma} - 1 \right) \quad (8.26)$$

Or, using Montgomery's notation (2013), we can express $C_{pnst}$ as

$$C_{pnst} = \frac{1}{\sqrt{1 + \xi^2}} \min \left( \frac{\mu - LSL}{3\sigma} - 1, \frac{USL - \mu}{3\sigma} - 1 \right)$$

$$C_{pnst} = \frac{1}{\sqrt{1 + \xi^2}} \left( C_{pk} - 1 \right) \quad (8.27)$$

where

$$\xi = \frac{\mu - T}{\sigma}$$

A point estimator of $C_{pnst}$ is given by

$$\hat{C}_{pnst} = \frac{1}{\sqrt{1 + \nu^2}} \left( \hat{C}_{pk} - 1 \right) \quad (8.28)$$

where

$$\nu = \hat{\xi} = \frac{\hat{\mu} - T}{\hat{\sigma}} \quad (8.29)$$

and

$$\hat{\mu} = \bar{x}, \hat{\sigma} = s \quad (8.30)$$

If 25 or more samples of size 4 or 5 are used, then we may use the following estimators of the process mean and process standard deviation:

$$\hat{\mu} = \overline{\overline{x}}, \hat{\sigma} = \frac{\overline{R}}{d_2} \ or \ \frac{\overline{s}}{c_4} \tag{8.31}$$

where values of $d_2$ and $c_4$ for various sample sizes are given in Table A.2 in the Appendix. Note that the second set of estimates is used only if the process is in a state of statistical control and the estimate based on sample range is used for samples of sizes less than 10.

The process is considered capable, barely capable, or incapable when $C_{pnst} > 0$, $C_{pnst} = 0$, or $C_{pnst} < 0$, respectively.

### 8.3.5.1 Comparing $C_{pnst}$ with $C_{pk}$ and $C_{pm}$

**Example 8.7   Bothe (2002)**
Consider two processes A and B with the following summary statistics:

$$\mu_A = 3.0, \mu_B = 0.0$$

$$\sigma_A = 1.143, \sigma_B = 2.66$$

Further, it is given that LSL = -5, USL = 5, and $T \neq M$.
Plugging these values into Eq. (8.7), we get

$$C_{pk}(A) = 0.58, C_{pk}(B) = 0.63$$

Thus, process A has a smaller $C_{pk}$ value than process B. However, process A is producing a higher percentage of conforming parts than process B, which is clearly a contradiction. If we choose $T = 1.5$ ($T \neq M$), so that $T$ is equidistant from the two means, then after substituting all the values in Eq. (8.26), we get

$$C_{pnst} = -0.2545 \text{ for process A}$$

$$= -0.3222 \text{ for process B}$$

which indicates that process A is better than process B. Moreover, both values of $C_{pnst}$ are negative, which indicates that both processes are incapable.

**Example 8.8   Bothe (2002)**
Consider two processes A and B with the following summary statistics:

$$\mu_A = 12, \mu_B = 15$$

$$\sigma_A = 2, \sigma_B = 0.667$$

Also, it is given that LSL = 6, USL = 18, and T = 12. Furthermore, it is known that process A is inferior to process B, in the sense that it is more likely to produce a higher percentage of nonconforming parts. From Bothe (2002), we have $C_{pm}$ for processes A and B equal to 1.0 and 0.65, respectively, and again these values are not indicative of the actual capability.

Using the summary statistics and the values of LSL, USL, and $T$, and plugging them into Eq. (8.26), we get

$$C_{pnst} = 0.0 \text{ for process A}$$
$$= 0.1085 \text{ for process B}$$

which are in line and indicate that process A is inferior to process B. Moreover, both values of $C_{pnst}$ are positive, which indicates that both processes are capable. However, process A is just barely capable. In other words, process A is more likely to produce a higher number of nonconforming parts than process B.

---

**Example 8.9   Pearn et al. (1992)**
Consider two processes A and B with the following summary statistics:

$$\mu_A = T - \frac{d}{2}, \quad \mu_B = T + \frac{d}{2}, \quad \sigma_A = \sigma_B = \frac{d}{3}, \quad T = \frac{1}{4}(3(\text{USL}) + \text{LSL})$$

where $d = \frac{1}{2}(\text{USL} - \text{LSL})$. Furthermore, under these conditions, processes A and B are expected to produce 0.27% and 50% nonconforming parts.

However, from Pearn et al. (1992), it is known that, for both processes, $C_{pm} = 0.555$, which is clearly, again, a contradiction. Now, substituting these values in Eq. (8.26), we get

$$C_{pnst} = 0.0 \text{ for process A}$$
$$= -0.55 \text{ for process B}$$

which are again in line with what we expected. Furthermore, the values of $C_{pnst}$ indicate that process A is barely capable, whereas process B is highly incapable.

These examples show that the capability index $C_{pnst}$ generally presents a clearer picture than the indices $C_{pk}$ and $C_{pm}$.

### 8.3.5.2   Other Features of $C_{pnst}$

- The index $C_{pnst}$ can take any real value: negative, zero, or positive.
- $C_{pnst} < 0$ implies that at least one of the NTLs does not fall within the specification limits, which implies that the process is no longer capable.
- $C_{pnst} = 0$ implies that at least one of the NTLs has coincided with the specification limits, and hence the process is barely capable. Under the normality assumptions, the process is producing either 1350 PPM or 2700 PPM nonconforming parts depending on whether one or both NTLs have coincided with the specification limits. Furthermore, any change in the process – a change in the process mean, process variation, or both – can make the process incapable.
- $0 < C_{pnst} < 1$ implies that the process is capable. The higher the value of the index, the better the process. As $C_{pnst}$ approaches 1, the process mean is approaching the target value, and the NTLs are moving away from the specification limits toward the target value.
- $C_{pnst} \geq 1$ implies a very desirable situation. The process standard deviation $\sigma$ has become very small, and the distance between the specification limits and the NTLs is at least $3\sigma$, that is, the distance between the target value $T$ and the specification limits is at least $6\sigma$. Thus the process is performing at a six-sigma quality level or better. The process mean may shift by as much as $1.5\sigma$ off the target

**Table 8.6** ($\sigma = 2$) values of $C_p$, $C_{pk}$, $C_{pm}$, $C_{pmk}$, and $C_{pnst}$ for $\mu = 20, 22, 24, 26, 28$; T = 24, LSL = 12, USL = 36.

| $\mu$ | $C_p$ | $C_{pk}$ | $C_{pm}$ | $C_{pmk}$ | $C_{pnst}$ |
|---|---|---|---|---|---|
| 20 | 2 | 1.333 | 0.894 | 0.596 | 0.149 |
| 22 | 2 | 1.667 | 1.414 | 1.179 | 0.471 |
| 24* | 2 | 2.000 | 2.000 | 2.000 | 1.000 |
| 26 | 2 | 1.667 | 1.414 | 1.179 | 0.471 |
| 28 | 2 | 1.333 | 0.894 | 0.596 | 0.149 |

**Table 8.7** ($\mu = 20$) Values of $C_p$, $C_{pk}$, $C_{pm}$, $C_{pmk}$, and $C_{pnst}$ for $\sigma = 2, 2.5, 3.0, 3.5, 4.0, 4.5$; T = 24, LSL = 12, USL = 36.

| $\sigma$ | $C_p$ | $C_{pk}$ | $C_{pm}$ | $C_{pmk}$ | $C_{pnst}$ |
|---|---|---|---|---|---|
| 2 | 2.000 | 1.333 | 0.894 | 0.596 | 0.149 |
| 2.5 | 1.600 | 1.067 | 0.848 | 0.565 | 0.035 |
| 3.0 | 1.333 | 0.889 | 0.800 | 0.533 | −0.067 |
| 3.5 | 1.143 | 0.762 | 0.753 | 0.502 | −0.157 |
| 4.0 | 1.000 | 0.663 | 0.707 | 0.471 | −0.236 |
| 4.5 | 0.889 | 0.593 | 0.664 | 0.443 | −0.305 |

value without causing any problems. This is because a shift in the process mean on either side of the target would lead to a rate of at most 3.4 PPM nonconforming; this is another characteristic of a process performing at a six-sigma quality level.

- It can easily be seen that as the process mean moves away from the target value, the value of the process standard deviation remains fixed (say, at $\sigma = 2$) and the distance between the NTLs and the USL changes; consequently, the value of $C_{pnst}$ changes (for example, see Table 8.6). The changes in $C_{pnst}$ are more drastic than those of the other indices ($C_p$, $C_{pk}$, $C_{pm}$, or $C_{pmk}$).
- Further, it can be seen that as the process variation increases while the value of process mean $\mu$ remains fixed (say $\mu = 20$), the distance between the NTLs and the specification limits and the value of the index $C_{pnst}$ change (for example, see Table 8.7). Again, the value $C_{pnst}$ changes at a more drastic rate than the values of any of the other indices.

### 8.3.6 Process Performance Indices: $P_p$ and $P_{pk}$

The PCIs we have studied so far have one characteristic in common: each of them is used when the process is in statistical control and hence stable. In 1991, the American Automotive Industry Group (AAIG) was founded, consisting of representatives from Ford Motor Company, General Motors Corporation, American Daimler/Chrysler Corporation, and the American Society for Quality. This group standardized the supplier quality reporting procedure for their industry, and they advocated the use of two sets of PCIs and process performance indices: one consisting of $C_p$ and $C_{pk}$, and the other consisting of $P_p$ and $P_{pk}$. Furthermore, they advised using $C_p$ and $C_{pk}$ when the process is stable, $P_p$ and $P_{pk}$ when the process is not stable, where the two sets of indices $P_p$, $P_{pk}$ and $C_p$, $C_{pk}$ are defined in exactly the same manner, differing only in how their estimates are defined. That is:

$$\hat{P}_p = \frac{USL - LSL}{6\hat{\sigma}_s} \tag{8.32}$$

$$\hat{C}_p = \frac{USL - LSL}{6\hat{\sigma}_{\bar{R}/d_2}} \tag{8.33}$$

and

$$\hat{P}_{pk} = \min\left(\frac{USL - \bar{\bar{X}}}{3\hat{\sigma}_s}, \frac{\bar{\bar{X}} - LSL}{3\hat{\sigma}_s}\right) \tag{8.34}$$

$$\hat{C}_{pk} = \min\left(\frac{USL - \bar{\bar{X}}}{3\hat{\sigma}_{\bar{R}/d_2}}, \frac{\bar{\bar{X}} - LSL}{3\hat{\sigma}_{\bar{R}/d_2}}\right) \tag{8.35}$$

In other words, they only differ in how the process standard deviation is defined and estimated. The estimate $\hat{\sigma}_s$ of the process standard deviation is obtained by using the overall sample standard deviation – in other words, the pooled standard deviation $s = \sqrt{\sum_{i=1}^{n}\sum_{j=1}^{m}\left(x_{ij} - \bar{\bar{x}}\right)^2/(mn-1)}$; whereas $\hat{\sigma}_{\bar{R}/d_2}$ is an estimate obtained by using the subgroup ranges $R_i$, $i = 1, 2, ..., m$ and the corresponding value of $d_2$. Note that, when the process is stable, the two sets of indices are essentially the same because the estimates $\hat{\sigma}_s$ and $\hat{\sigma}_{\bar{R}/d_2}$ are approximately equal.

Kotz and Lovelace (1998) strongly argued against the use of $P_p$ and $P_{pk}$. They wrote, "We highly recommend against using these indices when the process is not in statistical control. Under these conditions, the P-numbers are meaningless with regard to process capability, have no tractable statistical properties, and infer nothing about long-term capability of the process. Worse still, they provide no motivation to the user-companies to get their process in control. The P-numbers are a step backwards in the efforts to properly quantify process capability, and a step towards statistical terrorism in its undiluted form". Montgomery (2013) agrees with Kotz and Lovelace. He writes, "The process performance indices $P_p$ and $P_{pk}$ are more than a step backwards. They are a waste of engineering and management effort – they tell you nothing."

## 8.4 Pre-control

Pre-control was developed in 1954 by a team of consultants at Rath and Strong, as reported in "A Simple Effective Process Control Method," Report 54-1, Rath & Strong, Inc. (*1954*). The team included Dorin Shainin, Warren Purcell, Charlie Carter, and Frank Satterthwaite. *Pre-control* is more successful with processes that are inherently stable and not subject to rapid process drifts once they are set up. In other words, pre-control charts should be applied when a process meets the following conditions:

- It is in a state of statistical control.
- It is "capable" as determined by a process capability analysis.
- It exhibits a low defect rate.

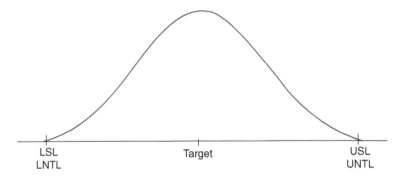

Figure 8.2  A barely capable process.

Earlier in this chapter, we discussed several PCIs. However, the most commonly used PCI is $C_{pk}$. In the case where $C_{pk}$ is equal to 1, the process variability equals the specifications or tolerance limits, as can be seen from Figure 8.2. This figure illustrates what is commonly referred to as a process that is *barely capable* of acceptable performance.

We see from Figure 8.2 that any process drift or shift to the left or the right will result in the production of products or services that are outside customer specifications. So, for pre-control to be a valid SQC tool, the process capability ratio must be strictly greater than 1, which means there is at least some room for the process to vary and yet still produce products or services that meet customer expectations.

Pre-control is used as a mechanism to reduce the amount of sampling and inspection required to validate that processes are producing products and services consistent with customer expectations, as defined by specifications. In many applications, if not most, SPC and process capability analysis are used to validate process performance by monitoring for statistically significant signals that indicate the presence/absence of variation due to assignable causes. However, SPC does require regular sampling of rational subgroups (typically, samples of size four or five or more), and sampling, measurements, and inspection can sometimes become expensive. As shown in Figure 8.3, pre-control follows the use of SPC and process capability analysis, and it requires much less sampling and fewer inspections and measurements than does SPC. Pre-control, therefore, is intended to provide an indication of process performance that is less expensive than SPC once a process has achieved long-term statistical control; however, pre-control is not intended to be a substitute for SPC, according to Ledolter and Swersey (1997). They also observe that such a use of pre-control will likely lead

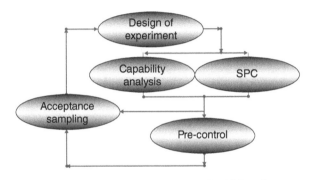

Figure 8.3  Relationships among the SQC tools.

to unnecessary tampering with the process, which can actually increase variability. Montgomery (2013) remarks that "Pre-control should only be considered in manufacturing process where the process capability ratio is much greater than one and where a near-zero defects environment has been achieved."

### 8.4.1 Global Perspective on the Use of Pre-control – Understanding the Color-Coding Scheme

Pre-control uses a color-coding scheme associated with each of the six zones, consistent with Figure 8.4.

Note in Figure 8.4 that there are actually two green zones, which are marked here as one large *green zone*; there are also two *yellow zones* and two *red zones*. Because pre-control is based on the standard normal distribution, which is a symmetrical distribution, there is one green, yellow, and red zone on each side of the process target (center). Whether originally intended by the inventors of pre-control, these zones apply a color-coding scheme used throughout the world and are analogous to the color scheme used in traffic control systems.

In virtually all countries, green, yellow, and red signify for vehicular and pedestrian traffic the operational conditions of *go*, *caution*, and *stop*, respectively. As will be discussed in the next section of this chapter, green indicates that the process is performing as expected and warrants continued production. Yellow indicates a low probability of observing process behavior if the process were performing as expected, which indicates that caution (and additional sampling/testing) is warranted. Red indicates that the product or service does not conform to customer expectations or specifications and the process should be stopped. This color-coding scheme is embedded in the mechanics of how we use pre-control.

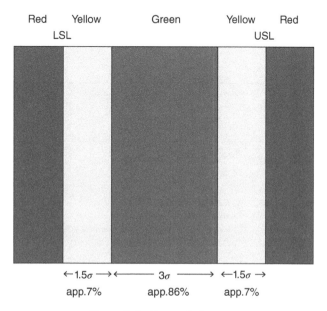

**Figure 8.4** Pre-control zones.

## 8.4.2 The Mechanics of Pre-control

The use of pre-control is accomplished in the following manner:

1) *Ensure that the process is sufficiently capable.* Several authors assert that, for pre-control to be valid, the process capability must be equal to 1. When the process capability is equal to 1, the process variability is exactly equal to the specifications or tolerances – in this case, there is no room for process drift without producing nonconforming products or services. To always be safely valid for pre-control, the process capability should be greater than 1 so that there is some room for process drift without producing nonconforming products or services. Montgomery (2013) recommends that the process capability ratio be at least 1.15.
2) *Establish the pre-control zones.* Zones for pre-control are established by dividing the specifications or tolerances zone by 4. Once we have defined four zones within the specification or tolerance limits, two of the zones are green (marked in Figure 8.4 as one large green zone) and two of the zones are yellow.
3) *Verify that the process is ready to begin pre-control.* A process is considered ready to begin pre-control when five consecutive pieces or service delivery iterations produce output that, when measured, fall within the green zone established in step 2. If the process is determined to be capable as defined in step 1 and yet does not produce five consecutive pieces in the green zone established in step 2, then the process is not yet a candidate for the use of pre-control, and so additional process improvements must be completed.
4) *Begin sampling.* Sampling for pre-control involves pulling samples of two consecutive pieces. The frequency of sampling is based on the amount of production completed without having to adjust the process. If 25 samples are taken without having to adjust the process, then sampling frequency is considered to be adequate. If the process requires adjustment prior to completing 25 samples, then sampling should be conducted more frequently. Likewise, if more than 25 samples are taken before a process adjustment is required, then sampling frequency can be decreased to let processes run longer before sampling.
5) *Apply the pre-control rules.* Rules for the use of pre-control are provided in Table 8.8.

**Table 8.8** Pre-control rules.

| Each sample (two consecutive pieces) | Pre-control zone | Measurement requirement | Pre-control step to follow |
| --- | --- | --- | --- |
| 1 | Green | Piece 2 not needed | Run process, step 4 |
| 1<br>2 | Yellow<br>Green | Measure piece 2 | Run process, step 4 |
| 1<br>2 | Yellow<br>Yellow | Measure piece 2<br>Reset process<br>Pieces 1 and 2 in the same yellow zone<br>Pieces 1 and 2 in opposite yellow zones | Step 3 |
| 1 | Red | Reset process | Step 1 |

### 8.4.3 The Statistical Basis for Pre-control

The statistical basis for pre-control is related to probability and the *standard normal distribution*: that is, a normal distribution with a mean of 0 and a standard deviation of 1. The statistical basis for pre-control begins by considering a process whose capability index is equal to 1, in which case the process variability is exactly equal to the specifications or tolerances. Dividing the specification or tolerance zone by 4 establishes the boundaries for four zones, the reasons for which are described in more detail later in this section. For now, the two central zones contained within the specification are referred to as green zones, and the other two are referred to as yellow zones. Dividing the specifications or tolerance zones by 4 creates the probability that 86% of the observations will fall within the green zones and 7% of the observations will fall into each of the yellow zones, as can be seen in Figure 8.5. With 7% of the process output falling into a single yellow zone, we have an approximately 1/14 chance of actually observing process output in a single yellow zone. With 7% of the process falling into each of two yellow zones (one zone on each side of the process target), we have an approximately $(1/14) \times (1/14) = (1/196) \cong 1/200$ chance of actually observing process output from the same sample that falls in either one of the yellow zones. It also follows from the rules given in Table 8.8 that when two consecutive units fall within the yellow zone, it is possible that the process has shifted to an out-of-control state. Similarly, the probability of one unit falling in one yellow zone and the second unit falling in the other yellow zone is very low.

We can also see from Figure 8.6 that, with a process capability equal to 1, we expect to see a very small number of observations beyond the specification limits.

### 8.4.4 Advantages and Disadvantages of Pre-control

The use of pre-control involves balancing advantages and disadvantages. As we noted earlier in this section, pre-control, if used at all, is used *after* SPC and process capability analysis. Pre-control has supporters and critics, which introduces the notion that we must consider and balance these advantages and disadvantages prior to using it.

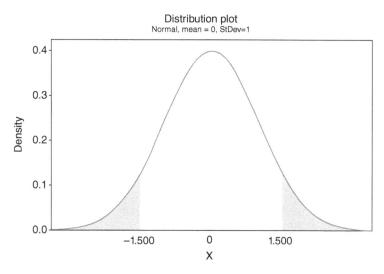

**Figure 8.5**  Standard normal curve with each yellow zone approximately 7%.

**Figure 8.6** Standard normal curve showing a barely capable process.

### 8.4.4.1  Advantages of Pre-control

If conditions warrant the use of pre-control, as has been discussed throughout this section, then it offers the following advantages:

- It is much easier for shop floor operators to use than SPC.
- There are cost savings from reduced measurements as compared to SPC.
- The measurement devices needed to support pre-control (go/no-go gauges) are less expensive than measurement devices needed for continuous variables.
- Pre-control does not require computation and charting as does SPC.
- Pre-control provides process performance data more quickly than SPC (remember that SPC requires 25 subgroups/samples to be drawn for control limits to be calculated, and that charting needs to be completed before any conclusions can be made about process performance).
- Pre-control is designed to ensure that defective products and services do not reach the customer (as compared to SPC, which is designed to detect assignable causes for variation).

### 8.4.4.2  Disadvantages of Pre-control

Using pre-control also presents certain disadvantages when compared to the use of other SQC tools such as SPC. Some of the disadvantages summarized here are also listed in Montgomery (2013):

- Since SPC is commonly discontinued when pre-control begins, diagnostic information contained as patterns of assignable cause variation in control charts is lost.
- Historical records of process performance over time are lost: that is, records of why assignable cause variations occurred.
- The small sample sizes involved with pre-control reduce our ability to detect "even moderate-to-large shifts."
- Pre-control provides no information helpful in establishing statistical control or in reducing variability.
- Pre-control will likely lead to unnecessary tampering with the process.
- Pre-control will have "more false alarms or more missed signals than the control chart" (Wheeler and Chambers 1992).

## 8.5   Measurement System Capability Analysis

Measurement system analysis (MSA) is used to understand and quantify the variability associated with measurements and measurement systems. The reference manual on MSA (2010) by the Automotive Industry Action Group (AIAG) defines measurement and measurement systems as follows:

---

**Definition 8.1**   *Measurement*: "the assignment of numbers (or values) to material things to represent the relations among them with respect to particular properties" (C. Eisenhart 1963).

---

**Definition 8.2**   *Gauge*: "any device used to obtain measurements; frequently used to refer specifically to the devices used on the shop floor; includes go/no-go devices."

---

**Definition 8.3**   *Measurement system*: "the collection of instruments or gauges, standards, operations, methods, fixtures, software, personnel, environment and assumptions used to quantify a unit of measure or fix assessment to the feature characteristic being measured; the complete process used to obtain measurements."

---

These definitions will serve as the basis of our discussion. Before we begin, however, we first must provide clarification. The term MSA is commonly used interchangeably with the term *gauge repeatability* and *reproducibility*. Sometimes they are referred to as the two R's of a measurement system or simply *Gauge R&R*. By *repeatability*, we mean that we get the same measured value if we measure the same unit several times under essentially the same conditions, whereas *reproducibility* means how the measured values vary when measurements are taken on the same units but under different conditions (including different operators, different shifts or times, and so on).

MSA is a more comprehensive analysis of quantifying variability components from gauge stability, gauge bias, gauge linearity, gauge repeatability, and reproducibility. Gauge repeatability and reproducibility is, in fact, a subset or a component of MSA. The reason underlying the mistaken interchange of the terms MSA and Gauge R&R is that many, if not most, problems with measurement systems are detected and corrected with the gauge R&R procedure without having to continue to the more comprehensive analysis of MSA. For a detailed discussion of analysis techniques for gauge stability, gauge bias, and gauge linearity, readers are referred to the AIAG Reference Manual for MSA (2010).

The point of interest is that variability is present in production and service delivery processes. Further, we continue to differentiate between common cause variation and assignable or special cause variation. Common cause variation is the type of variation that we expect in the natural behavior of a process, and it can be reduced but not completely eliminated. Assignable or special cause variation is the type of variation we would not expect to see in the natural behavior of a process, and it can and must be eliminated.

It is critically important that we understand that the methods, procedures, tools, and equipment we use to make measurements constitute an independent process that creates, and is susceptible to, its own variation. This means there are two sources or components of variation present in each measurement we take, as illustrated in Figure 8.7.

As we can see in the figure, the variability we observe in a single measurement taken from sample data has two sources or components of variability. It is important to ensure that the component of variability associated with the measurement system does not consume an excessive amount of the

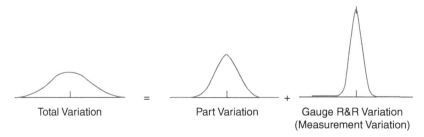

Total Variation = Part Variation + Gauge R&R Variation (Measurement Variation)

**Figure 8.7**   Components of total variation.

variability allowed in the process specification; and as mentioned earlier, quantifying measurement system variability is the purpose of MSA.

### 8.5.1   Evaluating Measurement System Performance

In our study so far, we have focused our full attention on bringing a process under statistical control. In other words, our efforts have been to reduce the overall variation in a process. In this section, we focus our entire attention on MSA, which constitutes an important part of the total process variation. The total process variation can be divided into three major categories: variations due to parts, measurement instruments or equipment (gauge), and operators. Part-to-part variation may be due to an environment, methods, materials, machines, or some combination thereof, as well as other factors. The variation due to a *measurement system* mainly consists of two major components: (i) the *instrument* being used for taking measurements and (ii) the *operators* (i.e. appraisers) who use the instrument. In addition, sometimes there is another component: the interaction between operators and instrument. In the industrial world, these components are usually referred to as *repeatability* and *reproducibility*, respectively. Thus, repeatability and reproducibility may be considered major indicators of measurement system performance. A little later, we discuss repeatability and reproducibility in more detail.

Since repeatability is related to the variation generated by the instrument (i.e. measurement equipment or gauge), it is referred to as *equipment variation* (EV). Reproducibility is related to variation generated by operators (i.e. appraisers) using measurement instruments, and it is referred to as *appraiser variation* (AV). As mentioned earlier, the study of gauge repeatability and reproducibility is also usually referred to as *Gauge R&R* study.

In the AIAG Reference Manual on MSA (2010), several methods for conducting gauge R&R are given. In this book, we study two methods: the range method and the analysis of variance (ANOVA) method. We discuss these methods using real-life data. In other words, we will focus our attention on explaining and interpreting the results of our example as it is worked out using one of the computer software tools discussed in this book.

### 8.5.2   The Range Method

The method discussed in this section has been presented by various authors, including IBM (1986) and Barrentine (2003). Before we discuss the details, we define specific terms: *measurement capability index, $K_1$ factor*, and *$K_2$ factor*. In addition, we define some other terms that are useful in understanding MSA.

MSA is a technique for collecting data and analyzing it to evaluate the effectiveness of a gauge. To collect the data, we randomly select some parts and select a certain number of operators (three or more; and as a general rule, the more, the better). Each operator then takes multiple measurements (at least two) on each part, and all the parts are measured in random order. These measurements are also known as *trials*. Using control chart terminology discussed in earlier chapters, the measurements on each part, or the *trials*, constitute a *rational subgroup*, and the number of *parts* times the number of *operators* constitutes the *number of subgroups or samples*. Then $\bar{R}$ is defined as the average of the ranges of trials for the same operator, and $\bar{\bar{R}}$ is defined as the average of the $\bar{R}$'s between the operators.

---

**Definition 8.4** A *measurement capability index* (MCI) is a measurement that quantifies our belief that the gauge is reliable enough to support the decisions that we will make under the existing conditions.

---

The MCI relates to four characteristics that are key to any measurement system:

- Precision
- Accuracy
- Stability
- Linearity

The *precision* characteristic is further subdivided into two categories: repeatability and reproducibility.

---

**Definition 8.5** *Repeatability* measures the preciseness of the observations taken under the same conditions, which is achieved by computing the variance of such observations.

---

For example, we say a gauge possesses the characteristic of repeatability if an operator obtains similar observations when measuring the same part again and again.

---

**Definition 8.6** *Reproducibility* measures the preciseness of the observations taken by different operators when measuring the same part.

---

For example, we say a gauge possesses the characteristic of reproducibility if various operators obtain similar observations when measuring the same part.

---

**Definition 8.7** The *accuracy* of a measurement system is the closeness between the average of the measurements taken and the true value.

---

The distinction between precision and accuracy is very well explained by the diagram shown in Figure 8.8.

---

**Definition 8.8** *Stability* is defined as the total variation in measurements obtained with a measurement system on the same master piece or the same part when measuring a single characteristic over an extended period.

---

The smaller the total variation, the more stable the measurement system is.

---

**Definition 8.9** *Linearity* is defined when the difference between a part's true value (master measurement) and the average of the observed measurements of the same part has the same distribution over the entire measurement range of the master part.

---

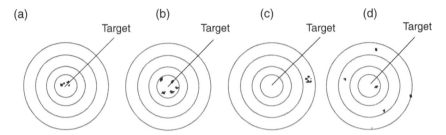

**Figure 8.8** (a) Accurate and precise; (b) accurate but not precise; (c) not accurate but precise; (d) neither accurate nor precise.

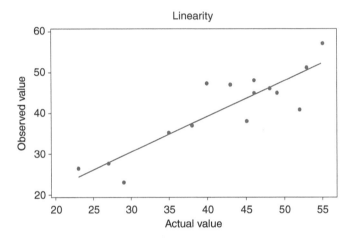

**Figure 8.9** Diagram showing the linear relationship between actual and observed values.

Linearity is best explained by the diagram in Figure 8.9.

As mentioned earlier, in any manufacturing process, the total variability consists of two components: one due to the variability among the parts and the other due to the variability in the measurement system. Thus, the MCI of a measurement system, which is directly related to the variability due to the measurement system, is a very pertinent factor in improving any process. It is difficult to effectively monitor or implement any process quality improvement project without an adequate measurement system. As mentioned, the total variability due to the measurement system consists of three components: (i) variability due to operators, (ii) variability due to the instrument, and (iii) the interaction between the operators and instrument. Statistically, these relationships can be expressed as follows:

$$\sigma_{Total}^2 = \sigma_{Parts}^2 + \sigma_{Meas.}^2 \tag{8.36}$$

$$\sigma_{Meas.}^2 = \sigma_{inst.}^2 + \sigma_{Operator}^2 + \sigma_{part. \times Operator}^2 \tag{8.37}$$

Note that an estimate $\sigma_{Total}^2$ in Eq. (8.36) is the total variability obtained from the observed data: that is,

$$\hat{\sigma}_{Total}^2 = S^2$$

The total variability due to the measurement $(\sigma^2_{Meas.})$ is also known as the total Gauge R&R variability $(\hat{\sigma}^2_{Gauge})$. The instrument variability is represented by the variability in the repeated measurements by the same operator, and for this reason, it is also known as *repeatability*. In the ANOVA method, the repeatability variance component is the error variance (i.e. $\sigma^2_{inst.} = \sigma^2_{EV} = \sigma^2$). The remainder of the variability in the measurement system comes from the various operators who use the instrument and the interactions between the instruments and the operators. Note that the interaction appears when any operator can measure one type of part better than another. This total variability from the operators and the interaction between the operators and the instruments is also known as *reproducibility*. Thus Eq. (8.37) can also be expressed as

$$\sigma^2_{Meas.} = \sigma^2_{Re\,peatability} + \sigma^2_{Re\,producibility} \qquad (8.38)$$

Using Barrentine's approach, *repeatability* is defined as

$$\text{Repeatability} = EV = 5.15\hat{\sigma}_{Re\,peatability} = 5.15\frac{\overline{\overline{R}}}{d_2} = K_1\overline{\overline{R}} \qquad (8.39)$$

where $K_1 = \dfrac{5.15}{d_2}$ and where the factor 5.15 represents the 99% range of the standard normal distribution, following from the fact that

$$P(-2.575 \leq Z \leq 2.575) = 0.99 \qquad (8.40)$$

Note that the AIAG recommends using the factor 6 instead of 5.15 since it covers almost 100% (to be exact, 99.74%) of the range. Values of $K_1$ for various sample sizes (i.e. the number of trials or the number of measurements taken on the same part by the same operator) are given in Table A.3 in the Appendix. As noted by Barrentine, Duncan (1986) points out that this estimation procedure should be slightly modified if $N = r \times n = (\text{\# operators}) \times (\text{\# parts})$ is less than 16. Thus, if $N$ is less than 16, then $K_1$ is defined as

$$K_1 = \frac{5.15}{d_2^*} \qquad (8.41)$$

The values of $d_2^*$ are listed in Table $D_3$ in Duncan (1986). The values of $K_1$ listed in Table A.3 in the Appendix are determined according to the value of $N$.

Again, using Barrentine's notation (2003, p. 57), reproducibility $= AV$, *ignoring the interaction term*, is defined as

$$\text{Reproducibility} = AV = \sqrt{(K_2 R_{\bar{x}})^2 - \frac{(EV)^2}{(r \times n)}} \qquad (8.42)$$

where $r$ is the number of trials, $n$ is the number of parts or samples, EV is as given in Eq. (8.39), $R_{\bar{x}}$ is the range of the operator's means, and the factor $K_2$ is defined as

$$K_2 = \frac{5.15}{d_2^{**}} \tag{8.43}$$

The value of $d_2^{**}$ can be found in Appendix C in the (AIAG) publication.
For example, if we have five operators and one part, then the value of $K_2$ is

$$K_2 = \frac{5.15}{2.48} = 2.077$$

We now illustrate the range and ANOVA methods with the following example.

**Example 8.10**
A manufacturer produces connect rods (con rods) used for gas turbine engine applications. These rods are machined out of a solid billet of metal. The manufacturer has installed a new measuring gauge to take measurements on these rods. To perform the MSA on the new gauge, the quality control manager selected randomly three Six Sigma Green Belt operators from the Department of Quality Control, who then decided to take a random sample of 10 rods. Each operator took three length measurements on each randomly selected rod. The data obtained are shown in Table 8.9.
  Analyze the data using the range method (later, we will also use the ANOVA method).

**Solution**

We discuss the gauge R&R using the approach that was originally presented by Barrentine (2003):

1) Verify if the gauge calibration is current.
2) Identify the operators. Three operators are typically used in gauge studies; however, the more operators, the better.
3) Select a random sample of parts, and have each operator measure all parts. One operator measures all parts, taking several measurements on each part; then the second operators

**Table 8.9**  Data for an experiment involving three operators, 10 parts (con rods), and 3 measurements on each con rod by each operator.

| Part # | Operator 1 | | | Operator 2 | | | Operator 3 | | |
|---|---|---|---|---|---|---|---|---|---|
| | Trial 1 | Trial 2 | Trial 3 | Trial 1 | Trial 2 | Trial 3 | Trial 1 | Trial 2 | Trial 3 |
| | 36 | 35 | 35 | 35 | 34 | 35 | 35 | 35 | 35 |
| | 38 | 38 | 38 | 36 | 38 | 36 | 34 | 35 | 34 |
| | 36 | 35 | 35 | 38 | 37 | 38 | 34 | 33 | 34 |
| | 35 | 36 | 35 | 32 | 32 | 32 | 34 | 35 | 34 |
| | 37 | 36 | 37 | 36 | 36 | 35 | 35 | 35 | 35 |
| | 41 | 40 | 40 | 40 | 39 | 40 | 36 | 37 | 37 |
| | 45 | 45 | 45 | 42 | 41 | 41 | 43 | 42 | 43 |
| | 44 | 45 | 45 | 42 | 43 | 42 | 43 | 44 | 44 |
| | 46 | 46 | 46 | 48 | 49 | 49 | 49 | 49 | 49 |
| | 35 | 34 | 35 | 38 | 37 | 37 | 38 | 38 | 38 |
| | $\bar{R}_1 = 0.7$ $\bar{x}_1 = 39.13$ | | | $\bar{R}_2 = 1.0$ $\bar{x}_2 = 38.60$ | | | $\bar{R}_3 = 0.7$ $\bar{x}_3 = 38.24$ | | |

takes measurements, then the third operators takes measurement, and so on. All parts are measured in random order.

4) Calculate the sample mean, inter-trial range for each sample, and average range for each operator. The sample means and average ranges are provided in Table 8.9.

5) Calculate the range of sample means ($R_{\bar{x}}$):

$$R_{\bar{x}} = Max(\bar{x}_i) - Min(\bar{x}_i), \ i = 1, 2, 3$$

$$= 39.13 - 38.24 = 0.89$$

6) Calculate the average range ($\bar{\bar{R}}$) for the operators:

$$\bar{\bar{R}} = \frac{\bar{R}_1 + \bar{R}_2 + \bar{R}_3}{3} = \frac{0.7 + 1.0 + 0.6}{3} = 0.767$$

7) Calculate repeatability (also referred to as equipment variation [EV]) and the estimate of its standard deviation ($\sigma_{\text{Re peatability}}$):

$$\text{Repeatability (EV)} = \bar{\bar{R}} \times K_1 = 0.767 \times 3.05 = 2.339$$

where, from Table A.3 in the Appendix, $K_1 = 3.05$.

$$\hat{\sigma}_{\text{Re peatability}} = \frac{EV}{5.15} = \frac{2.339}{5.15} = 0.454$$

8) Calculate reproducibility (also referred to as appraiser or operator variation [AV]) and the estimate of its standard deviation ($\sigma_{\text{Re producibility}}$):

$$\text{Reproducibility (AV)} = \sqrt{(R_{\bar{x}} \times K_2)^2 - \left[ (EV)^2 / (m \times r) \right]}$$

where, from Eq. (8.43), we have $K_2 = 3.04$, $m = 10 = $ # of parts or samples, and $r = 3 = $ # of trials so that the total number of samples is $n = m \times r = 10 \times 3 = 30$. Thus, we have

$$\text{Reproducibility (AV)} = \sqrt{(0.89 \times 3.04)^2 - \left[ (2.339)^2 / (10 \times 3) \right]}$$

$$= 2.671$$

Note that if the number under the radical is negative, then AV is zero. The estimate of the standard deviation of reproducibility ($\sigma_{\text{Reproducibility}}$) is

$$\hat{\sigma}_{\text{Re producibility}} = \frac{2.671}{5.15} = 0.518$$

Since, using this method, the reproducibility is calculated by ignoring the interaction term, the standard deviation of reproducibility may merely be looked on as the operator standard deviation.

9) Calculate the gauge R&R (i.e. repeatability and reproducibility) and the estimate of its standard deviation:

$$\sigma_{Gauge}^2 = \sigma_{\text{Re peatability}}^2 + \sigma_{\text{Re producibility}}^2$$

$$= (0.454)^2 + (0.518)^2$$

$$= 0.474$$

The estimate of the Gauge R&R standard deviation is given by

$$\hat{\sigma}_{Gauge} = 0.688$$

The gauge R&R standard deviation ($\hat{\sigma}_{Gauge}$) is a good estimate of gauge capability. Now, if we use USL = 60 and LSL = 25, then, using the notation of Montgomery (2013), an estimate of another *gauge capability* is the *precision-to-tolerance ratio* (P/T) given by

$$\frac{P}{T} = \frac{5.15 \times \hat{\sigma}_{Gauge}}{USL - LSL} = \frac{5.15 \times 0.688}{60 - 25} = 0.10$$

If the estimated value of the precision-to-tolerance ratio (P/T) is less than or equal to 0.1, then the gauge is considered to be adequately capable. Thus, in this case, we may say that the gauge is adequately capable. However, this rule should not be taken as a guarantee of the accuracy of the measurement, according to Montgomery (2013). The appraiser should make sure the measurements are sufficiently precise and accurate so that the experimenter can make the right decision. Note that the various estimates obtained by using the range method and the ANOVA method (discussed next) differ significantly from each other.

### 8.5.3 The ANOVA Method

Measurement system analysis using the ANOVA method is done by using two types of designs: crossed designs and nested designs. *Crossed designs* are used when each operator measures the same parts, whereas *nested designs* are used when each operator measures different parts, in which case we say that the parts are *nested within operators*. In this chapter, we discuss the case when each operator measures the same parts. In other words, we use crossed designs. Nested or hierarchical designs are beyond the scope of this book.

As an example, we will perform a measurement system analysis by using the data in Example 8.10. Obviously, we have two factors in Example 8.10, say factor A and factor B, where factor A denotes the operators and factor B denotes the parts. Then factor A has 3 levels, and factor B has 10 levels. Each level of factor A is replicated three times at each level of factor B. This example clearly satisfies the conditions of a crossed experimental design.

The gauge R&R study using the ANOVA method can be carried out by using one of the statistical packages discussed in this book. We will consider using Minitab, R, or JMP. Here we discuss how to use Minitab and R.

*Minitab*

1) Enter all of the data in column C1, identifiers of operators in column C2, and identifiers of part numbers in column C3.
2) From the menu bar, select **Stat > Quality Tools > Gage Study >Gage R&R Study (crossed)** to open the **Gage R&R study (crossed)** dialog box.
3) In the dialog box, enter the columns in which you have recorded the part numbers, operators, and measurement data. Now, if we select **Options**, another dialog box titled **Gage R&R Study (crossed): ANOVA Options** appears. In this dialog box, enter the value **6** into the box next to **Study variation**. You can enter a desired value (0.15 is a reasonable value) in the box next to **Alpha to remove interaction term**; then the interaction term will be removed only if the *p* value is greater than 0.15. Finally, check the box **Draw graphs on separate graphs, one graph per page**. Then click **OK** in both the dialog boxes. The output will appear in the **Session Window** as shown after the R code.

## R

To use R code to run a gauge R&R study using the ANOVA method, we first install the package "qualityTools" into your R library and then load it from the library ("qualityTools"). We explain using the data in Table 8.9. We entered all the data in columns C1–C3 of the Minitab worksheet and then saved the worksheet as Table 8.9.csv file in the **SQC-Data** folder on the Desktop. You may save it in any folder you like: **Documents**, etc. The data were entered in the following order:

36, 35, 35, 38, 36, 34, 36, 38, 34, ..., 35, 37, 38

The R code is as follows:

```
# first install the package "qualityTools"
install.packages("qualityTools")
# note that installing a package is only one-time event.
library (qualityTools)
GMA = gageRRDesign(Operators = 3, Parts = 10, Measurements = 3,
randomize = FALSE)
setwd("C:/Users/Person/Desktop/SQC-Data folder")
Y = read.csv("Table 8.9.csv")
response (GMA) = Y$Column_1
GMA = gageRR(GMA, method = "crossed", sigma = 6, alpha = 0.15)
plot(GMA)
```

All of the output, including the graphs, is identical to that produced using Minitab, which are presented in Table 8.10.

### Interpretation of Two-Way ANOVA Table with Interaction

In the ANOVA table, we test the following three hypotheses.

- $H_0$: All parts have similar variation vs. $H_1$: All parts do not have similar variation.
- $H_0$: All operators are equally good vs. $H_1$: All operators are not equally good.
- $H_0$: Interactions between parts and operators are negligible vs. $H_1$: Interactions between parts and operators are not negligible.

The decision whether to reject any of these hypotheses depends on the p-value (shown in the last column) and the corresponding value of the level of significance. If the p-value is less than or equal to the level of significance, then we reject the null hypothesis; otherwise, we do not reject the null hypothesis. Thus, for example, the p-value for parts is zero, which means we reject the null hypothesis that the parts have similar variation at any level of significance; in other words, the

**Table 8.10** Two-way ANOVA table with interaction.

| Source | DF | SS | MS | F | P |
|---|---|---|---|---|---|
| Parts | 9 | 1723.66 | 191.517 | 24.3263 | 0.000 |
| Operators | 2 | 12.29 | 6.144 | 0.7805 | 0.473 |
| Parts ∗ operators | 18 | 141.71 | 7.873 | 28.3422 | 0.000 |
| Repeatability | 60 | 16.67 | 0.278 | | |
| Total | 89 | 1894.32 | | | |

$\alpha$ *to remove interaction term* = *0.15.*

measurement system is capable of distinguishing the different parts. The p-value for operators is 0.473; therefore, at any level of significance, which is less than 0.473, we do not reject the null hypothesis that the operators are equally good (in most applications, an acceptable value of the level of significance is 0.05). Finally, the interaction in this example has a p-value of zero, which means that the interactions are not negligible at any level of significance. Usually, we assign a value to alpha so that if the p-value is greater than the assigned level of significance alpha (in this example, we assigned a value of 0.15), then we will have an ANOVA table without interaction. Thus, in this example, no matter what value is assigned to alpha, the interactions will not be removed. In other words, we cannot develop an ANOVA table in this example without interactions.

### Interpretation of Two-Way ANOVA Table without Interaction

If the interaction term is removed from the ANOVA table, then the SS (sum of squares) and DF (degrees of freedom) for interaction are merged with the corresponding terms of repeatability, which acts as an error due to uncontrollable factors. The interpretation for parts and operators is the same as in the two-way ANOVA table with interaction. However, it is important to note the p-values will usually change from one ANOVA table to another. This is because when the interaction term is removed from the ANOVA table, the SS and DF change; consequently, the F-values in Table 8.10 for operators and parts will change, and therefore the p-values will change.

Note that the variance components given in the Minitab printout in Table 8.11 may be determined manually by using MS values from the ANOVA table as follows:

$$\hat{\sigma}^2 = = MSE = MS\ (repeatability) = 0.278 = \hat{\sigma}^2_{repeatabilty}$$

$$\hat{\sigma}^2_{P*o} = \frac{MS_{p*o} - MSE}{n} = \frac{7.873 - 0.278}{3} = 2.518$$

$$\hat{\sigma}^2_{o} = \frac{MS_o - MS_{p*o}}{p \times n} = \frac{6.144 - 7.873}{10 \times 3} = 0.000$$

$$\hat{\sigma}^2_{p} = \frac{MS_p - MS_{p*o}}{o \times n} = \frac{191.517 - 7.873}{3 \times 3} = 20.4049$$

$$\hat{\sigma}^2_{reproducibility} = \hat{\sigma}^2_o + \hat{\sigma}^2_{p*o} = 0.000 + 2.518 = 2.518$$

$$\hat{\sigma}^2_{Gauge} = \hat{\sigma}^2_{reproducibility} + \hat{\sigma}^2_{repeatabilty} = 2.518 + 0.278 = 2.796$$

Note that these values match those given in the Minitab printout except for rounding errors. Furthermore, note that Python printout uses mean square (measurements) instead of mean square (repeatability).

**Table 8.11** Gauge R&R variance components.

| Source | Var comp | % contribution (of var comp) |
| --- | --- | --- |
| Total gauge R&R | 2.8095 | 12.10 |
| Repeatability | 0.2778 | 1.20 |
| Reproducibility | 2.5317 | 10.91 |
| Operators | 0.0000 | 0.00 |
| Operators * parts | 2.5317 | 10.91 |
| Part-to-part | 20.4049 | 87.90 |
| Total variation | 23.2144 | 100.00 |

The first column (var comp) in the Minitab printout provides the breakdown of the variance components (estimates of variances). The second column (% contribution) provides the percent contribution of the variance components, which becomes the basis of a gauge R&R study using the ANOVA method. Thus, for instance, the total variation due to the gauge is 12.10%, of which 1.20% is contributed by repeatability and the remaining 10.91% is contributed by reproducibility. The variation due to parts is 87.90% of the total variation. This indicates that the gauge is capable.

Note that the percent contributions are calculated simply by dividing the variance components by the total variation and then multiplying by 100%. Thus, for example, the percent contribution due to repeatability is given by

$$\frac{0.2778}{23.2144} \times 100\% = 1.20\%$$

The part of the Minitab printout shown in Table 8.12 provides various percent contributions using standard deviation estimates, which are obtained by taking the square root of the variance components. The comparison with standard deviation makes more sense because the standard deviation uses the same units as those of the measurements. The study variation (i.e. measurement system variation, which is equivalent to the process variation in the study of process control) is obtained by multiplying the standard deviation by 5.15 (some authors use this factor as 6 instead of 5.15; note that 5.15 gives a 99% spread, whereas 6 gives a 99.73% spread). The percent study variations are calculated by dividing the standard deviation by the total variation and then multiplying by 100%. Thus, for example, the percent contribution due to part-to-part is given by

$$\frac{4.51718}{4.81813} \times 100 = 93.75\%$$

The percent tolerance is obtained by dividing the $(6 \times SD)$ by the process tolerance (assuming that the process tolerance is 50) and then multiplying by 100%. Thus, for example, the process tolerance of total gauge R&R is given by

$$\frac{10.0569}{50} \times 100 = 20.1138\%$$

Note that the total percent tolerances in this example do not add to 100. Rather, the total sum is 57.82, which means the total variation is using 57.82% of the specification band.

**Table 8.12** Gauge evaluation.

| Source | Std. dev. (SD) | Study var (6 × SD) | % Study Var (% SV) |
| --- | --- | --- | --- |
| Total gauge R&R | 1.67615 | 10.0569 | 34.79 |
| Repeatability | 0.52705 | 3.1623 | 10.94 |
| Reproducibility | 1.59113 | 9.5468 | 33.02 |
| Operators | 0.00000 | 0.0000 | 0.00 |
| Operators ∗ parts | 1.59113 | 9.5468 | 33.02 |
| Part-to-part | 4.51718 | 27.1031 | 93.75 |
| Total variation | 4.81813 | 28.9088 | 100.00 |

*Number of distinct categories = 3.*

The last entry in the Minitab printout is the number of distinct categories, which in this case is three. The number of distinct categories can be determined as shown:

$$\text{Number of distinct categories} = \text{Integral part of } \left( \frac{part-to-part \ SD}{Total \ Gauge \ R \& R \ SD} \times 1.4142 \right)$$

$$= \text{Integral part of } \left( \frac{4.51718}{1.67615} \times 1.4142 \right)$$

$$= \text{Integral part of } (3.81123) = 3$$

Under the AIAG's recommendations, a measurement system is capable if the number of categories is greater than or equal to five. Thus, in this example, the measurement system is not capable of separating the parts into the different categories that they belong to. This quantity is equivalent to the one defined in AIAG (2002) and is referred to as the *signal-to-noise ratio* (SNR). That is,

$$\text{SNR} = \sqrt{\frac{2\hat{\rho}_p}{1 - \hat{\rho}_p}} \tag{8.44}$$

where

$$\hat{\rho}_p = \frac{\sigma_p^2}{\sigma_{Total}^2} \tag{8.45}$$

Montgomery (2013) suggest another measure of gauge capability, the *discrimination ratio*, defined as

$$\text{DR} = \frac{1 + \rho_p}{1 - \rho_p} \tag{8.46}$$

For a gauge to be capable, the DR value must exceed 4. Now, from the variance component table, we have

$$\sigma_p^2 = 20.4049, \quad \sigma_{Total}^2 = 23.2144$$

Substituting these values in Eq. (8.45), we obtain

$$\rho_p = 0.879$$

Thus, we have

$$\text{DR} = 15.53$$

which means that the gauge is capable.

### 8.5.4 Graphical Representation of a Gauge R&R Study

Figure 8.10 shows the various percent contributions of gauge R&R, repeatability, reproducibility, and part-to-part variations.

Gage R&R study-ANOVA method for Example 8.10

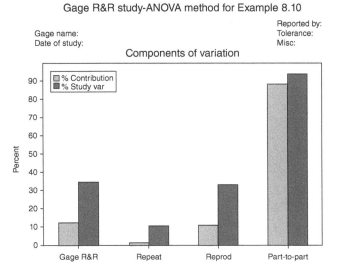

**Figure 8.10** Percent contribution of variance components for the data in Example 8.10.

Gauge R&R study using crossed design

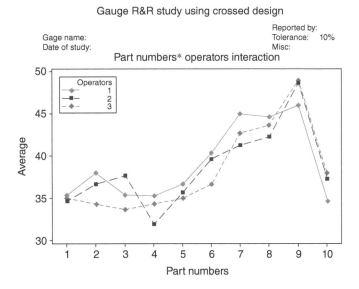

**Figure 8.11** Interaction between operators and parts for the data in Example 8.10.

The graph in Figure 8.11 plots the average measurements of each part for each of the operators. In this example, we have three line graphs corresponding to the three operators. These line graphs intersect each other; they also are not very close to each other. This implies that there is significant interaction between the operators and parts. In fact, the interactions in this example are significant at any level of significance.

In Figure 8.12, the shaded rectangles represent the spread of the measurements by each operator, and the circles inside the rectangles represent the means; the spread of measurements for operator 1 is notably the largest. The means fall almost on a horizontal line, which indicates that the average measurement for each operator is almost the same. Thus, in this example, we can say that the

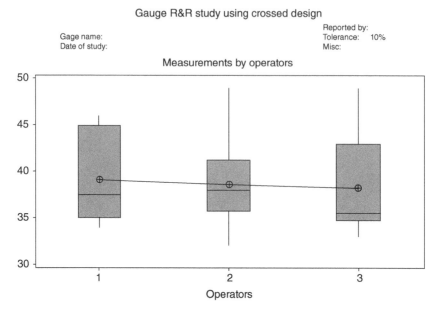

**Figure 8.12** Scatter plot for measurements vs. operators.

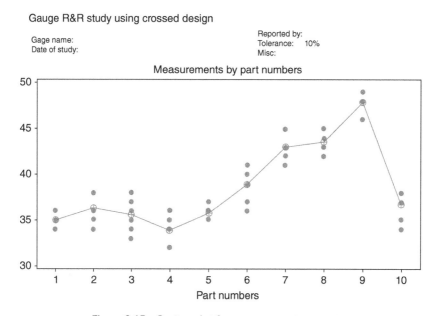

**Figure 8.13** Scatter plot for measurements vs. parts.

operators are measuring the parts consistently. In other words, the variation due to reproducibility is low. In Figure 8.13, the black circles represent the measurements for each part, while the clear circles marked with circled plus signs represent the mean for each part. In this case, the spread of measurements for each part is quite small. This means that each part is being measured with the same precision and accuracy. Combining the outcomes of Figures 8.12 and 8.13, we can conclude that overall the gauge R&R variability is not very significant.

### 8.5.5  Another Measurement Capability Index

Similar to a PCI, which quantifies the ability of a process to produce products of desired quality, a *measurement capability index* (MCI) quantifies the ability of a measurement system to provide accurate measurements. In other words, an MCI evaluates the adequacy of the measurement system. There are various MCIs in use. Two were defined earlier in Eqs. (8.44) and (8.46). We discuss here another MCI:

$$MCI_{pv} = 100 \times \frac{\hat{\sigma}_{Gauge\ R\&R}}{\hat{\sigma}_{Total}} \qquad (8.47)$$

The criteria for the assessment of this index, according to Barrentine (2003), are as follows:

1) $MCI_{pv} \leq 20\%$: good
2) $20\% < MCI_{pv} \leq 30\%$: marginal
3) $MCI_{pv} > 30\%$: unacceptable

In the example discussed earlier, $MCI_{pv} = 34.79\%$. Thus, according to this MCI, the measurement system is not capable.

## Review Practice Problems

**1**  A process is in statistical control, and 25 samples of 5 units each from this process produced $\overline{\overline{X}} = 18$ and $s = 1.3$. The customer's specification limits are LSL = 14 and 22.
   a) Estimate the PCI $C_p$.
   b) Assuming the unit values from this process are normally distributed, find the percentage of nonconforming units.

**2**  Suppose in Problem 1 that the process had some setback so that the new 25 samples of 5 units each produced $\overline{\overline{X}} = 19$ and $s = 1.3$; the specification limits are still the same.
   a) Estimate the PCI $C_p$, and find the percentage of nonconforming units.
   b) Compare the results of this problem with those of Problem 1, and comment on the reliability of the PCI $C_p$.

**3**  In Problems 1 and 2, calculate the values of the PCIs $\hat{C}_{pl}$ and $\hat{C}_{pu}$. Based on these two index estimates, comment on whether nonconforming units have increased below the LSL or above the USL. Justify your answer.

**4**  A process is producing copper wires of a given specification, and the process is in statistical control. The tensile strengths of these wires for a given length are measured producing $\overline{X} = 30$ lbs. and $s = 1.5$ lbs. The specification limits are LSL = 26 and USL = 34. Estimate the value of the PCI $C_{pk}$. Find the percentage of nonconforming wires that the given process is producing.

**5**  In Problem 4, find the percentage of nonconforming units below the LSL and above the USL. If the customer is willing to accept the wires with a tensile strength greater than 34 lb but not below the 26 lb limit, what advice about specific changes in the process would you give to the producer so that a smaller portion of wires are scrapped?

**6** A random sample of size 25 from a process produced $\overline{X} = 100$ and $s = 8$. The specification limits are given to be $100 \pm 20$. Estimate the PCIs $C_p$ and $C_{pk}$. Find the percentages of non-conforming products below the LSL and above the USL.

**7** Repeat Problem 6, assuming that the new specification limits for the same process are $103 \pm 20$. Find the percentages of nonconforming products below the LSL and above the USL. Compare your results with the ones you had obtained in Problem 6, and comment on the effect of moving the center point of the specifications from 100 to 103, specifically on how the estimates of $C_p$ and $C_{pk}$ changed. Assume that the values of the sample mean and sample standard deviation remain the same.

**8** A process has a controlled variability with $\sigma = 1$, but its mean is varying. The varying means are $\mu = 16, 18, 20, 22, 24$, and the specification limits are $20 \pm 6$. For each value of the mean, calculate the values of the PCIs $C_p$ and $C_{pk}$. Comment on your results.

**9** Suppose in Problem 8 that the process has a controlled mean $\mu = 20$, but its variance is changing. The five different values of its standard deviation are $\sigma = 1, 1.2, 1.4, 1.6, 1.8$, and the specification limits are still the same: $20 \pm 6$. For each value of the standard deviation, calculate the values of the PCIs $C_p$ and $C_{pk}$. Comment on your results.

**10** Refer to the values of $C_{pk}$ given in Table 8.2. Assuming that the process is producing units with a normally distributed quality characteristic, calculate the upper and lower bounds on the percentage of conforming units associated with each value of $C_{pk}$ given in Table 8.2.

**11** A process is in statistical control, and the process quality characteristic is normally distributed with mean $\mu = 20$ and $\sigma = 1$. The specification limits are LSL $= 14$ and USL $= 26$, and the target value is $T = 20$, which, as usual, is the center point or the mid-point of the specification limits. Calculate the values of PCIs $C_{pk}$ and $C_{pm}$.

**12** The process in Problem 11 is of six-sigma quality, so it can stand some variation in the process mean without affecting the percentage of nonconforming products. Suppose that the process mean suffers several upward shifts of one-half sigma – that is, $\mu = 20, 20.5, 21$, 21.5, and 22 – while the standard deviation and the target values remain the same as in Problem 11. For each value of the process mean, calculate the values of the PCIs $C_{pk}$ and $C_{pm}$. Compare the various values of these PCIs, and comment on what you learn about them. At what point does the process become incapable?

**13** Consider a process in statistical control with LSL $= 20$ and USL $= 40$ and the target value at $T = 30$. Suppose a random sample of size 25 from this process produced a sample mean of $\overline{x} = 36$ and sample standard deviation of $s = 2.5$. Determine the estimated values for the PCIs $C_p$, $C_{pk}$, and $C_{pm}$. Assume that the quality characteristic of the process is normally distributed.

**14** Refer to Problem 11. Calculate the value of the PCI $C_{pnst}$. Compare the value of the $C_{pnst}$ with the values of other PCIs you obtained in Problem 11, and state what you learn from the values of these indices.

**15** In each case given in Problem 12, calculate the value of the PCI $C_{pnst}$. Compare the values of the $C_{pnst}$ with the values of $C_{pk}$ and $C_{pm}$ you obtained in Problem 12. Comment on which PCI is reacting faster to the changes occurring in the process due to the half-sigma shifts in the process mean.

**16** Consider two processes, A and B, with the following summary statistics:

$$\mu_A = 18, \mu_B = 16$$
$$\sigma_A = 2.0, \sigma_B = 1.0$$

Assume LSL = 12 and USL = 24 and a target value of $T = 17$. Based on the percentage of specification band used, comment on which process is more likely to produce a higher percentage of nonconforming units if a minor shift in the process mean occurs.

**17** Refer to Problem 16. Calculate the values of the PCIs $C_{pm}$ and $C_{pnst}$. Comparing the information you obtained about the two processes in Problem 16 with the values of the indices $C_{pm}$ and $C_{pnst}$, which PCI would you consider to create a more representative picture?

**18** Refer to the data given in Example 8.1, and suppose that $T = 272$. Using the information given in this example, do the following:
a) Estimate the PCIs $C_{pm}$ and $C_{pnst}$.
b) Find the percentage of nonconforming rods below the LSL and above the USL.

**19** Refer to the information obtained in Problem 18. Based on the values of the PCIs $C_{pm}$ and $C_{pnst}$ obtained in part (a), examine whether the process is capable. Now, based on the percentage of nonconforming rods found in part (b), examine whether the process is capable. Does your conclusion based on part (a) match with your conclusion based on part (b)?

**20** Using the information given in Example 8.1, and assuming that $T = 272$, find an estimate of the PCI $C_{pmk}$ and determine whether the process is capable.

**21** Suppose that 30 samples of size $n = 5$ from a process that is considered under statistical control yielded $\overline{\overline{X}} = 45$ and $\overline{R} = 6.90$. Additionally, suppose that the customer's specification limits are LSL = 36, USL = 54, and $T = 45$. Estimate the PCIs $C_{pm}$, $C_{pmk}$, and $C_{pnst}$.

**22** Using the information given in Problem 21, estimate the value of the PCI $C_{pk}$. From your estimated value of $C_{pk}$, determine the lower and upper bounds on the percentage of conforming units associated with the value of $C_{pk}$.

**23** A quality engineer in the semiconductor industry wants to maintain control charts for an important quality characteristic. After collecting each sample, $\overline{X}$ and $R$ are computed. Suppose that 30 samples of size $n = 5$ produce the following summary statistics:

$$\sum_{i=1}^{30} \overline{X}_i = 450 \text{ and } \sum_{i=1}^{30} R_i = 80$$

a) Assuming that the quality characteristic is normally distributed and the specification limits are $15 \pm 3.5$, determine whether the process is capable.

b) With the target value set at T = 15, estimate the PCIs $C_{pm}$, $C_{pmk}$, and $C_{pnst}$.

**24** Consider a process with mean $\mu = 20$ and $\sigma = 2$. Five consecutive observations from the process are collected: 16, 18, 22, 24, and 25. Determine whether the process is ready for pre-control, assuming that the process is in statistical control.

**25** Consider a process with mean $\mu = 45$ and $\sigma = 4$. A random sample of size 10 consecutive observations is taken from this process, which produces the following data:

| 52 | 29 | 32 | 42 | 34 | 57 | 26 | 63 | 39 | 47 |
|----|----|----|----|----|----|----|----|----|----|

Determine whether the process was ready for pre-control at any stage during the collection of these data, assuming that the process is in statistical control.

**26** A random sample of size 30 from a process produced $\overline{X} = 100$ and $s = 6$. The specification limits are given to be $100 \pm 20$.

a) Estimate the PCIs $C_p$ and $C_{pk}$.

b) Construct a 95% confidence interval for $C_p$ and $C_{pk}$.

**27** A process is in statistical control, and the quality characteristic of interest is normally distributed with specification limits $60 \pm 10$. A random sample of 36 units produced a sample mean of $\overline{X} = 60$ and a sample standard deviation of $s = 2.5$.

a) Find a point estimate for the PCI $C_{pk}$.

b) Construct a 95% confidence interval for $C_{pk}$.

**28** Refer to Problem 27. Suppose that the process suffered an upward shift so that a new sample of 36 units produced a sample mean of $\overline{X} = 62$ and sample standard deviation of $s = 3$. Assume that the specifications have not changed.

a) Find a point estimate of $C_{pk}$.

b) Find a 95% confidence interval for $C_{pk}$.

**29** A process is in statistical control, and 20 samples of size 5 each produced the summary statistics $\overline{\overline{X}} = 45, \overline{R} = 4.65$. Suppose that the customer's specifications are $45 \pm 6$.

a) Find point estimates for the PCIs $C_p$ and $C_{pk}$.

b) Find a 95% confidence interval for $C_p$ and $C_{pk}$.

**30** Consider two processes A and B with the following summary statistics:

$$\mu_A = 6.0, \mu_B = 2.0$$
$$\sigma_A = 2.5, \sigma_B = 1.8$$

Further, suppose that LSL = −2, USL = 10. For each process, find the percentage of non-conforming units below the LSL and above the USL. Which process is producing (more) non-conforming units?

**31** In Problem 30, suppose that $T = 4.0$. Find the values of the PCIs $C_{pm}$ and $C_{pnst}$ for the two processes. Compare the percentage of nonconforming units produced by the two processes in Problem 30 with their PCIs. Do you believe that the two PCIs represent the true picture?

**32** A process in statistical control with quality characteristic specification limits given as LSL = 30 and USL = 36. Furthermore, the quality characteristic is normally distributed with $\mu = 33$, $\sigma = 1$, and the target value $T = 33$. Find the values of the PCIs $C_{pm}$ and $C_{pnst}$ for the process.

**33** The process in Problem 32 had a setback due to the shipment of bad raw material. Consequently, the process mean experienced a downward shift of one standard deviation, so that the new mean and standard deviation are $\mu = 32$, $\sigma = 1$. The specification limit and target value remained the same.
a) Find the values of the PCIs $C_{pm}$ and $C_{pnst}$ for the affected processes, and comment.
b) Find the percentage of nonconforming units below the LSL and above the USL.

**34** A manufacturing company got a gauge system to use in the implementation of its new SPC program. An operator from the quality control department who will operate the gauge wants to assess its capability. To achieve his goal, he received 16 parts from the production department. He measured the dimension of each part twice. The data collected are given in Table 8.13. Estimate the gauge capability, and then find the precision-to-tolerance ratio (P/T) if the specification limits of the part used by the operator are LSL = 10 and USL = 70.

**Table 8.13** Measurement data for Problem 8.34.

| Sample # | $X_1$ | $X_2$ | $\bar{X}$ | R |
|----------|-------|-------|-----------|---|
| 1 | 25 | 25 | 25.0 | 0 |
| 2 | 23 | 23 | 23.0 | 0 |
| 3 | 25 | 24 | 24.5 | 1 |
| 4 | 26 | 24 | 25.0 | 2 |
| 5 | 25 | 23 | 24.0 | 2 |
| 6 | 24 | 25 | 24.5 | 1 |
| 7 | 25 | 25 | 25.0 | 0 |
| 8 | 26 | 24 | 25.0 | 2 |
| 9 | 24 | 22 | 23.0 | 2 |
| 10 | 24 | 24 | 24.0 | 0 |
| 11 | 23 | 24 | 23.5 | 1 |
| 12 | 24 | 22 | 23.0 | 2 |
| 13 | 24 | 24 | 24.0 | 0 |
| 14 | 24 | 24 | 24.0 | 0 |
| 15 | 25 | 24 | 24.5 | 1 |
| 16 | 24 | 24 | 24.0 | 0 |
| | | | $\bar{\bar{X}} = 24.375$ | $\bar{R} = 0.875$ |

**35** Refer to the data in Problem 34.
   a) Determine the total, gauge, and parts standard deviations.
   b) Calculate the signal-to-noise ratio (SNR), and check whether the gauge is capable.
   c) Check the gauge capability using the discrimination ratio (DR).

**36** An operator from the quality control department, who will operate a new gauge that the company bought recently, wants to assess the gauge adequacy. The operator measures eight parts four times each. The data obtained are shown in Table 8.14.
   a) Find an estimate of the total variability $\sigma^2_{Total}$.
   b) Find an estimate of the parts variability $\sigma^2_{Parts}$.
   c) Find an estimate of the gauge variability $\sigma^2_{Gauge}$.
   d) If the part specifications are LSL = 30 and USL = 80, evaluate the gauge adequacy by determining the value of P/T.

Table 8.14   Measurement data for Problem 8.36.

| Sample no. | $X_1$ | $X_2$ | $X_3$ | $X_4$ | $\overline{X}$ | R |
|---|---|---|---|---|---|---|
| | | | Measurements | | | |
| 1 | 78 | 78 | 79 | 79 | 78.50 | 1 |
| 2 | 74 | 75 | 74 | 75 | 74.50 | 1 |
| 3 | 72 | 73 | 72 | 73 | 72.50 | 1 |
| 4 | 69 | 68 | 68 | 69 | 68.50 | 1 |
| 5 | 62 | 63 | 62 | 63 | 62.50 | 1 |
| 6 | 52 | 52 | 51 | 52 | 51.75 | 1 |
| 7 | 65 | 66 | 65 | 66 | 65.50 | 1 |
| 8 | 61 | 61 | 62 | 61 | 61.00 | 0 |

**37** A company manufactures small tubes used in heavy equipment. The company has installed a new gauge to take measurements of these tubes. To perform the measurement system analysis on the new gauge, three operators are selected randomly from the quality control department. The operators decided to take a random sample of 10 tubes. Each operator took three measurements on each tube selected randomly. The data obtained are shown in Table 8.15.

Table 8.15   Measurement data for Problem 8.37.

| Tube no. | Operator 1 Meas. 1 | Meas. 2 | Meas.3 | Operator 2 Meas. 1 | Meas. 2 | Meas.3 | Operator 3 Meas. 1 | Meas. 2 | Meas. 3 |
|---|---|---|---|---|---|---|---|---|---|
| 1 | 59 | 59 | 61 | 59 | 59 | 61 | 59 | 59 | 60 |
| 2 | 44 | 45 | 44 | 44 | 45 | 44 | 45 | 45 | 44 |
| 3 | 63 | 65 | 64 | 63 | 65 | 64 | 63 | 63 | 64 |
| 4 | 62 | 63 | 62 | 62 | 63 | 62 | 62 | 63 | 62 |
| 5 | 63 | 62 | 63 | 63 | 62 | 63 | 63 | 62 | 62 |
| 6 | 52 | 51 | 52 | 52 | 51 | 51 | 51 | 51 | 52 |
| 7 | 41 | 42 | 41 | 42 | 42 | 41 | 41 | 42 | 42 |
| 8 | 58 | 59 | 58 | 58 | 59 | 58 | 58 | 59 | 58 |
| 9 | 56 | 57 | 57 | 56 | 57 | 56 | 56 | 56 | 57 |
| 10 | 55 | 56 | 56 | 56 | 56 | 56 | 55 | 55 | 56 |
| | $\overline{X}_1 = 55.6667$ | $\overline{R}_1 = 1.2$ | | $\overline{X}_2 = 55.6667$ | $\overline{R}_2 = 1.1$ | | $\overline{X}_3 = 55.5$ | $\overline{R}_3 = 1.0$ | |

Using the range method (discussed in Section 8.5.2), do the following:

a) Estimate the gauge repeatability, $\sigma_{repeatability}$.

b) Estimate the gauge reproducibility, $\sigma_{reproducibility}$.

c) Estimate the gauge R&R, $\sigma_{Gauge}$.

d) If the specification limits are LSL = 30 and USL = 80, find the gauge capability by finding P/T.

**38** Repeat Problem 37 using the ANOVA method discussed in Section 8.5.3.

**39** Refer to Problem 37.

a) Estimate the SNR.

b) Estimate the DR.

c) Using the results obtained in parts (a) and (b), comment on the gauge capability.

# 9

# Acceptance Sampling Plans

## 9.1   Introduction

Acceptance sampling plans are one of the oldest methods of quality control. An *acceptance sampling* plan is the process of inspecting a portion of the product in a lot to decide whether to classify the entire lot as either conforming or nonconforming to quality specifications and thus decide whether to accept or reject the lot. Inspection is done either with sampling or with 100% inspection, in which case all units are inspected.

This method is also called a *lot-by-lot* acceptance sampling plan. This method of quality control started before World War II and became very popular during and after the war when there was great demand for certain products and manufacturers started mass production; the main goal was to have tight control over incoming products and raw materials. In the early 1950s and 1960s, pioneering statisticians such as W. Edwards Deming, Joseph M. Juran, and others introduced the modern techniques of statistical process control (SPC). These techniques quickly became very popular, particularly with Japanese industries. As we saw in earlier chapters, the purpose of modern techniques of statistical process control is to improve process performance throughout the production process rather than depending primarily on acceptance sampling.

Today, acceptance sampling is used in conjunction with SPC methods when it is assumed that the process is under statistical control. In this chapter, we discuss acceptance sampling plans for attributes and variables.

## 9.2   The Intent of Acceptance Sampling Plans

As we noted in the introduction, the primary goal of acceptance sampling plans is to make sure units or raw materials to be used in the manufacturing process are of good quality. The process consists of taking a sample from the shipment or lot of units or raw material received from the supplier, and inspecting some desired characteristics of the units in the sample. Based on the result of that inspection, and using the guidelines of the sampling plan, a decision is made regarding the lot (that is, the lot is accepted or rejected). An accepted lot will be used for manufacturing, whereas a rejected lot can be handled in various ways, including simply sending it back to the supplier, conducting a 100% inspection to sort out bad units and replace them with good units, or introducing a more rigorous acceptance sampling plan on the rejected lots.

Note that acceptance sampling plans are sometimes used as a deterrent against a supplier that is known to ship bad units or raw material. But if a supplier always strictly uses SPC with a very high

*Statistical Quality Control: Using Minitab, R, JMP, and Python*, First Edition. Bhisham C. Gupta.
© 2021 John Wiley & Sons, Inc. Published 2021 by John Wiley & Sons, Inc.
Companion website: www.wiley.com/go/college/gupta/SQC

process capability index – say, two or higher – and has attained excellent quality of all units or materials, then an acceptance sampling plan may not be necessary.

Another point to note is that an acceptance sampling plan is not a guarantee of quality control. Often, all lots may be of the same quality, but some lots may be accepted while others are rejected outright and put through 100% inspection or simply put through a screening process, also called a *rectifying inspection*. In this process, all the nonconforming units are sorted out from the rejected lot. The sorted bad units are either sent back to the supplier, reworked, or simply replaced with good units. Finally, we may remark here that an acceptance sampling plan is not a substitute for SPC.

## 9.3 Sampling Inspection vs. 100% Inspection

The following are some of the scenarios when acceptance sampling is preferred over 100% inspection:

- When human error is a concern. Quite often, 100% inspection may overlook nonconforming units due to human fatigue and the monotonous nature of the process.
- When testing is destructive. For example, if a part is put through a test until it fails then, the 100% inspection would mean destroying all the units in the lot.
- When testing is costly and/or time-consuming but the cost of passing a nonconforming part is not as high.
- When the supplier is known to have a fully implemented SPC and has attained very high quality for their product.
- When we wish to evaluate suppliers in terms of their products' quality. Suppliers that consistently deliver high-quality products can receive preferred status such as reduced inspection, priority for contracts, etc.
- When suppliers do not meet quality requirements. Decisions based on sampling will be much cheaper than those based on 100% inspections.
- When sampling provides an economic advantage due to fewer units being inspected. In addition, the time required to inspect a sample is substantially less than that required for the entire lot, and there is less damage to the product due to reduced handling.
- When it helps the supplier/customer relationship. By inspecting a small fraction of the lot and forcing the supplier to screen 100% of rejected lots (which is the case for rectifying inspection), the customer emphasizes that the supplier must be concerned about quality.

However, in certain situations, it is preferable to inspect 100% of the product. For example, when we are testing very expensive or complicated products, the cost of making the wrong decision may be too high. Screening is also appropriate when the fraction of nonconforming units is extremely high. In such cases, the majority of lots would be rejected under an acceptance sampling plan, and the acceptance of lots would be a result of statistical variations rather than better quality. Lastly, screening is appropriate when the fraction of nonconforming units is not known, and an estimate based on a large sample is needed.

## 9.4 Classification of Sampling Plans

Acceptance sampling plans are typically classified into three broad categories: sampling plans for attributes, sampling plans for variables, and special sampling plans. It should be noted that acceptance sampling is not advised for processes in continuous production or those in a state of statistical control. For these processes, Deming (1986) provides decision rules for selecting either 100% inspection or no inspection at all.

Sampling plan for attributes is based on quality characteristics that are either conforming or nonconforming. Sampling plan for variables is based on quality characteristics measured on a numerical scale. Sampling plan for attributes has an advantage over sampling plans for variables when a part is nonconforming due to several quality characteristics. In such cases, a single-sampling plan for attributes will suffice, whereas if we use sampling plans for variables, then we must develop an acceptance sampling plan for each quality characteristic. That said, sample sizes for acceptance sampling plans for attributes are usually much larger than for sampling plans for variables. For special sampling plans, Montgomery (2013), for example, suggests starting with 100% inspection; when a predefined number of units is found to be conforming, sampling inspection is started.

### 9.4.1 Formation of Lots for Acceptance Sampling Plans

Lots for an acceptance sampling plan should be carefully formed so that they are as homogeneous as possible. A given lot should not contain units produced by more than one machine, from different shifts, or from different production lines. A lot should also be as large as possible.

Lots should be defined to facilitate the handling, shipping, and selection of samples. Normally, this is achieved by selecting proper or manageable packaging, which can make shipping and selection of samples easier and more economical. As noted earlier, larger lots are beneficial: they may let us select large samples at low cost. Moreover, they may allow us to develop more informative operating characteristic (OC) curves.

### 9.4.2 The Operating Characteristic (OC) Curve

If lots are used for an acceptance sampling plan, then the probability of accepting a lot is defined by the proportion of lots accepted from a given process over some time. The OC curve for the acceptance sampling plan is a curve that gives the probability of accepting lots at varying levels of percent defective.

No matter which type of sampling plan is being considered, an OC curve is an important evaluation tool. It illustrates the risks involved in an acceptance sampling plan. In the following example, we discuss OC curves and other curves under different sampling plans.

**Example 9.1**

Using a statistical package, construct OC curves for the following sampling plans:

a) $N = 5000, n = 100, 75, 50, 25; c = 1$

b) $N = 5000, n = 200, c = 4; n = 100, c = 2; n = 50, c = 1$

c) $N = 5000, n = 100, c = 1; n = 100, c = 2; n = 100, c = 3$

**Solution**

a) Figure 9.1 shows OC curves with lot size 5000 and four different sampling plans. (We constructed these curves using the statistical package Minitab. The instructions for using Minitab are presented later after we introduce some required terminology.) Note that in Figure 9.1a–c, the constant $c$ represents a number such that if the number of nonconforming units is less than or equal to $c$, the lot is accepted; otherwise, the lot is rejected. The value of $c$ is also called the *acceptance number*.

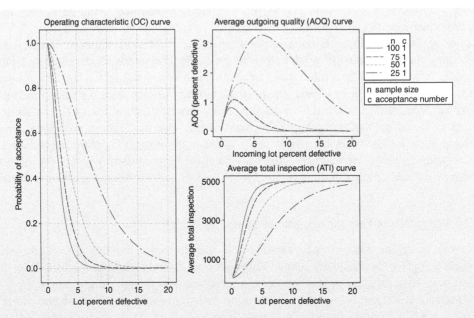

**Figure 9.1** OC, average outgoing quality (AOQ), and average total inspection (ATI) curves with $N = 5000$; $n = 100, 75, 50, 25$; and $c = 1$.

If we assume, for example, that the lot is 5% defective, then the OC curves in Figure 9.1 show that for samples of sizes 100, 75, 50, and 25, the probabilities of accepting the lot are approximately 4%, 10%, 30%, and 65%, respectively. We observe that, if the value of $c$ remains constant, then as the sample size decreases, the probability of accepting the lot increases. However, the rate of increase in probability will depend on the percent defectives in the lot.

b) In part (a), the value of $c$ was held constant, whereas here, the value of $c$ changes with the sample size. For these sampling plans, the corresponding OC curves are shown in Figure 9.2. If we assume that the lot is 5% defective, then under the sampling plans

$$n = 200, c = 4; n = 100, c = 2; n = 50, c = 1$$

the probabilities of accepting the lot are approximately 2%, 15%, and 30%, respectively.

c) Here we hold the sample size constant at 100, while the value of $c$ changes. The sampling plans and corresponding OC curves are shown in Figure 9.3. We again assume that the lot is 5% defective, and under the sampling plans

$$n = 100, c = 1; n = 100, c = 2; n = 100, c = 3$$

the probabilities of accepting the lot are approximately 2%, 12%, and 25%, respectively.

If we assume that the lot is 2% defective, then the OC curves in Figure 9.2 give probabilities of acceptance with $n = 200, c = 4; n = 100, c = 2$; and $n = 50, c = 1$ are approximately equal to 95%, 93%, 90%, respectively. If, however, $c$ increases and the sample size is held constant, these probabilities are approximately equal to 90%, 95%, and 99%, respectively (see Figure 9.3).

Figures 9.1, 9.2, and 9.3 show other curves as well: the AOQ and ATI curves. We will comment on these curves after introducing some requisite terminology.

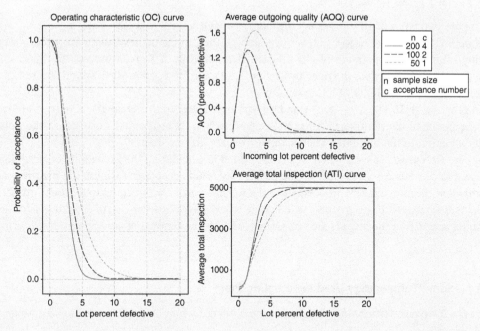

**Figure 9.2** OC, AOQ, and ATI curves with $n$ = 200, c = 4; $n$ = 100, $c$ = 2; $n$ = 50, $c$ = 1.

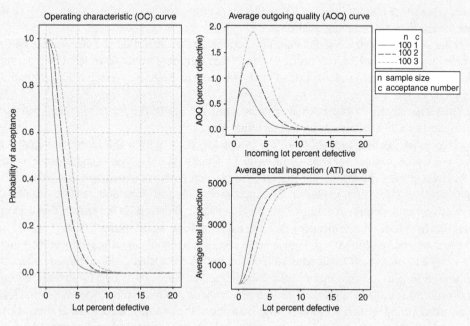

**Figure 9.3** OC, AOQ, and ATI curves with $n$ = 100, $c$ = 1; $n$ = 100, $c$ = 2; $n$ = 100, $c$ = 3.

### 9.4.3 Two Types of OC Curves

There are two types of OC curves to consider: *type B OC curves* and *type A OC curves*.

*Type B OC curves* are used when we can assume the sample comes from a large lot or we are taking samples from a stream of lots selected at random from a given process. In this case, we use the *binomial distribution* (refer to Section 3.5.1) and/or Poisson distribution (Section 3.5.3) to calculate the probability of lot acceptance.

In general, ANSI/ASQ Z1.4-2003 standard (discussed later in this chapter) OC curves are based on the binomial distribution when sample sizes are less than or equal to 80, or on the Poisson approximation to the binomial when sample sizes are greater than 80.

*Type A OC curves* are used when we are interested in calculating the probability of accepting an isolated lot. In this case, to calculate the probability of lot acceptance, we use the *hypergeometric distribution* (Section 3.5.2) with $N$ equal to the lot size and $M = Np$ equal to the total number of defectives or nonconforming products in the lot, where $p$ is the percentage of defectives or nonconforming products in the lot, $n$ is the sample size, and $c$ is the number of acceptable defectives in the sample.

### 9.4.4 Some Terminology Used in Sampling Plans

We now introduce terminology commonly encountered when discussing specific accepting sampling plans:

- *Acceptable quality level* (AQL): The poorest quality level of the supplier's process that the customer deems acceptable as a process average. Products at AQL have a high probability (generally around 95%, although it may vary) of being accepted.
- *Average outgoing quality limit* (AOQL): The maximum possible number of defectives or defect rate for the average outgoing quality.
- *Lot tolerance percent defective* (LTPD): The poorest quality level that the customer is willing to accept in an individual lot. It is the quality level that routinely would be rejected by the sampling plan. In many sampling plans, the LTPD is the percent defective having a 10% probability of acceptance.
- *Rejectable quality level* (RQL): An alternative name for LTPD; the poorest quality level that the customer is willing to tolerate in an individual lot.
- *Producer's risk* ($\alpha$): The probability of rejecting a lot that is within the acceptable quality level. This means the producer or supplier faces the possibility (at the $\alpha$ level of significance) of having a lot rejected even though the lot has met the requirements stipulated by the AQL level.
- *Consumer's risk* ($\beta$): The probability of acceptance (usually 10%) for a designated numerical value of relatively poor submitted quality. The consumer's risk, therefore, is the probability of accepting a lot (at the $\beta$ level of significance) that has a quality level equal to the LTPD.
- *Average outgoing quality* (AOQ): The expected average quality of outgoing products. This includes all accepted lots, as well as all rejected lots that have been sorted 100% and have had all of the nonconforming units replaced by conforming units. There is a given AOQ for specific fractions of nonconforming units of submitted lots sampled under a given sampling plan. When the fraction of nonconforming units is very low, a large majority of the lots will be accepted as submitted. The few lots that are rejected will be sorted 100% and have all nonconforming units replaced with conforming units. Thus, the AOQ will always be less than the submitted quality. As the quality of submitted lots declines in relation to the AQL, the percent of lots rejected increases in proportion to accepted lots. As these rejected lots are sorted and combined with accepted lots, an AOQ lower than the average fraction of nonconformances of submitted lots emerges. So, when the

level of quality of incoming lots is good, the AOQ is good; when the incoming quality is bad, and most lots are rejected and sorted, the AOQ is also good.

In Figures 9.1, 9.2, and 9.3, the AQL and ATI curves give the values of AQL and the total number of units that are inspected at the varying level of defectives in the incoming lots. Note that if the lot fraction of defectives increases, the AQL level decreases. This is because more lots are rejected, and consequently, more lots are going through the screening or 100% inspection process, and all bad units are either replaced or reworked. However, in this case, ATI increases again because of the screening process. The general formula for calculating ATI is

$$\text{ATI} = n + (1 - p_a)(N - n) \tag{9.1}$$

where $p_a$ is the probability of accepting a lot.

**Example 9.2**
Find AOQ and ATI for an acceptance sampling by attributes when $n = 100$, $c = 2$, and $N = 5000$.

**Solution**

We find the AOQ by using Minitab and following these steps:

1) From the menu bar, select **Stat > Quality Tools > Acceptance Sampling by Attributes**.
2) A dialog box titled **Acceptable Sampling by Attributes** appears. In this dialog box, from the dropdown menu, select **Compare User Defined Sampling Plans**.
3) Enter the following values in the boxes: Next to **AQL: 2** (the quality level of the manufacturer's process that the customer would consider acceptable as a process average); next to **RQL or LTPD: 10** (the poorest level of quality that the customer is willing to accept in an individual lot); next to **Sample size: 100**; next to **Acceptance numbers: 2**; and next to **Lot size: 5000**. Click **OK**. The results will appear in the **Session Window**.

The Minitab output shows that if the incoming lot percent defective is 10%, then the ATI is equal to 4990.5. This is because most of the lots will be rejected and, consequently, will go through the screening process.

Note that you can enter more than one sample size and more than one acceptance number. More detailed instructions are given shortly.

**Method**

| | |
|---|---|
| **Acceptable quality level (AQL)** | 2 |
| Rejectable quality level (RQL or LTPD) | 10 |

**Compare user-defined plan(s)**
Sample size 100
Acceptance number 2
*Accept lot if defective items in 100 sampled items ≤2, otherwise reject.*

| Percent defective | Probability accepting | Probability rejecting | AOQ | ATI |
|---|---|---|---|---|
| 2 | 0.677 | 0.323 | 1.326 | 1684.2 |
| 10 | 0.002 | 0.998 | 0.019 | 4990.5 |

Using the formula given in Eq. (9.1), the ATI of an incoming lot at a given percent defectives is given by

$$ATI = n + (1 - p_a)(N - n)$$

Now, suppose the percent defective is 2%. If we substitute the values of $n = 100$ and $N = 5000$ into the previous equation and use $(1 - p_a) = 0.323$, we have

$$ATI = 100 + 4900 \times 0.323 = 1682.7$$

That is approximately equal to the ATI value given in Minitab (a rounding error occurs with the $p_a$ value).

As remarked earlier, the AOQ level is 1.326%, which is even lower than the incoming lot fraction defective of 2%. The ATI value increases as the lot percent defective increases. This is because as the lot percent defective increases, the number of rejected lots increases, and the majority of the lots must go through the screening process. These values can also be found using Figure 9.4.

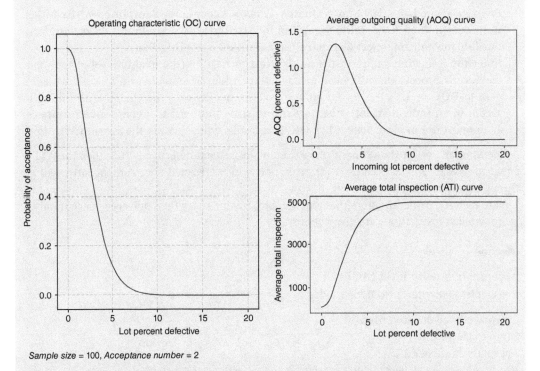

*Sample size = 100, Acceptance number = 2*

**Figure 9.4** The OC, AOQ, and ATI curves with $N = 5000$, $n = 100$, and $c = 2$.

Note that the OC curves and other curves such in those in Figures 9.1–9.4 can be constructed using either of the software packages discussed in this book. For instance, we constructed Figures 9.1–9.4 for various cases in Example 9.1 using Minitab by taking the following steps:

**Minitab**
1) From the menu bar, select **Stat > Quality Tools > Acceptable Sampling by Attributes**.
2) A dialog box titled **Acceptable Sampling by Attributes** appears. In this dialog box, select **Compare User Defined Sampling Plans** from the dropdown menu.
3) In the box next to **AQL**, enter the desired level of AQL: say, **2**.
4) In the box next to **RQL or LTPD**, enter the desired level of RQL: say, **10**.
5) In the box next to **Sample size**, enter **200 100 50**. Do not put any commas next to these numbers – just leave one space between them.
6) In the box next to **Acceptance numbers**, enter **4 2 1**. Again, leave one space between these numbers.
7) In the box next to **Lot size**, enter **5000**. Click **OK**. All the results will appear in the **Session Window**.

**R**

The OC curves can be generated by entering the following in the R console. However, currently no R packages are available to generate ATI and AOQ. If and when they become available, we will include them in Chapter 10, which is available for download on the book's website: www.wiley.com/college/gupta/SQC.

a)

```
install.packages(AcceptanceSampling)
library(AcceptanceSampling)
x=mapply(function(nsamp,cmax)OC2c(nsamp,cmax,type="hypergeom",N=5000,pd=seq
(0,0.2,0.001)),nsamp=c(100,75,50,25),cmax=c(1,1,1,1),SIMPLIFY = FALSE)
pt_code=c(1,10,19,20)
plot(x[[1]]);for(i in 2:4)points(x[[i]]@pd, x[[i]]@paccept, pch=pt_code[i])
```

Note that the "install.packages()" function only needs to be called once on a computer to have a package installed in R's local packages library. For any R session, the "library ()" function has to be called once during the session to load a local package's functions. You may investigate the other functions demonstrated here on your own.

b)

```
library(AcceptanceSampling)
x=mapply(function(nsamp,cmax)OC2c(nsamp,cmax,

type="hypergeom",N=5000,pd=seq(0,0.2,0.001)),nsamp=c(200,100,50),cmax=c(4,2,1),SIMPLIFY
= FALSE)
pt_code=c(1, 10, 19)
plot(x[[1]]); for (i in 2:3)points(x[[i]]@pd,x[[i]]@paccept,
pch = pt_code[i])
legend("topright", pch = pt code,, c("n=200, c=4", "n=100, c=2", "n=50, c=1"))
```

c)

```
library(AcceptanceSampling)
x=mapply(function(nsamp,cmax)OC2c(nsamp,cmax,type="hypergeom",N=5000,pd=seq
(0,0.2,0.001)),nsamp=c(100,100,100),cmax=c(3,2,1),SIMPLIFY = FALSE)
pt_code=c(1, 10, 19)
plot(x[[1]]); for (i in 2:3)points(x[[i]]@pd,x[[i]]@paccept,
pch = pt_code[i])
legend("topright", pch = pt code,, c("n=100, c=3", "n=100, c=2", "n=100, c=1"))
```

## 9.5 Acceptance Sampling by Attributes

Acceptance sampling by attributes is generally used for two purposes: (i) protection against accepting lots from a continuing process whose average quality deteriorates beyond an AQL and (ii) protection against isolated lots that may have levels of nonconformance greater than can be considered acceptable. This is the most commonly used form of acceptance sampling plan; and the most widely used standard of all attribute plans is ANSI/ASQ Z1.4-2003. The following sections provide more details on the characteristics of acceptance sampling as well as a discussion of military standards in acceptance sampling. We again review some of the terminology discussed earlier in this chapter.

### 9.5.1 Acceptable Quality Limit (AQL)

As part of the revision of ANSI/ASQ Z1.4-1993, the acceptable quality *level* was changed to the acceptable quality *limit* in ANSI/ASQ Z1.4-2003. As mentioned earlier, it is defined as the quality level that is the worst tolerable process average when a continuing series of lots is submitted for acceptance sampling. This means a lot that has a fraction defective equal to the AQL has a high probability (generally in the neighborhood of 95%, although it may vary) of being accepted. As a result, plans that are based on the AQL, such as ANSI/ASQ Z1.4-2003, favor the producer in terms of getting lots accepted that are in the general neighborhood of the AQL for fraction defective in a lot.

### 9.5.2 Average Outgoing Quality (AOQ)

The AOQ is the expected average quality of outgoing products obtained over a long sequence of lots from a process with fraction defective *p*. This includes all accepted lots as well as all rejected lots that have been 100% sorted and have had all of the nonconforming units replaced by conforming units.

There is a given AOQ for a specific fraction of nonconforming units of submitted lots sampled under a given sampling plan. When the fraction of nonconforming units is very low, a large majority of the lots will be accepted as submitted. The few lots that are rejected will be 100% sorted and have all nonconforming units replaced with conforming units. Thus, the AOQ will always be less than the submitted quality. As the quality of submitted lots declines in relation to the AQL, the percent of lots rejected increases in proportion to accepted lots. As these rejected lots are sorted and combined with accepted lots, an AOQ lower than the average fraction of nonconformances of submitted lots emerges. Therefore, when the level of quality of incoming lots is good, the AOQ is good; when the incoming quality is bad, most lots are rejected, and sorted and bad units are replaced by the good ones, the quality of outgoing lots becomes good again (see Figure 9.1–9.4). Except when the incoming quality is bad, the value of ATI increases.

Earlier, we saw how to find the value of AOQ for varying incoming lot fractions defective by using one of the statistical software packages discussed in this book. Now we show how to manually calculate the AOQ for a specific fraction of nonconforming units in a sampling plan. The first step is to calculate the probability of accepting the lot at that level of fraction of nonconforming units; then we multiply the probability of acceptance by the fraction of nonconforming units for the AOQ. Thus,

$$\text{AOQ} = p_a \times p\,[1 - (n/N)] = \frac{p_a \times p(N - n)}{N} \tag{9.2}$$

where $n$ is the sample size and, after inspection, the lot contains no defective units, since all encountered defective units are replaced with good units; there are $(N - n)$ units that are not defective units, since if the lot was rejected and therefore screened, all of the defective units would have been replaced with good units. If the lot is accepted, then it contains $p(N - n)$ defectives; and, finally, $p_a$ is the probability of accepting the lot with defectives fraction $p$. (If desired, the AOQ can be expressed as a percentage by multiplying by 100.)

### 9.5.3 Average Outgoing Quality Limit (AOQL)

The AOQL is the maximum AOQ for all possible levels of incoming quality. The AOQ is a variable dependent on the quality level of incoming lots. When the AOQ is plotted for all possible levels of incoming quality, as shown in the AOQ curves in Figures 9.1–9.4, the AOQL is the highest value on the AOQ curve. For example, AOQL = 1.326 in Figure 9.4.

### 9.5.4 Average Total Inspection (ATI)

As discussed earlier in this chapter, the ATI is the total number of units inspected under the given acceptance sampling plan and is given by

$$\text{ATI} = n + (1 - p_a)(N - n) \tag{9.3a}$$

where, $p_a$ is the probability of accepting the lot.

**Example 9.3**

Suppose we have an acceptance sampling plan with sample size $n = 100$ and acceptance number $c = 3$ and that the sample is taken from a continuous process. Find $p_a$, the probability of acceptance with fraction of lot defective $p = 0.01, 0.02, 0.03, 0.04, \cdots, 0.06, 0.07$. Then construct the OC curve for the sampling plan in this example.

**Solution**

Since the samples in this example are taken from a continuous process, we can use the binomial distribution to find the probabilities $p_a$ for various values of $p$ as follows.

At $p = 0.01$:

$$P_a = P(x \leq c) = P(x \leq 3)$$

$$= \binom{100}{0}(0.01)^0(0.99)^{100} + \binom{100}{1}(0.01)^1(0.99)^{99}$$

$$+ \binom{100}{2}(0.01)^2(0.99)^{98}$$

$$+ \binom{100}{3}(0.01)^3(0.99)^{97}$$

$$= 0.3660 + 0.3697 + 0.1849 + 0.0610$$

$$= 0.9816$$

Similarly, we can obtain probabilities $p_a$ for other values of $p$. Table 9.1a lists all the probabilities $p_a$ obtained at $p = 0.01, 0.02, 0.03, 0.04, \cdots, 0.06$, and $0.07$. By plotting $p$ vs. $P_a$, we get the OC curve, which is shown in Figure 9.5.

**Table 9.1a** Probability of acceptance with various fraction of lot defective.

| $p$ | 0.01 | 0.02 | 0.03 | 0.04 | 0.05 | 0.06 | 0.07 |
|-----|------|------|------|------|------|------|------|
| $P_a$ | 0.9816 | 0.8590 | 0.6473 | 0.4295 | 0.2578 | 0.1430 | 0.0744 |

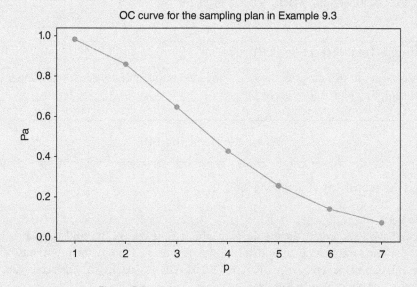

**Figure 9.5** OC curves with $n = 100$ and $c = 3$.

**Example 9.4**

Refer to Example 9.3. Construct the AOQ curve. Then find the value of the average outgoing quality limit for the plan in Example 9.3.

**Solution**

From Eq. (9.2), we know that for a continuous process, i.e. when $N$ is very large, the AOQ is given by

$$AOQ = p_a \times p \qquad (9.3b)$$

For very large $N$, the quantity $(N - n) / N$ approaches 1. Thus, using the values of $p_a$ and $p$, the values of AOQ are computed as in Table 9.1b. The values of AOQ in the table are given as a percentage (that is, each value of AOQ is multiplied by 100).

By plotting AOQ vs. $p_a$, we get the AOQ curve, which is shown in Figure 9.6.

As can be seen, the AOQ rises until the incoming quality level of 0.03 nonconforming is reached. The maximum AOQ point is 1.942%, which is called the AOQL. This is the AOQL for an infinite (very large) lot size, sample size = 100, and acceptance number $c = 3$.

**Table 9.1b** Acceptance probabilities and AOQ% vs. lot percent defective.

| $p$ | 0.01 | 0.02 | 0.03 | 0.04 | 0.05 | 0.06 | 0.07 |
|---|---|---|---|---|---|---|---|
| $p_a$ | 0.9816 | 0.8590 | 0.6473 | 0.4295 | 0.2578 | 0.1430 | 0.0744 |
| AOQ% | 0.9816 | 1.718 | 1.942 | 1.718 | 1.289 | 0.8580 | 0.5208 |

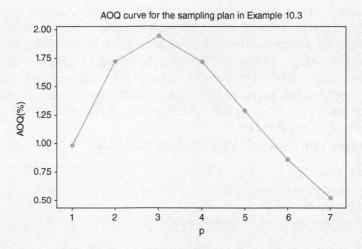

**Figure 9.6** AOQ curves with $n = 100$ and $c = 3$.

## 9.6 Single Sampling Plans for Attributes

Suppose that we have to make a decision about a lot of size $N$. A *single sampling plan* is defined by the sample size ($n$) and the acceptance number ($c$). Thus if we have a lot of size 8000 (say) and the single-sampling plan is defined by

$$n = 100, c = 3$$

from this lot of size 8000, a random sample of size $n = 100$ units will be selected and inspected for the quality characteristic(s) of interest, and the number of defectives or nonconforming units, $m$, will be is determined. If $m \leq 3$, then the lot is accepted; otherwise, the lot is rejected. Note that we are dealing with a quality characteristic that is an attribute, so we may observe units for one or more quality characteristics on go/no-go basis since a part could be defective or nonconforming due to one or more quality characteristics. This procedure is called a single-sampling plan because the fate of the lot is decided based on the result of just one sample. In Figures 9.1–9.4, we saw how the OC curves change as the sample size $n$ changes, the acceptance number $c$ changes, or both sample size $n$ and acceptance number $c$ change. Hence, $p_a$, the probability of acceptance, changes considerably as the sample size $n$ and/or acceptance number $c$ change.

## 9.7 Other Types of Sampling Plans for Attributes

Several types of attribute sampling plans are in use, with the most common being single, double, multiple, and sequential sampling plans. The type of sampling plan used is determined by its ease of use and administration, the general quality level of incoming lots, the average sample number, and so on. In Section 9.6, we discussed single sampling plans for attributes. In this and the following sections, we briefly discuss the other sampling plans for attributes.

### 9.7.1 Double-Sampling Plans for Attributes

The use of double-sampling plans arises under conditions when a second sample is needed before the final decision about the fate of a lot is made. Usually, a smaller first sample is drawn from the submitted lot, and one of three decisions is made: (i) accept the lot, (ii) reject the lot, or (iii) draw another sample. If, after the first sample, condition (iii) prevails, then a second sample is drawn from the submitted lot and inspected. After the inspection of the second sample, the lot is either accepted or rejected. Double-sampling plans have the advantage of smaller total sample size when the incoming quality is either excellent or poor because the lot is either accepted or rejected on the first sample.

A double-sampling plan is defined by the following parameters:

$n_1$ = size of the first sample
$c_1$ = acceptance number for the first sample
$n_2$ = size of the second sample
$c_2$ = acceptance number for the combined samples

We illustrate here how a double-sampling plan is executed with the help of an example. Suppose we assign the following values to four parameters defined earlier:

$$n_1 = 50, c_1 = 1, n_2 = 100, c_2 = 3$$

We draw a random sample of size 50 units from the submitted lot, say of size $N = 5000$, and record the number of defectives $m_1$. If $m_1 \leq c_1 = 1$, then the lot is accepted on the first sample. If $m_1 > c_2 = 3$, then the lot is rejected on the first sample. If $c_1 < m_1 \leq c_2 = 3$, then a second sample is drawn from the submitted lot, and the number of defectives or nonconforming products $m_2$ is observed. If the combined defectives or nonconforming products is such that $m_1 + m_2 \leq c_2 = 3$, then the lot is accepted; otherwise, the lot is rejected.

The major benefit of a double-sampling plan over a single-sampling plan is that it may result in a reduced total number of inspections. As mentioned before, in double-sampling plans, the size of the first sample is usually significantly smaller than the sample size in a single-sampling plan. If the quality of the products is either excellent or bad, then there are significant chances of deciding about the fate of the lot after the first sample: if the quality is excellent, then we might accept the lot after inspecting the first sample; and if it is quite bad, then we might reject it. So, in these cases, we will have fewer inspections, which will curtail the cost. Other times, we may reject a lot during the inspection of the second sample if the combined defectives become greater than $c_2$ before completing the inspection.

## 9.7.2  The OC Curve

The performance of double-sampling plans, as in the case of single-sampling plans, can be explained with the help of the OC curve. The OC curve for double sampling plans is somewhat more complex than for single samplingplans. In double-sampling plans, the OC curve has three curves: one curve for the probability of lot acceptance for combined samples, a second showing the probability of lot acceptance after the first sample, and a third showing the probability of rejecting the lot. We illustrate the derivation of the OC curve for a double-sampling plan with the help of the following example.

### Example 9.5
Consider the following double-sample acceptance plan:

$$n_1 = 50, c_1 = 1, n_2 = 100, c_2 = 3$$

Suppose lot percent or fraction defectives are $p = 0.01, 0.02, 0.03, 0.04, 0.05, 0.06, 0.07$. Do the following:

a) Compute the probabilities of acceptance on the first sample.
b) Compute the probabilities of rejection on the first sample.
c) Compute the probabilities of acceptance on the second sample.
d) Compute the probabilities of acceptance on combined samples.
e) Calculate the AOQ percentage points, assuming the process is continuous.
f) Construct the OC curves for parts (a)–(d).

### Solution

a) Let $p_a^1$ denote the probability of accepting the lot on the first sample. Then $p_a^1$ at $p = 0.01$ is given by

$$p_a^1 = \sum_{m_1=0}^{1} \frac{50!}{m_1! \times (50-m_1)!} p^{m_1}(1-p)^{(50-m_1)}$$

$$= \sum_{m_1=0}^{1} \frac{50!}{m_1! \times (50-m_1)!} (0.01)^{m_1}(1-0.01)^{(50-m_1)}$$

$$= 0.9106$$

Other probabilities when $p = 0.02, 0.03, 0.04, 0.05, 0.06, 0.07$ are calculated similarly.

b) Let $p_r^1$ denote the probability of rejecting the lot on the first sample. Then $p_r^1$ at $p = 0.01$ is given by

$$p_r^1 = \sum_{m_1>3} \frac{50!}{m_1! \times (50-m_1)!} p^{m_1}(1-p)^{(50-m_1)}$$

$$= 1 - \sum_{m_1=0}^{3} \frac{50!}{m_1! \times (50-m_1)!} p^{m_1}(1-p)^{(50-m_1)}$$

$$= 1 - \sum_{m_1=0}^{3} \frac{50!}{m_1! \times (50-m_1)!} (0.01)^{m_1}(1-0.01)^{(50-m_1)}$$

$$= 0.0016$$

Other probabilities when $p = 0.02, 0.03, 0.04, 0.05, 0.06, 0.07$ are calculated similarly.

c) Let $p_a^2$ denote the probability of accepting the lot on the second sample. Then $p_a^2$ at $p = 0.01$ is given by

$$p_a^2 = P(m_1 = 2, m_2 \le 1) + P(m_1 = 3, m_2 = 0)$$
$$= P(m_1 = 2) \times P(m_2 \le 1) + P(m_1 = 3) \times P(m_2 = 0)$$

$$= \frac{50!}{2! \times 48!} p^2(1-p)^{48} \times \sum_{m_2=0}^{1} \frac{100!}{m_2! \times (100-m_2)!} p^{m_2}(1-p)^{(100-m_2)}$$

$$+ \frac{50!}{3! \times 47!} p^3(1-p)^{47} \times \frac{100!}{0! \times 100!} p^0(1-p)^{100}$$

$$= \frac{50!}{2! \times 48!} (0.01)^2(1-0.01)^{48}$$

$$\times \sum_{m_2=0}^{1} \frac{100!}{m_2! \times (100-m_2)!} (0.01)^{m_2}(1-0.01)^{(100-m_2)}$$

$$+ \frac{50!}{3! \times 47!} (0.01)^3(1-0.01)^{47} \times \frac{100!}{0! \times 100!} (0.01)^0(1-0.01)^{100}$$

$$= 0.0756 \times 0.7358 + 0.0122 \times 0.3660$$
$$= 0.0556 + 0.0045$$
$$= 0.0601$$

d) Let $p_a$ denote the probability of acceptance on the combined samples, and let $p_a^1$ and $p_a^2$ denote the probabilities of acceptance on the first and second samples, respectively. Then it is obvious that we have

$$p_a = p_a^1 + p_a^2$$

Thus, for example, $p_a$ at $p = 0.01$ is given by

$$p_a = 0.9106 + 0.0601$$
$$= 0.9707$$

Other probabilities when $p = 0.02, 0.03, 0.04, 0.05, 0.06, 0.07$ are calculated similarly.

e) In acceptance sampling plans, as discussed in Section 9.5, rejected lots usually go through a screening process where all the defective units are sorted out and reworked or replaced with good units. In such cases, the average fraction defective in the lots that customers receive is usually quite low, or we may say the AOQ is even lower than the incoming lot fraction defective. The AOQ points in double acceptance sampling at a given lot fraction (percent) defective $p$ are given by

$$AOQ = \frac{\left(p_a^1(N - n_1) + p_a^2(N - n_1 - n_2)\right)p}{N} \quad (9.4)$$

If we assume the process is continuous or we assume that the lot size $N$ is very large, then Eq. (9.4) reduces to

$$AOQ = (p_a^1 + p_a^2)p = p_a \times p \quad (9.5)$$

Thus, in the case where $N$ is very large, we can determine the AOQ points easily by multiplying the probability of acceptance $p_a$ with the lot fraction defective $p$.

All the probabilities, including the AOQ points and the probabilities of accepting and rejecting a lot on the first sample, accepting a lot on the second sample, and accepting a lot on a combined sample, are summarized in the Table 9.2 when the lot percents or fractions defective are $p = 0.01, 0.02, 0.03, 0.04, \cdots, 0.06, 0.07$.

f) The various OC curves, including the AOQ curve, are as shown in Figure 9.7. As discussed earlier in this chapter, the **AOQL** is the highest value on the AOQ curve. Thus, for example, from the AOQ curve given (or from the previous table), we see that the AOQL for the sampling plan in this example is 1.839%, which is less than 3% of the level of incoming lot percent defective.

**Table 9.2** $p_a^1, p_r^1, p_a^2, p_a$ probabilities and AOQ% vs. lot percent defective $p$

| $p$ | 0.01 | 0.02 | 0.03 | 0.04 | 0.05 | 0.06 | 0.07 |
|---|---|---|---|---|---|---|---|
| $p_a^1$ | 0.9106 | 0.7358 | 0.5553 | 0.4005 | 0.2794 | 0.1900 | 0.1265 |
| $p_r^1$ | 0.0016 | 0.0178 | 0.0628 | 0.1391 | 0.2396 | 0.3527 | 0.4673 |
| $p_a^2$ | 0.0.0601 | 0.0830 | 0.0577 | 0.0272 | 0.0110 | 0.0039 | 0.0013 |
| $p_a = p_a^1 + p_a^2$ | 0.9707 | 0.8188 | 0.6130 | 0.4277 | 0.2904 | 0.1939 | 0.1278 |
| AOQ% | 0.9707 | 1.6376 | 1.8390 | 1.7108 | 1.4520 | 1.1634 | 0.8946 |

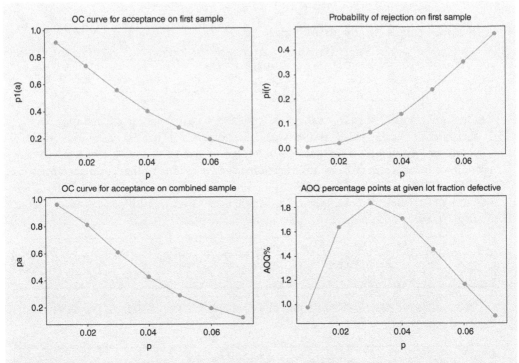

**Figure 9.7** OC curves for acceptance and rejection on the first sample and for acceptance on a combined sample, and AOQ percentage points at a given lot fraction defective.

**Example 9.6**
Refer to the acceptance sampling plan given in Example 9.5. Under this plan, find the value of the ATI for lots with fractions defective of $p = 0.01, 0.02, 0.03, 0.04, \cdots, 0.06, 0.07$. Assume that the lot size $N = 10,000$.

**Solution**

The ATI is given by

$$\text{ATI} = = n_1 p_a^1 + (n_1 + n_2)p_a^2 + N(1 - p_a)$$

Thus, for example, when $p = 0.01$, the ATI is equal to (refer to Table 9.2 in Example 9.5)

$$\text{ATI} = 50(0.9106) + (50 + 100)(0.0601) + 10\,000(1 - 0.9707)$$

$$= 45.53 + 9.02 + 293 = 347.55 = 348$$

Other ATI values at lot fractions defective $p = 0.02, 0.03, 0.04, \cdots, 0.06, 0.07$ are calculated similarly. Thus, we have

| $p$ | 0.01 | 0.02 | 0.03 | 0.04 | 0.05 | 0.06 | 0.07 |
|-----|------|------|------|------|------|------|------|
| ATI | 348 | 955 | 1971 | 2886 | 3564 | 4041 | 4368 |

Note that as the incoming lot fraction defective increases, the ATI value also increases.

### 9.7.3  Multiple-Sampling Plans

Multiple sampling plans work the same way as double sampling with an extension of the number of samples to be taken up to *seven*, according to ANSI/ASQ Z1.4-2003. In the same manner that double sampling is performed, acceptance or rejection of submitted lots may be reached before the seventh sample, depending on the acceptance/rejection criteria established for the plan.

### 9.7.4  Average Sample Number

The average sample number (ASN) is the expected average number of inspections per lot for a given sampling plan. The ASN for single sampling plans is a constant value that is equal to the single sample size for the plan. The ASN for double sampling plans is the sum of the first sample size plus the second sample size times the probability that a second sample will be required. The ASN is also a function of incoming lot fraction of nonconforming units when working with a double-sampling plan. In a double-sampling plan, the ASN is found by using the following formula.

$$\text{ASN} = n_1 + n_2\left(1 - p_a^1 - p_r^1\right) \tag{9.6}$$

where:

$n_1$ = size of the first sample
$n_2$ = size of the second sample
$p_a^1$ = probability of accepting a lot on the first sample
$p_r^1$ = probability of rejecting a lot on the first sample
$(1 - p_a^1 - p_r^1)$ = probability of requiring a second lot

**Example 9.7**
Refer to the double-sampling plan considered in Example 9.5. That is, $n_1 = 50$, $c_1 = 1$, $n_2 = 100$, $c_2 = 3$. Find the ASN when the incoming lot fraction defective is 2%.

**Solution**

Using the formula given in Eq. (9.6) and the values of $p_a^1$ and $p_r^1$ from Table 9.2, we have

$$\text{ASN} = 50 + 100\,(1 - 0.7358 - 0.0178) = 50 + 24.64 = 75$$

The ASN values at other values of incoming lot fraction defective rate $p$ are calculated in the same manner.

By calculating the ASN at other values of incoming lot fraction defective rate $p$, it can easily be seen that, when comparing sampling plans with equal protection, double-sampling plans will generally result in smaller average sample sizes when the quality is either excellent or poor. When quality is near the indifference level, double sampling could result in a greater ASN, but only rarely.

### 9.7.5  Sequential-Sampling Plans

A sequential-sampling acceptance plan is an extension of double/multiple-sample acceptance plans. Sequential-sampling plans were popularized by Wald (1973). In such a plan, a sequence of samples is taken until the decision about the fate of the lot is made. In other words, sampling could continue until all the units in the lot are inspected. In practice, however, sampling is terminated when the number of units inspected is equal to three times the number of units that would have been inspected in a single-sampling plan. In sequential-sampling plans, the sample size could be either one or greater than one. Sequential plans with sample size equal to one are called *item-by-item* sequential-sampling plans. These sampling plans are based on Wald's sequential probability ratio test.

In sequential sampling plan, the total number of units inspected is plotted against the total number of defective units observed at the time of the last inspected unit. As shown in Figure 9.8, the graph consists of two parallel lines, called the *acceptance zone line* and the *rejection zone line*. They divide the x-y plane into three zones, called the *acceptance zone*, the *rejection zone*, and the *continue-sampling zone*. If the plotted point falls below the acceptance line or in the acceptance zone, then the lot is accepted; if it falls above the rejection zone line or in the rejection zone, then the lot is rejected; otherwise, sampling continues.

The y-axis scales the number of nonconforming units in the total number of units inspected, while the x-axis scales the total number of units inspected. The equations for the acceptance zone line and the rejection zone line are

$$\text{Acceptance zone line: } y_A = -h_1 + sn$$

$$\text{Rejection zone line: } y_R = h_2 + sn$$

where $n$ is the sample size, and $h_1$, $h_2$, and $s$ are defined as follows:

$$h_1 = \frac{\log((1-\alpha)/\beta)}{d}$$

$$h_2 = \frac{\log((1-\beta)/\alpha)}{d}$$

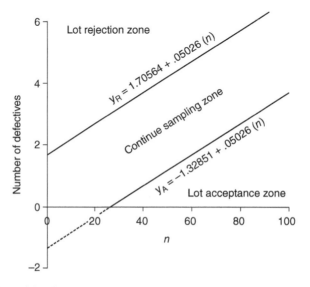

**Figure 9.8** Graphical device for the sequential plan in Example 9.8.

$$s = \frac{\log\left((1-p_1)/(1-p_2)\right)}{d}$$

$$d = \log\left(p_2(1-p_1)/p_1(1-p_2)\right)$$

A sequential-sampling plan is determined by following four parameters:

$\alpha$ = producer's risk
AQL = acceptable quality level = $p_1$
$\beta$ = consumer's risk
RQL = rejectable (or unacceptable) quality level = $p_2$

In the following example, we use $\alpha = 0.05$, AQL = $p_1 = 0.02$, $\beta = 0.10$, and RQL = $p_2 = 0.10$. This results in a plan that will have a 5% chance of rejecting a lot whose lot fraction defective is 2% and a 10% chance of accepting a lot whose lot fraction defective is 10%.

**Example 9.8**
Suppose we would like to develop a sequential sampling plan with the following parameter values: $\alpha = 0.05$, AQL = $p_1 = 0.02$, $\beta = 0.10$, and RQL = $p_2 = 0.10$.

**Solution**

Using the parameter values given in this example, we have

$$\begin{aligned} d &= \log\left((0.1(1-0.02))/(0.02(1-0.1))\right) \\ &= \log\left(0.098/0.018\right) \\ &= 0.735954 \end{aligned}$$

$$\begin{aligned} h_1 &= \log\left(0.95/0.10\right)/0.735954 \\ &= \log\left(9.5\right)/0.735954 \\ &= 1.32851 \end{aligned}$$

$$\begin{aligned} h_2 &= \log\left(0.90/0.05\right)/0.735954 \\ &= \log\left(18\right)/0.735954 \\ &= 1.70564 \end{aligned}$$

$$\begin{aligned} s &= \log\left(0.98/0.90\right)/0.735954 \\ &= \log\left(1.088888\right)/0.735954 \\ &= 0.05025 \end{aligned}$$

Thus, the acceptance and rejection zone lines are

$$y_A = -1.32851 + 0.05026(n) \tag{9.7}$$

$$y_R = 1.70564 + 0.05026(n) \tag{9.8}$$

Note here that to make any decision about the fate of a lot, we can use either the graph shown in Figure 9.8 as a device or the table format (*item-by-item*) given in Table 9.3. Note that in

Table 9.3, A denotes the acceptance number, R denotes the rejection number, and $n$ is the number of items inspected. Furthermore, note that A and R should be greater than or equal to zero and that $n = 1, 2, 3, \cdots$. To construct Table 9.3, we proceed as follows:

**Table 9.3** Item-by-item sequential-sampling plan with $\alpha = 0.05$, $p_1 = .02$, $\beta = 0.10$, $p_2 = .10$.

| Number of items inspected, n | A | R | Number of items inspected, n | A | R | Number of items inspected, n | A | R |
|---|---|---|---|---|---|---|---|---|
| 1 | a | 2 | 21 | a | 3 | 41 | 0 | 4 |
| 2 | a | 2 | 22 | a | 3 | 42 | 0 | 4 |
| 3 | a | 2 | 23 | a | 3 | 43 | 0 | 4 |
| 4 | a | 2 | 24 | a | 3 | 44 | 0 | 4 |
| 5 | a | 2 | 25 | a | 3 | 45 | 0 | 4 |
| 6 | a | 3 | 26 | a | 4 | 46 | 0 | 5 |
| 7 | a | 3 | 27 | 0 | 4 | 47 | 1 | 5 |
| 8 | a | 3 | 28 | 0 | 4 | 48 | 1 | 5 |
| 9 | a | 3 | 29 | 0 | 4 | 49 | 1 | 5 |
| 10 | a | 3 | 30 | 0 | 4 | 50 | 1 | 5 |
| 11 | a | 3 | 31 | 0 | 4 | 51 | 1 | 5 |
| 12 | a | 3 | 32 | 0 | 4 | 52 | 1 | 5 |
| 13 | a | 3 | 33 | 0 | 4 | 53 | 1 | 5 |
| 14 | a | 3 | 34 | 0 | 4 | 54 | 1 | 5 |
| 15 | a | 3 | 35 | 0 | 4 | 55 | 1 | 5 |
| 16 | a | 3 | 36 | 0 | 4 | 56 | 1 | 5 |
| 17 | a | 3 | 37 | 0 | 4 | 57 | 1 | 5 |
| 18 | a | 3 | 38 | 0 | 4 | 58 | 1 | 5 |
| 19 | a | 3 | 39 | 0 | 4 | 59 | 1 | 5 |
| 20 | a | 3 | 40 | 0 | 4 | 60 | 1 | 5 |

1) In Eqs. (9.7) and (9.8), plug in the values of $n = 1, 2, 3, \cdots$.
2) Equations 9.7 and 9.8 give the values of A and R, respectively. The values of A are rounded to the next integer less than or equal to $y_A$, whereas values of R are rounded to the next integer greater than or equal to $y_R$. The concept of rounding these numbers becomes very clear by referring to the acceptance and rejection zones in the graph in Figure 9.8. If the value of A is negative, then we replace it with the letter $a$ and declare that acceptance is not possible since we cannot have a negative number of defectives.

As can be seen from the table, acceptance of the lot is not possible until the 27th sample unit is inspected, and rejection of the lot is not possible until the 2nd unit is inspected. The probability of acceptance $p_a$ at $p = p_1$, $p = p_2$ is equal to $1 - \alpha$ and $\beta$, respectively.

Clearly, these two points would fall on the OC curve. The third point on the OC curve can be found by plotting the value of $p_a$ at $p = s$, where $p_a = h_2/(h_1 + h_2)$ (Statistical Research Group 1945). Thus, by joining these three points, we obtain the OC curve. The values of ASN, AOQ, and ATI for a sequential plan are not discussed here. However, the interested reader may refer to Montgomery (2013) and Duncan (1986).

## 9.8 ANSI/ASQ Z1.4-2003 Sampling Standard and Plans

In this section, we discuss sampling standard ANSI/ASQ Z1.4-2003. This standard is an acceptance sampling system to be used with switching rules on a continuing stream of lots for the AQL specified. It provides tightened, normal, and reduced sampling plans to be applied for attributes inspection for percent nonconforming or nonconformities per 100 units. Another standard plan similar to ANSI/ASQ Z1.4-2003 is Military Standard 105E. Note that the name of the standard was changed from ANSI/ASQC Z1.4 to ANSI/ASQ Z1.4: the letter *C* was dropped to reflect the 1997 name change of the American Society for Quality Control to its current name, the American Society for Quality.

> Note: ANSI/ASQ Z1.4-2003 Tables I, VIII, II-A, III-A, and IV-A and ANSI/ASQ Z1.9-2003 Tables A-2, C-1, B-3, and B-5 of the standards are reproduced with permission in Tables 9.4–9.12, respectively, at the end of this chapter.

ANSI/ASQ Z1.4-2003 is a revision of ANSI/ASQC Z1.4-1993. It consists of sampling procedures and tables for inspection by attributes. As mentioned, it is quite similar to Military Standard 105E. ANSI/ASQ Z1.4-2003 is a collection of sampling schemes, whereas the acceptance sampling plans discussed in earlier sections were individual acceptance sampling plans. Because ANSI/ASQ Z1.4-2003 is a collection of acceptance sampling schemes, it is usually known as an acceptance sampling system that is used with switching rules on a continuous process or a continuous stream of lots with a specified AQL. It allows tightened, normal, and reduced plans to be applied for attributes inspection for percent nonconforming or nonconformities per 100 units.

We define again here the acceptance number, acceptance quality level (AQL), and acceptance quality limit (AQL):

- The *acceptance number* (c) is the maximum number of nonconforming units in the sample that will permit the acceptance of a lot.
- The *acceptance quality level* (AQL) is defined as the maximum percent nonconforming products (or the maximum number of nonconformities per 100 units) that for sampling inspection can be considered satisfactory as a process average.
- The *acceptance quality limit* (AQL) is defined as the quality limit that is the worst tolerance process average when a continuing series of lots is submitted for acceptance sampling plan.

Note that both the acceptance quality *level* and the acceptance quality *limit* are abbreviated AQL.

The concept of an AQL applies only when an acceptance sampling scheme with rules for switching between normal, tightened, reduced, and discontinuance of sampling inspection is used. These rules are designed to encourage suppliers to have process averages consistently better than the AQL. If suppliers fail to do so, there is a high probability of being switched from normal inspection to tightened inspection where lot acceptance becomes more difficult. Once on tightened inspection, and unless corrective action is taken to improve product quality, it is very likely that the rule requiring discontinuance of sampling inspection will be invoked (see this rule here).

As noted earlier, ANSI/ASQ Z1.4-2003 is a revision of ANSI/ASQC Z1.4-1993 and includes the following changes:

- The definition of AQL was changed from acceptance quality *level* to acceptance quality *limit*.
- The discontinuation-of-inspection rule has been changed from 10 consecutive lots on tightened inspection to 5 consecutive lots not accepted on tightened inspection.

- ANSI/ASQ Z1.4-2003 makes it optional to use the limit numbers for switching to reduced. In addition, ANSI/ASQ Z1.4-2003 contains additional OC curves called *scheme OC curves* that describe the protection provided by the switching procedure during periods of constant quality.

Other than the previous changes and some revisions in the footnotes of some tables, all tables, table numbers, and procedures used in MIL-STD-105E (which was cancelled in 1995) and ANSI/ASQC Z1.4-1993 have been retained.

ANSI/ASQ Z1.4-2003 is probably the most commonly used standard for attribute sampling plans. Its wide recognition and acceptance of the plan could be due to government contracts stipulating the standard, rather than its statistical importance. Producers submitting products at a nonconformance level within AQL have a high probability of having the lot accepted by the customer.

When using ANSI/ASQ Z1.4-2003, the characteristics under consideration should be classified. The general classifications are critical, major, and minor nonconformities or defects:

- *Critical defect*: A defect that judgment and experience indicate is likely to result in hazardous or unsafe conditions for the individuals using, maintaining, or depending on the product; or a defect that judgment and experience indicate is likely to prevent the performance of the unit. In practice, critical characteristics are commonly inspected to an AQL level of 0.40–0.65% if not 100% inspected. 100% inspection is recommended for critical characteristics if possible. Acceptance numbers are always zero for critical defects.
- *Major defect*: A defect, other than critical, that is likely to result in failure or to reduce the usability of the unit of product materially for its intended purpose. In practice, AQL levels for major defects are generally about 1%.
- *Minor defect*: A defect that is not likely to reduce the usability of the unit of product materially for its intended purpose. In practice, AQL levels for minor defects generally range from 1.5 to 2.5%.

### 9.8.1 Levels of Inspection

Seven levels of inspection are used in ANSI/ASQ Z1.4-2003: reduced inspection, normal inspection, tightened inspection, and four levels of special inspection. The special inspection levels should only be used when small sample sizes are necessary and large risks can be tolerated. When using ANSI/ASQ Z1.4-2003, a set of *switching rules* must be followed as to the use of reduced, normal, and tightened inspection.

The following guidelines are taken from ANSI/ASQ Z1.4-2003:

- *Initiation of inspection*: Normal inspection Level II (**normal inspection level II** is the default **inspection level**). ISO 2859 states for **normal inspection**: "**normal inspection** is used when there is no reason to suspect that the quality **level** differs from an acceptable **level**." Normal inspection will be used at the start of inspection unless otherwise directed by the responsible authority.
- *Continuation of inspection*: Normal, tightened, or reduced inspection shall continue unchanged for each class of defects or defectives on successive lots or batches except where the following switching procedures require change. The switching procedures shall be applied to each class of defects or defectives independently.
- *Switching procedures*: Switching rules are shown graphically in Figure 9.9.
- *Normal to tightened*: When normal inspection is in effect, tightened inspection shall be instituted when two out of five consecutive lots or batches have been rejected on original inspection (that is, ignoring resubmitted lots or batches for this procedure).

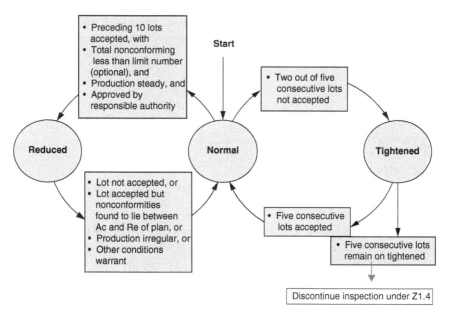

**Figure 9.9**   Switching rules for normal, tightened, and reduced inspection.

- *Tightened to normal:* When tightened inspection is in effect, normal inspection shall be instituted when five consecutive lots or batches have been considered acceptable on original inspection.
- *Normal to reduced*: When normal inspection is in effect, reduced inspection shall be instituted provided all of the following conditions are satisfied:
  1) The preceding 10 lots or batch (or more), as indicated by the note on ANSI/ASQ Z1.4-2003 Table VIII (Table 9.4), have been on normal inspection, and none has been rejected on original inspection.
  2) The total number of defectives (or defects) in the sample from the preceding 10 lots or batches (or such other number as was used for condition (1)) is equal to or less than the applicable number given in Table VIII of ANSI/ASQ Z1.4-2003. If a double or multiple sampling plan is in use, all samples inspected should be included, not "first" samples only.
  3) Production is at a steady rate. That is, there is no interruption in production due to any machine failure or any other such cause.
  4) Reduced inspection is considered desirable by the responsible authority.

- *Reduced to normal*: When reduced inspection is in effect, normal inspection shall be instituted if any of the following occur on original inspection:
  1) A lot or batch is rejected.
  2) A lot or batch is considered acceptable under reduced inspection, but the sampling procedures terminated without either acceptance or rejection criteria having been met. In these circumstances, the lot or batch will be considered acceptable, but normal inspection will be reinstated, starting with the new lot or batch.
  3) Production becomes irregular or delayed.
  4) Other conditions warrant that normal inspection be instituted.

- *Discontinuation of inspection*: If the cumulative number of lots not accepted in a sequence of consecutive lots on tightened inspection reaches five, the acceptance procedures of this standard shall be discontinued. Inspection under the provisions of this standard shall not be resumed until corrective action has been taken by the supplier. Tightened inspection shall then be used as "normal to tightened."

## 9.8.2 Types of Sampling

ANSI/ASQ Z1.4-2003 allows three types of sampling:

- Single sampling
- Double sampling
- Multiple sampling

The choice of the type of plan depends on many variables. Single sampling is the easiest to administer and perform but usually results in the largest ATI. Double sampling in ANSI/ASQ Z1.4-2003 results in a lower ATI than single sampling. However, as noted earlier in this chapter, a double-sampling plan requires more decisions to be made, such as:

- Accept the lot after the first sample
- Reject the lot after the first sample
- Take a second sample
- Accept the lot after the second sample
- Reject the lot after the second sample

Multiple sampling further reduces the ATI but also increases the number of decisions to be made. As many as seven samples may be required before a decision to accept or reject the lot can be made. This type of plan requires the most administration.

A general procedure for selecting plans from ANSI/ASQ Z1.4-2003 is as follows:

1) Select the AQL.
2) Select the inspection level.
3) Determine the lot size.
4) Find the appropriate sample size code letter. See ANSI/ASQ Z1.4-2003 Table 1 (Table 9.5).
5) Determine the type of sampling plan to be used: single, double, or multiple.
6) Using the selected AQL and sample size code letter; enter the appropriate table to find the desired plan to be used.
7) Determine the normal, tightened, and reduced plans as required from the corresponding tables.

**Example 9.9**
Suppose that a product is submitted in lots of size $N = 2000$ units. Suppose they are to be inspected at an AQL level of 1.5%. Determine the appropriate single, double, and multiple sampling plans for general inspection Level II.

**Solution**

To define the ANSI/ASQ Z1.4-2003 plans are as follows:

1) Table I (Table 9.5) on page 10 of ANSI/ASQ Z1.4-2003, under general inspection level II, indicates that an appropriate sample size code letter is K.

2) From Table II-A (Table 9.6) of ANSI/ASQ Z1.4-2003 on page 11 of the standard, for a *single-sampling* inspection plan under general inspection level II, we have $n = 125$, $c = 5$.
3) From Table III-A (Table 9.7) of ANSI/ASQ Z1.4-2003 on page 14 of the standard, for a *double-sampling* inspection plan under general inspection level, 2 samples of size 80 will be required.
4) From Table IV-A (Table 9.8) of ANSI/ASQ Z1.4-2003 on page 17 of the standard, for a *multiple-sampling* inspection plan under general inspection level II, at least 2 samples of 32 are required, and it may take up to 7 samples of 32 before an acceptance or rejection decision is made.

Complete single, double, and multiple sampling plans under normal inspection level II are as follows:

| Sampling plan | | Sample(s) size | AC | RE |
|---|---|---|---|---|
| Single sampling | | 125 | 5 | 6 |
| Double sampling | First | 80 | 2 | 5 |
| | Second | 80 | 6 | 7 |
| Multiple sampling | First | 32 | * | 4 |
| | Second | 32 | 1 | 5 |
| | Third | 32 | 2 | 6 |
| | Fourth | 32 | 3 | 7 |
| | Fifth | 32 | 5 | 8 |
| | Sixth | 32 | 7 | 9 |
| | Seventh | 32 | 9 | 10 |

*AC = acceptance number (AC), RE = rejection number (RE).*
*Acceptance not permitted at this sample size.*

## 9.9 Dodge-Romig Tables

Dodge-Romig tables were designed as sampling plans to minimize ATI. These plans require an accurate estimate of the process average nonconforming when selecting the sampling plan to be used. Dodge-Romig tables use the AOQL and LTPD values for plan selection, rather than AQL as in ANSI/ASQ Z1.4-2003. For single and double sampling plans using each of these approaches, tables are available in Dodge and Romig (1959). When the process average nonconforming is controlled to requirements, Dodge-Romig tables result in lower ATI, but rejection of lots and sorting tend to minimize the gains if process quality deteriorates. We do not cover any further details of Dodge-Romig tables in this book.

Note that if the process average nonconforming shows statistical control, then acceptance sampling should not be used. The most economical course of action in this situation is to implement either no inspection or 100% inspection, according to Deming (1986).

## 9.10 ANSI/ASQ Z1.9-2003 Acceptance Sampling Plans by Variables

*Variables sampling plans* use the actual measurements of sample products for decision-making rather than classifying products as conforming or nonconforming, as in the case of attribute sampling plans. Variables sampling plans are more complex to administer than attribute plans; thus they require more skill. The cost of observations in variable sampling is greater than that of attribute sampling plans, but this cost is usually offset by the reduced sample sizes, which are less than 60% of the sample size for an attribute-sampling plan with equal protection. This, of course, depends greatly on whether the population standard deviation is known or unknown. Variable sampling plans are used to control the process fraction defective as well as to control the process mean.

The most common standard for variables sampling, which we discuss in this chapter, is ANSI/ASQ Z1.9-2003, which has plans for when: (i) variability is known, (ii) variability is unknown but estimated using the standard deviation method, and (iii) variability is unknown but estimated using the range method. Such a sampling plan can be used to test for a single specification limit, a double (or bilateral) specification limit, the estimation of the process average, and the estimation of the dispersion of the parent population.

Here, we give two of the benefits that variable sampling plans offer over attribute sampling plans as discussed by Neubauer and Luko (2013a, 2013b):

- Variables sampling plans for protection equal to attribute sampling requires smaller sample sizes. As mentioned, several types of variables sampling plans are in use, three of which are for when (i) $\sigma$ is known, (ii) $\sigma$ is unknown but can be estimated using the sample standard deviation s, and (iii) $\sigma$ is unknown but can be estimated using the range method. Furthermore, in these plans, the sample size is smaller when $\sigma$ is known than when it is unknown.
- Variables sampling allows the determination of how close to a specification limit the process is performing. Attribute sampling plans either accept or reject a lot, whereas variable sampling plans give information about how well or how poorly the process is performing.

Variables sampling plans such as ANSI/ASQ Z1.9-2003 have some disadvantages and limitations:

- They rely on the assumption of normality of the population from which the samples are being drawn.
- Unlike attribute sampling plans, separate characteristics on the same units will have different averages and dispersions, resulting in a separate sampling plan for each characteristic.
- Variables plans are more complex to administer.
- Variables gauging is generally more expensive than attributes gauging.

### 9.10.1 ANSI/ASQ Z1.9-2003 – Variability Known

ANSI/ASQ Z1.9-2003 is a revision of ANSI/ASQC Z1.9-1993. Included in the revision was a change of the term acceptable quality *level* to acceptable quality *limit*, a change in the definition and explanation of AQL, and a change to the discontinuation of the inspection rule, as explained previously for ANSI/ASQ Z1.4-2003.

As in ANSI/ASQ Z1.4-2003, several AQL levels are used, and specific switching procedures for normal, reduced, and tightened inspection are followed. ANSI/ASQ Z1.9-2003 allows for the same AQL value for each specification limit plan or double specification limits plan or the use of different AQL values for each specification limit. The AQL values are designated $ML$ for the lower specification limit and $MU$ for the upper specification limit.

Two forms are used for every specification limit in an ANSI/ASQ Z1.9-2003 plan: Form 1 and Form 2. Form 1 provides only acceptance or rejection criteria, whereas Form 2 estimates the percent below the lower specification and the percent above the upper specification limit. These percentages are compared to the AQL for acceptance/rejection criteria. Figure 9.10 summarizes the structure and organization of ANSI/ASQ Z1.9-2003.

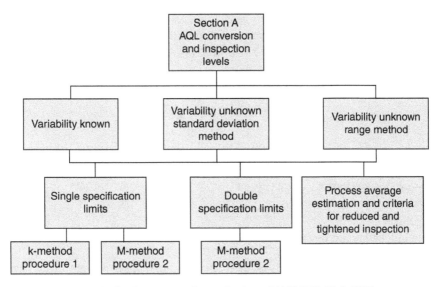

**Figure 9.10**  Structure and organization of ANSI/ASQ Z1.9-2003.

Various AQL levels are used in ANSI/ASQ Z1.9-2003 that are consistent with the AQL levels used in ANSI/ASQ Z1.4-2003. Section A of ANSI/ASQ Z1.9-2003 contains both an AQL conversion table and a table for selecting the desired inspection level. General Inspection Level II should be used unless otherwise specified. See Section A7.1 of the standard for further information about levels.

Table A-3 on page 7 of ANSI/ASQ Z1.9-2003 contains the OC curves for the sampling plans in Sections B, C, and D:

- *Section B* contains sampling plans used when the variability is unknown but estimated using the standard deviation method. Part I is used for a single specification limit; Part II is used for a double specification limit; and Part III is used for the estimation of a process average and criteria for reduced and tightened inspection.
- *Section C* contains sampling plans used when the variability is unknown but estimated using the range method. Parts I, II, and III are the same as Parts I, II, and III in Section B.
- *Section D* contains sampling plans used when variability is known. Parts I, II, and III are the same as Parts I, II, and III in Section B.

As previously mentioned, if the standard deviation, $\sigma$, is unknown, then two methods for acceptance sampling plan designs are detailed in ANSI/ASQ Z1.9-2003. One method is to design the sampling plan by using sample standard deviation $s$ to estimate $\sigma$; the other method is to design the sampling plan based on the sample range R method. In this book, we will briefly discuss both of these methods.

### 9.10.2   Variability Unknown – Standard Deviation Method

In this section, a sampling plan is shown for the situation where the variability is not known and the sample standard deviation $s$ is used to estimate the population standard deviation $\sigma$. The sampling plan for a double specification limit is found in Section B of the standard with one AQL value for both upper and lower specification limits combined.

The acceptability criterion is based on comparing an estimated percent of nonconforming products to a maximum allowable percent nonconforming products for the given AQL level. The estimated percent nonconforming is found in ANSI/ASQ Z1.9-2003 Table B-5 (Table 9.12).

The quality indices for this sampling plan are

$$Q_u = \frac{U - \overline{X}}{S} \text{ and } Q_L = \frac{\overline{X} - L}{S}$$

where

$U$ = upper specification limit
$L$ = lower specification limit
$\overline{X}$ = sample mean
$S$ = estimate of lot standard deviation

The quality level of the lot is in terms of the lot percent nonconforming products. Three values are calculated: $P_U$, $P_L$, and $P$. $P_U$ is an estimate of nonconformance with the upper specification limit; $P_L$ is an estimate of nonconformance with the lower specification limit; and $P$ is the sum of $P_U$ and $P_L$.

The value of $P$ is then compared with the maximum allowable percent nonconforming products. If $P$ is greater than or equal to $M$ in Table B-5 of ANSI/ASQ Z1.9-2003 (Table 9.12), or if either $Q_U$ or $Q_L$ is negative, then the lot is rejected. Note that if either $Q_U$ or $Q_L$ is negative, that means $\overline{X}$ falls beyond the specification limits. The following example illustrates this procedure.

### Example 9.10 Operation Temperature Data

The minimum temperature of operation for a certain component is specified as 150 °F. The maximum temperature is 185 °F. A lot of 100 items is submitted for inspection using General Normal Inspection Level II with AQL = 1%. ANSI/ASQ Z1.9-2003 Table A-2 (Table 9.9) gives code letter F, which results from ANSI/ASQC Z1.9-2003 Table B-3 (Table 9.11), in a sample size of 10. Note that Table B-3 is for normal and tightened inspection for plans based on unknown variability (double specification limits Form 2 and single specification limit Form 1). In this example, we have double specification limits. The resulting 10 measurements in degrees Fahrenheit are as follows:

| 182 | 168 | 167 | 185 | 166 | 167 | 164 | 153 | 160 | 164 |
|-----|-----|-----|-----|-----|-----|-----|-----|-----|-----|

Determine whether the lot meets the acceptance criteria.

### Solution

Using the following formula (see Eq. (3.9) in Chapter 3),

$$S^2 = \frac{1}{n-1}\left(\sum X_i^2 - \frac{(\sum X_i)^2}{n}\right)$$

we can easily determine that the sample mean and sample standard deviation for the given data are $\overline{X} = 167.60$ and $S = 9.49$, respectively. Now we are given the upper specification limit U = 185 and the lower specification limit $L = 150$. These specification limits result in the following quality indices:

$$Q_U = \frac{185 - 167.60}{9.49} = 1.83$$

$$Q_L = \frac{167.60 - 150}{9.49} = 1.85$$

Now, using Table B-5 of the standard ANSI/ASQ Z1.9 (Table 9.12), estimates for the lot percent nonconforming products above $U$ and below $L$ are equal to $P_U = 2.25$ and $P_L = 2.09$, respectively. Thus, the combined estimate percent nonconforming in the lot is equal to

$$P = P_U + P_L = 2.25 + 2.09 = 4.34$$

Now we can find that the *allowable lot percent nonconforming products*, given in Table B-3 of the standard ANSI/ASQ Z1.9 (Table 9.11) is $M = 3.27$. Using the acceptable criterion, the lot obviously does not meet the acceptable criterion since $P = 4.34$ is greater than 3.27.

Note that ANSI/ASQ Z1.9-2003 provides a variety of other examples for variables sampling plans.

### 9.10.3 Variability Unknown – Range Method

In this section, a sampling plan is discussed for when the variability is not known and the range method is used to estimate the unknown variability. To illustrate this plan for a single specification limit, we use an example from Section C of the standard ANSI/ASQ Z1.9.

However, before we work on this example, it is important to note that the tables used are the same as when sample range $R$ is used in place of standard deviation $S$. For samples of size seven or less, the statistics $\frac{U - \overline{X}}{R}$ and $\frac{\overline{X} - L}{R}$ are used in place of $\frac{U - \overline{X}}{S}$ and $\frac{\overline{X} - L}{S}$. For samples of size 10 or greater, the statistics $\frac{U - \overline{X}}{\overline{\overline{R}}}$ and $\frac{\overline{X} - L}{\overline{\overline{R}}}$ are used in place of the statistics $\frac{U - \overline{X}}{S}$ and $\frac{\overline{X} - L}{S}$, where $\overline{R}$ is the averages of the subgroups of five units each taken in the order of production. For more details, see Duncan (1986).

The quality indices for a single specification limit depending on the sample size are

$$\frac{U - \overline{X}}{R} \text{ or } \frac{\overline{X} - L}{R} ; \frac{U - \overline{X}}{\overline{\overline{R}}} \text{ or } \frac{\overline{X} - L}{\overline{\overline{R}}}$$

where

$U$ = upper specification limit
$L$ = lower specification limit
$\overline{X}$ = sample average
$R$ = range when the sample size is seven or less
$\overline{R}$ = average range of the subgroups of size five

The acceptance criterion is a comparison of the quality indices $(U - X) / R$ or $(X - L) / R$, or $(U - \overline{X})/\overline{R}$ or $(\overline{X} - L)/\overline{R}$ to the acceptability constant $k$ shown in Table C-1 (Table 9.10) of the standard. If the calculated quantity is equal to or greater than $k$, then the lot is accepted; if the calculated quantity is negative or less than $k$, then the lot is rejected.

The following example illustrates the use of a plan when the variability is unknown and the range method is employed. The Form I variables sampling plan is similar to the examples from Section C of ANSI/ASQ Z1.9-2003.

**Example 9.11 Electrical Resistance Data**
The lower specification limit for electrical resistance of a certain electrical component is 620 ohms. A lot of 100 items is submitted for inspection. General Inspection Level II, normal inspection, with AQL = 0.4%, is to be used. From ANSI/ASQ Z1.9-2003 Table A-2 (Table 9.9) and Table C-1 (Table 9.10), it is seen that a sample of size 10 is required.

Suppose that values of the sample resistances (reading from left to right or in order of production) are

$$645, 651, 621, 625, 658 \ (R = 658 - 621 = 37)$$
$$670, 673, 641, 638, 650 \ (R = 673 - 638 = 35)$$

Determine compliance with the acceptability criterion.

**Solution**

| Line | Information needed | Explanation | Value |
|---|---|---|---|
| 1 | Sample size: $n$ | 10 | |
| 2 | Sum of measurement: $\Sigma X_i$ | 6472 | |
| 3 | Sample mean $\overline{X} = (\Sigma X_i)/n$ | 6472/10 | 647.2 |
| 4 | Average range $\overline{R} = (\Sigma R_i)/m$, where $m$ is the no. of subgroups | (37 + 35) / 2 | 36 |
| 5 | Specification limit (lower): L | 620 | |
| 6 | Quantity $(\overline{X}-L)/\overline{R}$ | (647.2 − 620) / 36 | 0.756 |
| 7 | Acceptability constant: $k$* | See Table C-1 | 0.811 |
| 8 | Acceptability criterion: Compare $(\overline{X}-L)/\overline{R}$ with $k$ | 0.756 < 0.811 | |

*The table value is found from Table C-1 of ANSI/ASQ Z1.9 (Table 9.10).

The lot does not meet the acceptability criterion, since $(\overline{X}-L)/\overline{R}$ is less than $k$.

Note: If a single upper specification limit, $U$, is given, then compute the quantity $(U-\overline{X})/\overline{R}$ in line 6, and compare it with the value of $k$. The lot meets the acceptability criterion if $(U-\overline{X})/\overline{R}$ is greater than or equal to $k$.

## 9.11 Continuous-Sampling Plans

All of the acceptance sampling plans we have discussed so far are *lot-by-lot* plans. Many production processes do not produce lots, and thus the lot-by-lot acceptance sampling plans discussed earlier cannot be applied. Continuous-sampling plans have been developed for such cases. Such plans usually start with 100% inspection; when a stated number of units are found to be free of nonconformities, 100% inspection is replaced with a sampling inspection plan. Sampling plans remain current until a stated number of nonconforming units are found, at which time 100% inspection is resumed.

In brief, 100% inspection and sampling inspections are alternately applied. The most recent standard for developing continuous sampling is MIL-STD-1235B.

Continuous sampling is characterized by two parameters: $i$ is the *clearance number* or the number of conforming units under 100% inspection, and $f$ is the *fraction inspected*, which is a ratio of units inspected and the total number of units produced passing through the inspection station.

### 9.11.1   Types of Continuous-Sampling Plans

There are two different standards for continuous sampling plans:

- *Dodge's continuous sampling plans*: These include CSP-1 and CSP-2. These plans take the AOQL (average outgoing quality limit) as an index. That is, for every AOQL value, there are different combinations of $i$ and $f$.
- *MIL-STD-1235B*: These plans are selected using a sample size code letter and an AQL value. The standard includes CSP-1, CSP-2, CSP-F, CSP-T, and CSP-V plans.

### 9.11.2   Dodge's Continuous Sampling Plans

As mentioned earlier, these plans include CSP-1 and CSP-2 sampling plans, and they use the AOQL as a quality index.

Dodge's *CSP-1* continuous sampling plan operates as follows for a selected AOQL value. Start with 100% inspection. When $i$ (clearance number) consecutive number of units are found free from nonconformities, then 100% inspection is discontinued, and a fraction $f$ of the units are inspected. In other words, 100% inspection is replaced with a sampling inspection plan whereby only a fraction $f$ of units are selected randomly for inspection. Furthermore, note that the sample units are selected one at a time at randomly from a continuous flow of production. As soon as one sample unit is found nonconforming, 100% inspection is resumed.

Dodge's *CSP-2* continuous sampling plan operates as follows for a selected AOQL value. Start with 100% inspection. If $i$ consecutive units are found free from nonconformities, then 100% inspection is replaced with a sampling inspection whereby only a fraction $f$ of units is randomly selected for inspection. CPS-2 is a modification of the CPS-1 plan in that 100% inspection resumes only when the number of conforming sampled units between two nonconforming sampled units is less than a specified number $m$. It is quite common to use $m$ equal to the clearance number $i$. If the number of conforming units between two nonconforming units is greater than or equal to $m$, then the inspection plan continues.

### 9.11.3   MIL-STD-1235B

This standard uses the same parameters, $i$ and $f$, as defined in Section 9.11.2. The standard includes CSP-1, CSP-2, CSP-F, CSP-V, and CSP-T plans.

The *CSP-1* and *CSP-2* plans operate in the same way as Dodge's CSP-1 and CSP-2 plans, but they are selected based on a sample size code letter and an AQL value as a quality index. The sample size code letter is selected based on the number of units in the production interval.

The *CSP-F* plan works the same way as CSP-1 plan, providing alternate sequences of 100% and sampling inspection procedures, but the difference in CSP-F is that the AOQL and number of units in the production interval are used to characterize the plan. CSP-F is a single-level continuous sampling scheme. Moreover, CSP-F plan is applied to a relatively short run of production and therefore permit smaller clearance numbers to be used.

The *CSP-T* plan is multi-level continuous sampling inspection plan, which provides a reduced sampling fraction once the product shows superior quality. A CSP-T plan works as follows:

1) Start with 100% inspection. When $i$ (clearance number) consecutive number of units are found free from nonconformities, 100% inspection is then replaced with sampling inspection. At that point, a fraction $f$ of units is randomly selected for inspection. If one sampled unit is found nonconforming, then 100% inspection resumes. If $i$ consecutive sampled units are found free from nonconformities, then the fraction $f$ is reduced to $f/2$.
2) If one sampled unit is found nonconforming, then 100% inspection resumes. If $i$ consecutive sampled units are found free from nonconformities, then the fraction $f/2$ is reduced to $f/4$.
3) If one sampled unit is found nonconforming, then 100% inspection resumes.

The *CSP-V* plan works the same way as CSP-T, but with a reduced $i$ instead of a reduced $f$. The procedure is as follows:

1) Start with 100% inspection. When $i$ (clearance number) consecutive number of units are found free from nonconformities, 100% inspection is replaced with sampling inspection. A fraction $f$ of units is randomly selected for inspection.
2) If one sampled unit is found nonconforming, then 100% inspection resumes. If $i$ consecutive units are found free from nonconformities, the sampling inspection continues with the same fraction $f$.
3) If one sampled unit is found nonconforming, then 100% inspector resumes. If $i / 3$ sampled units are found free from nonconformities, then sampling inspection continues with the same fraction $f$.

## Review Practice Problems

**1**  Use a statistical package to construct OC curves for the following sampling plans:

a)  $N = 5000, n = 200, c = 4$

b)  $N = 5000, n = 100, c = 1$

**2**  Construct AOQ and ATI curves for the sampling plans in Problem 1.

**3**  A single-sampling plan for a process in statistical control is $n = 150$ with acceptance number $c = 3$. Find $p_a$, the probability of acceptance, with lot fractions defective of $p = 0.005, 0.01, 0.02, 0.03, 0.04$, and 0.05; then construct the OC curve for the sampling plan in this example.

**4**  Using the acceptance probabilities found in Problem 3, construct the AOQ curve. Then, using the AOQ curve you just constructed, find the average outgoing quality limit (AOQL). Assume that the lot size is 5000.

**5**  Construct a type B OC curve for the single-sampling plan $n = 100, c = 2$. Assume that the lot size is 5000.

**6**  A single-sampling plan $n = 50, c = 1$ is being used by a customer that received the shipment of a lot of size $N = 5000$. Construct type A and type B OC curves for the sampling plan being used by the customer. If the customer asks for advice on which OC curve to use, what will your recommendation be?

**7** Suppose a single-sampling plan $n = 150$, $c = 3$ is being proposed to a customer that receives shipments of lots of size 4000.
   a) Construct an AOQ curve for the single-sampling plan proposed in this problem, and find the AOQL value.
   b) Construct an ATI curve for the single-sampling plan proposed in this problem.

**8** Consider the following double-sampling acceptance plan:

$$n_1 = 50, c_1 = 1; n_2 = 150, c_2 = 3$$

   Suppose there are lot fractions defective of $p = 0.01, 0.02, 0.03, 0.04, 0.05$, and $0.06$. Find the following:
   a) Probabilities of acceptance on the first sample
   b) Probabilities of rejection on the first sample
   c) Probabilities of acceptance on the second sample
   d) Probabilities of acceptance on the combined samples

**9** Refer to Problem 8, and do the following:
   a) Calculate the AOQ percent points, assuming the process is continuous. Assume that the lot size is $N = 8000$.
   b) Construct the OC curves for parts (a) and (d) in Problem 8.

**10** Refer to the double-sampling acceptance plan in Problem 8. Find the value of the average total number of units inspected (ATI) for lot fractions defective of $p = 0.01, 0.02, 0.03, 0.04, \cdots$, and $0.06$. Assume that the lot size is $N = 8000$.

**11** Consider the following double-sampling plan:

$$n_1 = 50, c_1 = 1; n_2 = 75, c_2 = 3$$

   Find value of the average sample number (ASN) when the incoming lot fractions defective are 1%, 2%, 3%, 4%, 5%, 6%; and construct the ASN curve.

**12** Develop an item-by-item sequential-sampling plan for $\alpha = 0.05$, AQL $= p_1 = 0.02$, $\beta = 0.10$, RQL $= p_2 = 0.15$.

**13** Develop an item-by-item sequential-sampling plan for $\alpha = 0.05$, AQL $= p_1 = 0.01$, $\beta = 0.10$, RQL $= p_2 = 0.10$.

**14** Draw the graphical device for the sequential acceptance sampling plan considered in Problem 12.

**15** A supplier submits products in lots of size $N = 5000$ units. Suppose they are to be inspected at an AQL level of 1.5% using ANSI/ASQ Z1.4. Determine the appropriate single, double, and multiple sampling for General Inspection Level II from ANSI/ASQ Z1.4.

**16** A supplier submits products in lots of size $N = 8000$ units. Suppose they are to be inspected at an AQL level of 2.5% using ANSI/ASQ Z1.4. Determine the appropriate single, double, and multiple sampling plans for
   a) Special Inspection Level S-3 from ANSI/ASQ Z1.4
   b) General Inspection Level III from ANSI/ASQ Z1.4

**17** In your own words, describe the switching rules used in the acceptance sampling plans in sampling standard ANSI/ASQ Z1.4:
a) Normal to tightened
b) Tightened to normal
c) Normal to reduced

**18** A supplier ships products in a lot of 750 units to a customer. The customer is interested in setting up a sampling plan from sampling standard ANSI/ASQ Z1.9 using Special Inspection Level S3. If the AQL is 2.5%, determine the normal and tightened sampling plans, assuming that variability is unknown. Assume that the lower specification limit (LSL) of the units is given to be 250 °F.

**19** A supplier ships products in a lot of 450 units to the customer. The customer is interested in finding a sampling plan from sampling standard ANSI/ASQ Z1.9 using General Inspection level II. If the AQL is 1.5%, find the normal and tightened sampling plans, assuming that variability is unknown. Assume that the lower specification limit (LSL) of the units is given to 100 mg.

**20** Suppose that the lower specification limit of a quality characteristic of a copper plate used in an electronic application is LSL = 400. A lot of 100 plates is submitted for inspection, and General Normal Inspection level II with AQL = 0.65% is to be used. Using sampling standard ANSI/ASQ Z1.9, it can be seen that a sample of size 10 is needed. Suppose that values of the quality characteristic in the order of production are

| 475 | 409 | 401 | 413 | 451 | (R = 475 − 401 = 76) |
| 454 | 436 | 479 | 421 | 437 | (R = 479 − 421 = 58) |

Determine whether the lot acceptability criterion is met.

**21** Now suppose in Problem 20 that we are given USL = 480 instead of a lower specification limit. Use the data in Problem 20 to determine whether the lot acceptability criterion is met.

**22** The specification limits for the resistivity (ohm/cm) of individual silicon wafers are LSL = 75, USL = 76. A lot of 100 wafers is submitted for inspection, and General Normal Inspection Level II from ANSI/ASQ Z1.9 with AQL = 1% is to be used. Using Table 9.9, we find code letter F, which results in a sample size of 10 from ANSI/ASQC Z1.9-2003 Table B-3 (Table 9.11). The results of 10 measurements are as follows.

| 75.76 | 75.26 | 75.35 | 75.93 | 75.49 | 75.48 | 75.39 | 75.30 | 75.65 | 75.88 |

Determine whether the lot meets acceptance criteria.

**23** Define side-by-side the CSP-T and CSP-V sampling plans from MIL-STD 1235B and then discuss, with the help of an example, how the two plans differ from each other.

Table 9.4  ANSI/ASQ Z1.4-2003 Table VIII: Limit numbers for reduced inspection. *Source:* ANSI/ASQ (2003a).

| Number of Sample Units from Last 10 Lots or Batches | AQLs in Percent Nonconforming Items and Nonconformities per 100 Items (normal inspection) | | | | | | | | | | | | | | | | | | | | | | | | | |
|---|---|---|---|---|---|---|---|---|---|---|---|---|---|---|---|---|---|---|---|---|---|---|---|---|---|---|
| | 0.010 | 0.015 | 0.025 | 0.040 | 0.065 | 0.10 | 0.15 | 0.25 | 0.40 | 0.65 | 1.0 | 1.5 | 2.5 | 4.0 | 6.5 | 10 | 15 | 25 | 40 | 65 | 100 | 150 | 250 | 400 | 650 | 1000 |
| 20–29 | * | * | * | * | * | * | * | * | * | * | * | * | * | * | * | 0 | 0 | 2 | 4 | 8 | 14 | 22 | 40 | 68 | 115 | 181 |
| 30–49 | * | * | * | * | * | * | * | * | * | * | * | * | * | * | 0 | 0 | 1 | 3 | 7 | 13 | 22 | 36 | 63 | 105 | 178 | 277 |
| 50–79 | * | * | * | * | * | * | * | * | * | * | * | * | * | 0 | 0 | 2 | 3 | 7 | 14 | 25 | 40 | 63 | 110 | 181 | 301 | |
| 80–129 | * | * | * | * | * | * | * | * | * | * | * | * | 0 | 0 | 2 | 4 | 7 | 14 | 24 | 42 | 68 | 105 | 181 | 297 | | |
| 130–199 | * | * | * | * | * | * | * | * | * | * | * | 0 | 0 | 2 | 4 | 7 | 13 | 25 | 42 | 72 | 115 | 177 | 301 | 490 | | |
| 200–319 | * | * | * | * | * | * | * | * | * | * | 0 | 0 | 2 | 4 | 8 | 14 | 22 | 40 | 68 | 115 | 181 | 277 | 471 | | | |
| 320–499 | * | * | * | * | * | * | * | * | * | 0 | 0 | 1 | 4 | 8 | 14 | 24 | 39 | 68 | 113 | 189 | | | | | | |
| 500–799 | * | * | * | * | * | * | * | * | 0 | 0 | 2 | 3 | 7 | 14 | 25 | 40 | 63 | 110 | 181 | | | | | | | |
| 800–1249 | * | * | * | * | * | * | * | 0 | 0 | 2 | 4 | 7 | 14 | 24 | 42 | 68 | 105 | 181 | | | | | | | | |
| 1250–1999 | * | * | * | * | * | * | 0 | 0 | 2 | 4 | 7 | 13 | 24 | 49 | 69 | 110 | 169 | | | | | | | | | |
| 2000–3149 | * | * | * | * | * | 0 | 0 | 2 | 4 | 8 | 14 | 22 | 40 | 68 | 115 | 181 | | | | | | | | | | |
| 3150–4999 | * | * | * | * | 0 | 0 | 1 | 4 | 8 | 14 | 24 | 38 | 67 | 111 | 186 | | | | | | | | | | | |
| 5000–7999 | * | * | * | 0 | 0 | 2 | 3 | 7 | 14 | 25 | 40 | 63 | 110 | 181 | | | | | | | | | | | | |
| 8000–12,499 | * | * | 0 | 0 | 2 | 4 | 7 | 14 | 24 | 42 | 68 | 105 | 181 | | | | | | | | | | | | | |
| 12,500–19,999 | * | 0 | 0 | 2 | 4 | 7 | 13 | 24 | 40 | 69 | 110 | 169 | | | | | | | | | | | | | | |
| 20,000–31,499 | 0 | 0 | 2 | 4 | 8 | 14 | 22 | 40 | 68 | 115 | 181 | | | | | | | | | | | | | | | |
| 31,500 and over | 0 | 1 | 4 | 8 | 14 | 24 | 38 | 67 | 111 | 186 | | | | | | | | | | | | | | | | |

* = Denotes that the number of sample units from the last 10 lots or batches is not sufficient for reduced inspection for this AQL. In this instance, more than 10 lots or batches may be used for the calculation, provided that the lots or batches used are the most recent ones in sequence, that they have all been on normal inspection, and that none have been rejected while on original inspection.

**Table 9.5** ANSI/ASQ Z1.4-2003 Table I: Sample size code letters. *Source:* ANSI/ASQ (2003a).

| Lot or Batch Size | | | Special Inspection Levels | | | | General Inspection Levels | | |
|---|---|---|---|---|---|---|---|---|---|
| | | | S-1 | S-2 | S-3 | S-4 | I | II | III |
| 2 | to | 8 | A | A | A | A | A | A | B |
| 9 | to | 15 | A | A | A | A | A | B | C |
| 16 | to | 25 | A | A | B | B | B | C | D |
| 26 | to | 50 | A | B | B | C | C | D | E |
| 51 | to | 90 | B | B | C | C | C | E | F |
| 91 | to | 150 | B | B | C | D | D | F | G |
| 151 | to | 280 | B | C | D | E | E | G | H |
| 281 | to | 500 | B | C | D | E | F | H | J |
| 501 | to | 1200 | C | C | E | F | G | J | K |
| 1201 | to | 3200 | C | D | E | G | H | K | L |
| 3201 | to | 10,000 | C | D | F | G | J | L | M |
| 10,001 | to | 35,000 | C | D | F | H | K | M | N |
| 35,001 | to | 150,000 | D | E | G | J | L | N | P |
| 150,001 | to | 500,000 | D | E | G | J | M | P | Q |
| 500,001 | and | over | D | E | H | K | N | Q | R |

**Table 9.6** ANSI/ASQ Z1.4-2003 Table II-A: Single-sampling plans for normal inspection. *Source:* ANSI/ASQ (2003a).

AQLs in percent nonconforming items and nonconformities per 100 items (normal inspection). Each AQL cell shows the acceptance and rejection numbers as "Ac Re". ↓ = use first sampling plan below arrow; ▲ = use first sampling plan above arrow.

| Code | Sample size | 0.010 | 0.015 | 0.025 | 0.040 | 0.065 | 0.10 | 0.15 | 0.25 | 0.40 | 0.65 | 1.0 | 1.5 | 2.5 | 4.0 | 6.5 | 10 | 15 | 25 | 40 | 65 | 100 | 150 | 250 | 400 | 650 | 1000 |
|---|---|---|---|---|---|---|---|---|---|---|---|---|---|---|---|---|---|---|---|---|---|---|---|---|---|---|---|
| A | 2 | ↓ | ↓ | ↓ | ↓ | ↓ | ↓ | ↓ | ↓ | ↓ | ↓ | ↓ | ↓ | ↓ | ↓ | ↓ | ↓ | 0 1 | 1 2 | 2 3 | 3 4 | 5 6 | 7 8 | 10 11 | 14 15 | 21 22 | 30 31 |
| B | 3 | ↓ | ↓ | ↓ | ↓ | ↓ | ↓ | ↓ | ↓ | ↓ | ↓ | ↓ | ↓ | ↓ | ↓ | ↓ | 0 1 | 1 2 | 2 3 | 3 4 | 5 6 | 7 8 | 10 11 | 14 15 | 21 22 | 30 31 | 44 45 |
| C | 5 | ↓ | ↓ | ↓ | ↓ | ↓ | ↓ | ↓ | ↓ | ↓ | ↓ | ↓ | ↓ | ↓ | ↓ | 0 1 | 1 2 | 2 3 | 3 4 | 5 6 | 7 8 | 10 11 | 14 15 | 21 22 | 30 31 | 44 45 | ▲ |
| D | 8 | ↓ | ↓ | ↓ | ↓ | ↓ | ↓ | ↓ | ↓ | ↓ | ↓ | ↓ | ↓ | ↓ | 0 1 | 1 2 | 2 3 | 3 4 | 5 6 | 7 8 | 10 11 | 14 15 | 21 22 | 30 31 | 44 45 | ▲ | |
| E | 13 | ↓ | ↓ | ↓ | ↓ | ↓ | ↓ | ↓ | ↓ | ↓ | ↓ | ↓ | ↓ | 0 1 | 1 2 | 2 3 | 3 4 | 5 6 | 7 8 | 10 11 | 14 15 | 21 22 | 30 31 | 44 45 | ▲ | | |
| F | 20 | ↓ | ↓ | ↓ | ↓ | ↓ | ↓ | ↓ | ↓ | ↓ | ↓ | ↓ | 0 1 | 1 2 | 2 3 | 3 4 | 5 6 | 7 8 | 10 11 | 14 15 | 21 22 | 30 31 | 44 45 | ▲ | | | |
| G | 32 | ↓ | ↓ | ↓ | ↓ | ↓ | ↓ | ↓ | ↓ | ↓ | ↓ | 0 1 | 1 2 | 2 3 | 3 4 | 5 6 | 7 8 | 10 11 | 14 15 | 21 22 | 30 31 | 44 45 | ▲ | | | | |
| H | 50 | ↓ | ↓ | ↓ | ↓ | ↓ | ↓ | ↓ | ↓ | ↓ | 0 1 | 1 2 | 2 3 | 3 4 | 5 6 | 7 8 | 10 11 | 14 15 | 21 22 | 30 31 | 44 45 | ▲ | | | | | |
| J | 80 | ↓ | ↓ | ↓ | ↓ | ↓ | ↓ | ↓ | ↓ | 0 1 | 1 2 | 2 3 | 3 4 | 5 6 | 7 8 | 10 11 | 14 15 | 21 22 | 30 31 | 44 45 | ▲ | | | | | | |
| K | 125 | ↓ | ↓ | ↓ | ↓ | ↓ | ↓ | ↓ | 0 1 | 1 2 | 2 3 | 3 4 | 5 6 | 7 8 | 10 11 | 14 15 | 21 22 | 30 31 | 44 45 | ▲ | | | | | | | |
| L | 200 | ↓ | ↓ | ↓ | ↓ | ↓ | ↓ | 0 1 | 1 2 | 2 3 | 3 4 | 5 6 | 7 8 | 10 11 | 14 15 | 21 22 | 30 31 | 44 45 | ▲ | | | | | | | | |
| M | 315 | ↓ | ↓ | ↓ | ↓ | ↓ | 0 1 | 1 2 | 2 3 | 3 4 | 5 6 | 7 8 | 10 11 | 14 15 | 21 22 | 30 31 | 44 45 | ▲ | | | | | | | | | |
| N | 500 | ↓ | ↓ | ↓ | ↓ | 0 1 | 1 2 | 2 3 | 3 4 | 5 6 | 7 8 | 10 11 | 14 15 | 21 22 | 30 31 | 44 45 | ▲ | | | | | | | | | | |
| P | 800 | ↓ | ↓ | ↓ | 0 1 | 1 2 | 2 3 | 3 4 | 5 6 | 7 8 | 10 11 | 14 15 | 21 22 | 30 31 | 44 45 | ▲ | | | | | | | | | | | |
| Q | 1250 | ↓ | ↓ | 0 1 | 1 2 | 2 3 | 3 4 | 5 6 | 7 8 | 10 11 | 14 15 | 21 22 | 30 31 | 44 45 | ▲ | | | | | | | | | | | | |
| R | 2000 | ↓ | 0 1 | 1 2 | 2 3 | 3 4 | 5 6 | 7 8 | 10 11 | 14 15 | 21 22 | 30 31 | 44 45 | ▲ | | | | | | | | | | | | | |

**Table 9.7** ANSI/ASQ Z1.4-2003 Table III-A: Double-sampling plans for normal inspection. *Source:* ANSI/ASQ (2003a).

### AQLS (reduced inspection)

| Sample size code letter | Sample | Sample size | Cumulative sample size | 0.010 Ac Re | 0.015 Ac Re | 0.025 Ac Re | 0.040 Ac Re | 0.065 Ac Re | 0.10 Ac Re | 0.15 Ac Re | 0.25 Ac Re | 0.40 Ac Re | 0.65 Ac Re | 1.0 Ac Re | 1.5 Ac Re | 2.5 Ac Re | 4.0 Ac Re | 6.5 Ac Re | 10 Ac Re | 15 Ac Re | 25 Ac Re | 40 Ac Re | 65 Ac Re | 100 Ac Re | 150 Ac Re | 250 Ac Re | 400 Ac Re | 650 Ac Re | 1000 Ac Re |
|---|---|---|---|---|---|---|---|---|---|---|---|---|---|---|---|---|---|---|---|---|---|---|---|---|---|---|---|---|---|
| A | | | | | | | | | | | | | | | | | | | | | | | | | | | | |
| B | First / Second | 2 / 2 | 2 / 4 | ↓ | ↓ | ↓ | ↓ | ↓ | ↓ | ↓ | ↓ | ↓ | ↓ | ↓ | ↓ | ↓ | ↓ | 8 | ↑ | 0 2 / 1 2 | 0 3 / 3 4 | 1 4 / 4 5 | 2 5 / 6 7 | 3 7 / 8 9 | 5 9 / 12 13 | 7 11 / 18 19 | 11 16 / 26 27 | + | + | + |
| C | First / Second | 3 / 3 | 3 / 6 | ↓ | ↓ | ↓ | ↓ | ↓ | ↓ | ↓ | ↓ | ↓ | ↓ | ↓ | ↓ | ↓ | 8 | ↑ | 0 2 / 1 2 | 0 3 / 3 4 | 1 4 / 4 5 | 2 5 / 6 7 | 3 7 / 8 9 | 5 9 / 12 13 | 7 11 / 18 19 | 11 16 / 26 27 | 11 16 / 26 27 | 17 22 / 27 37 | 25 31 / 56 57 |
| D | First / Second | 5 / 5 | 5 / 10 | ↓ | ↓ | ↓ | ↓ | ↓ | ↓ | ↓ | ↓ | ↓ | ↓ | ↓ | ↓ | 8 | ↑ | ↓ | 0 2 / 1 2 | 0 3 / 3 4 | 1 4 / 4 5 | 2 5 / 6 7 | 3 7 / 8 9 | 5 9 / 12 13 | 7 11 / 18 19 | 11 16 / 26 27 | 17 22 / 37 38 | 25 31 / 56 57 | |
| E | First / Second | 8 / 8 | 8 / 16 | ↓ | ↓ | ↓ | ↓ | ↓ | ↓ | ↓ | ↓ | ↓ | ↓ | ↓ | 8 | ↑ | ↓ | 0 2 / 1 2 | 0 3 / 3 4 | 1 4 / 4 5 | 2 5 / 6 7 | 3 7 / 8 9 | 5 9 / 12 13 | 7 11 / 18 19 | 11 16 / 26 27 | 17 22 / 37 38 | 25 31 / 56 57 | | |
| F | First / Second | 13 / 13 | 13 / 26 | ↓ | ↓ | ↓ | ↓ | ↓ | ↓ | ↓ | ↓ | ↓ | ↓ | 8 | ↑ | ↓ | 0 2 / 1 2 | 0 3 / 3 4 | 1 4 / 4 5 | 2 5 / 6 7 | 3 7 / 8 9 | 5 9 / 12 13 | 7 11 / 18 19 | 11 16 / 26 27 | | | | | |
| G | First / Second | 20 / 20 | 20 / 40 | ↓ | ↓ | ↓ | ↓ | ↓ | ↓ | ↓ | ↓ | ↓ | 8 | ↑ | ↓ | 0 2 / 1 2 | 0 3 / 3 4 | 1 4 / 4 5 | 2 5 / 6 7 | 3 7 / 8 9 | 5 9 / 12 13 | 7 11 / 18 19 | 11 16 / 26 27 | | | | | | |
| H | First / Second | 32 / 32 | 32 / 64 | ↓ | ↓ | ↓ | ↓ | ↓ | ↓ | ↓ | ↓ | 8 | ↑ | ↓ | 0 2 / 1 2 | 0 3 / 3 4 | 1 4 / 4 5 | 2 5 / 6 7 | 3 7 / 8 9 | 5 9 / 12 13 | 7 11 / 18 19 | 11 16 / 26 27 | | | | | | | |
| J | First / Second | 50 / 50 | 50 / 100 | ↓ | ↓ | ↓ | ↓ | ↓ | ↓ | ↓ | 8 | ↑ | ↓ | 0 2 / 1 2 | 0 3 / 3 4 | 1 4 / 4 5 | 2 5 / 6 7 | 3 7 / 8 9 | 5 9 / 12 13 | 7 11 / 18 19 | 11 16 / 26 27 | | | | | | | | |
| K | First / Second | 80 / 80 | 80 / 160 | ↓ | ↓ | ↓ | ↓ | ↓ | ↓ | 8 | ↑ | ↓ | 0 2 / 1 2 | 0 3 / 3 4 | 1 4 / 4 5 | 2 5 / 6 7 | 3 7 / 8 9 | 5 9 / 12 13 | 7 11 / 18 19 | 11 16 / 26 27 | | | | | | | | | |
| L | First / Second | 125 / 125 | 125 / 250 | ↓ | ↓ | ↓ | ↓ | ↓ | 8 | ↑ | ↓ | 0 2 / 1 2 | 0 3 / 3 4 | 1 4 / 4 5 | 2 5 / 6 7 | 3 7 / 8 9 | 5 9 / 12 13 | 7 11 / 18 19 | 11 16 / 26 27 | | | | | | | | | | |
| M | First / Second | 200 / 200 | 200 / 400 | ↓ | ↓ | ↓ | ↓ | 8 | ↑ | ↓ | 0 2 / 1 2 | 0 3 / 3 4 | 1 4 / 4 5 | 2 5 / 6 7 | 3 7 / 8 9 | 5 9 / 12 13 | 7 11 / 18 19 | 11 16 / 26 27 | | | | | | | | | | | |
| N | First / Second | 315 / 315 | 315 / 630 | ↓ | ↓ | ↓ | 8 | ↑ | ↓ | 0 2 / 1 2 | 0 3 / 3 4 | 1 4 / 4 5 | 2 5 / 6 7 | 3 7 / 8 9 | 5 9 / 12 13 | 7 11 / 18 19 | 11 16 / 26 27 | | | | | | | | | | | | |
| P | First / Second | 500 / 500 | 500 / 1000 | ↓ | ↓ | 8 | ↑ | ↓ | 0 2 / 1 2 | 0 3 / 3 4 | 1 4 / 4 5 | 2 5 / 6 7 | 3 7 / 8 9 | 5 9 / 12 13 | 7 11 / 18 19 | 11 16 / 26 27 | | | | | | | | | | | | | |
| Q | First / Second | 800 / 800 | 800 / 1600 | ↓ | 8 | ↑ | ↓ | 0 2 / 1 2 | 0 3 / 3 4 | 1 4 / 4 5 | 2 5 / 6 7 | 3 7 / 8 9 | 5 9 / 12 13 | 7 11 / 18 19 | 11 16 / 26 27 | | | | | | | | | | | | | | |
| R | First / Second | 1250 / 1250 | 1250 / 2500 | 8 | ↑ | ↓ | 0 2 / 1 2 | 0 3 / 3 4 | 1 4 / 4 5 | 2 5 / 6 7 | 3 7 / 8 9 | 5 9 / 12 13 | 7 11 / 18 19 | 11 16 / 26 27 | | | | | | | | | | | | | | | |

↓ = Use first sampling plan below arrow. If sample size equals or exceeds lot or batch size, do percent inspection.

↑ = Use first sampling plan above arrow.

AC = Acceptance number.

Re = Rejection number.

8 = Use corresponding single sampling plan.

+ = Use corresponding single sampling plan or double sampling plan for code letter B.

**Table 9.8** ANSI/ASQ Z1.4-2003 Table IV-A: Multiple sampling plans for normal inspection. *Source:* ANSI/ASQ (2003a).

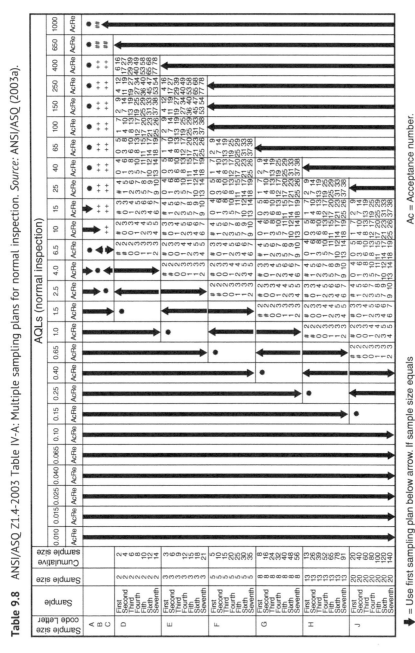

➡ = Use first sampling plan below arrow. If sample size equals or exceeds lot or batch size, do 100 percent inspection.

⬅ = Use first sampling plan below arrow.

✱ = Use corresponding single sampling plan.

++ = Use corresponding double sampling plan or multiple sampling plan for code letter D.

Ac = Acceptance number.

Re = Rejection number.

# = Acceptance not permitted at this sample size.

## = Use corresponding double sampling plan.

**Table 9.8** (Continued)

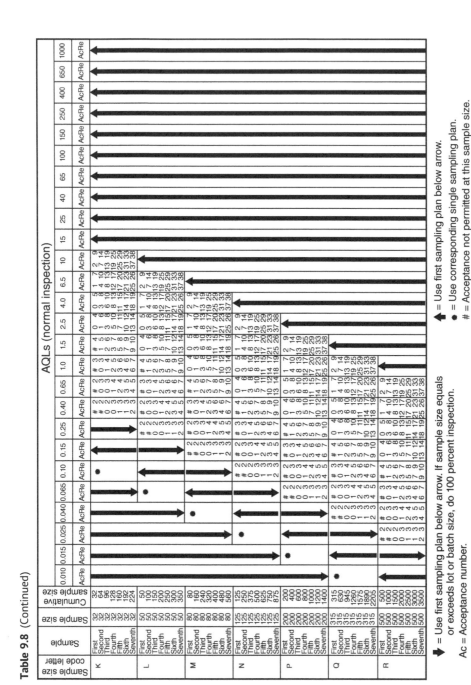

AQLs (normal inspection)

Sample size code letters: K, L, M, N, P, Q, R

| Sample size code letter | Sample | Sample size | Cumulative sample size |
|---|---|---|---|
| K | First | 32 | 32 |
| | Second | 32 | 64 |
| | Third | 32 | 96 |
| | Fourth | 32 | 128 |
| | Fifth | 32 | 160 |
| | Sixth | 32 | 192 |
| | Seventh | 32 | 224 |
| L | First | 50 | 50 |
| | Second | 50 | 100 |
| | Third | 50 | 150 |
| | Fourth | 50 | 200 |
| | Fifth | 50 | 250 |
| | Sixth | 50 | 300 |
| | Seventh | 50 | 350 |
| M | First | 80 | 80 |
| | Second | 80 | 160 |
| | Third | 80 | 240 |
| | Fourth | 80 | 320 |
| | Fifth | 80 | 400 |
| | Sixth | 80 | 480 |
| | Seventh | 80 | 560 |
| N | First | 125 | 125 |
| | Second | 125 | 250 |
| | Third | 125 | 375 |
| | Fourth | 125 | 500 |
| | Fifth | 125 | 625 |
| | Sixth | 125 | 750 |
| | Seventh | 125 | 875 |
| P | First | 200 | 200 |
| | Second | 200 | 400 |
| | Third | 200 | 600 |
| | Fourth | 200 | 800 |
| | Fifth | 200 | 1000 |
| | Sixth | 200 | 1200 |
| | Seventh | 200 | 1400 |
| Q | First | 315 | 315 |
| | Second | 315 | 630 |
| | Third | 315 | 945 |
| | Fourth | 315 | 1260 |
| | Fifth | 315 | 1575 |
| | Sixth | 315 | 1890 |
| | Seventh | 315 | 2205 |
| R | First | 500 | 500 |
| | Second | 500 | 1000 |
| | Third | 500 | 1500 |
| | Fourth | 500 | 2000 |
| | Fifth | 500 | 2500 |
| | Sixth | 500 | 3000 |
| | Seventh | 500 | 3500 |

AQL column headings: 0.010, 0.015, 0.025, 0.040, 0.065, 0.10, 0.15, 0.25, 0.40, 0.65, 1.0, 1.5, 2.5, 4.0, 6.5, 10, 15, 25, 40, 65, 100, 150, 250, 400, 650, 1000 — each with Ac and Re sub-columns.

↓ = Use first sampling plan below arrow. If sample size equals or exceeds lot or batch size, do 100 percent inspection.

↑ = Use first sampling plan above arrow.

• = Use corresponding single sampling plan.

\# = Acceptance not permitted at this sample size.

Ac = Acceptance number.

Re = Rejection number.

**Table 9.9**   ANSI/ASQ Z1.9-2003 Table A-2[*]: Sample size code letters[**].
*Source:* ANSI/ASQ (2003b).

| Lot Size | | | Inspection Levels | | | | |
|---|---|---|---|---|---|---|---|
| | | | Special | | General | | |
| | | | S3 | S4 | I | II | III |
| 2 | to | 8 | B | B | B | B | C |
| 9 | to | 15 | B | B | B | B | D |
| 16 | to | 25 | B | B | B | C | E |
| 26 | to | 50 | B | B | C | D | F |
| 51 | to | 90 | B | B | D | E | G |
| 91 | to | 150 | B | C | E | F | H |
| 151 | to | 280 | B | D | F | G | I |
| 281 | to | 400 | C | E | G | H | J |
| 401 | to | 500 | C | E | G | I | J |
| 501 | to | 1200 | D | F | H | J | K |
| 1201 | to | 3200 | E | G | I | K | L |
| 3201 | to | 10,000 | F | H | J | L | M |
| 10,001 | to | 35,000 | G | I | K | M | N |
| 35,001 | to | 150,000 | H | J | L | N | P |
| 150,001 | to | 500,000 | H | K | M | P | P |
| 500,001 | and | over | H | K | N | P | P |

\* The theory governing inspection by variables depends on the properties of the normal distribution; therefore, this method of inspection is applicable only when there is reason to believe that the frequency distribution is normal.

\*\* Sample size code letters given in the body of the table are applicable when the indicated inspection levels are to be used.

**Table 9.10**  ANSI/ASQ Z1.9-2003 Table C-1: Master table for normal and tightened inspection for plans based on variability unknown (single specification limit – Form 1). *Source:* ANSI/ASQ (2003b).

| Sample size code letter | Sample size | AQLs (normal inspection) | | | | | | | | | | | |
|---|---|---|---|---|---|---|---|---|---|---|---|---|---|
| | | T | .10 | .15 | .25 | .40 | .65 | 1.00 | 1.50 | 2.50 | 4.00 | 6.50 | 10.00 |
| | | k | k | k | k | k | k | k | k | k | k | k | k |
| B | 3 | ↓ | ↓ | ↓ | ↓ | ↓ | ↓ | ↓ | ↓ | .587 | .502 | .401 | .296 |
| C | 4 | ↓ | ↓ | ↓ | ↓ | ↓ | ↓ | .651 | .598 | .525 | .450 | .364 | .276 |
| D | 5 | ↓ | ↓ | ↓ | ↓ | ↓ | .663 | .614 | .565 | .498 | .431 | .352 | .272 |
| E | 7 | ↓ | ↓ | ↓ | .702 | .659 | .613 | .569 | .525 | .465 | .405 | .336 | .266 |
| F | 10 | ↓ | ↓ | .916 | .863 | .811 | .755 | .703 | .650 | .579 | .507 | .424 | .341 |
| G | 15 | 1.04 | .999 | .958 | .903 | .850 | .792 | .738 | .684 | .610 | .536 | .452 | .368 |
| H | 25 | 1.10 | 1.05 | 1.01 | .951 | .896 | .835 | .779 | .723 | .647 | .571 | .484 | .398 |
| I | 30 | 1.10 | 1.06 | 1.02 | .959 | .904 | .843 | .787 | .730 | .654 | .577 | .490 | .403 |
| J | 40 | 1.13 | 1.08 | 1.04 | .978 | .921 | .860 | .803 | .746 | .668 | .591 | .503 | .415 |
| K | 60 | 1.16 | 1.11 | 1.06 | 1.00 | .948 | .885 | .826 | .768 | .689 | .610 | .521 | .432 |
| L | 85 | 1.17 | 1.13 | 1.08 | 1.02 | .962 | .899 | .839 | .780 | .701 | .621 | .530 | .441 |
| M | 115 | 1.19 | 1.14 | 1.09 | 1.03 | .975 | .911 | .851 | .791 | .711 | .631 | .539 | .449 |
| N | 175 | 1.21 | 1.16 | 1.11 | 1.05 | .994 | .929 | .868 | .807 | .726 | .644 | .552 | .460 |
| P | 230 | 1.21 | 1.16 | 1.12 | 1.06 | .996 | .931 | .870 | .809 | .728 | .646 | .553 | .462 |
| | | .10 | .15 | .25 | .40 | .65 | 1.00 | 1.50 | 2.50 | 4.00 | 6.50 | 10.00 | |
| | | AQLs (tightened inspection) | | | | | | | | | | | |

All AQL values are in percent nonconforming. T denotes plan used exclusively on tightened inspection and provides symbol for identification of appropriate OC curve.

↓ = Use first sampling plan below arrow, that is, both sample size and *k* value. When sample size equals or exceeds lot size, every item in the lot must be inspected.

**Table 9.11** ANSI/ASQ Z1.9-2003 Table B-3: Master table for normal and tightened inspection for plans based on variability unknown (double specification limit and Form 2 – single specification limit). *Source:* ANSI/ASQ (2003b).

| Sample size code letter | Sample size | T | .10 | .15 | .25 | .40 | .65 | 1.00 | 1.50 | 2.50 | 4.00 | 6.50 | 10.00 |
|---|---|---|---|---|---|---|---|---|---|---|---|---|---|
| | | *M* | *M* | *M* | *M* | *M* | *M* | *M* | *M* | *M* | *M* | *M* | *M* |
| B | 3 | | | | | | | 7.59 | 18.86 | 26.94 | 33.69 |
| C | 4 | | | | | | | 1.49 | 5.46 | 10.88 | 16.41 | 22.84 | 29.43 |
| D | 5 | | | | | 0.041 | 1.34 | 3.33 | 5.82 | 9.80 | 14.37 | 20.19 | 26.55 |
| E | 7 | | 0.005 | 0.087 | 0.421 | 1.05 | 2.13 | 3.54 | 5.34 | 8.40 | 12.19 | 17.34 | 23.30 |
| F | 10 | 0.077 | 0.179 | 0.349 | 0.714 | 1.27 | 2.14 | 3.27 | 4.72 | 7.26 | 10.53 | 15.17 | 20.73 |
| G | 15 | 0.186 | 0.311 | 0.491 | 0.839 | 1.33 | 2.09 | 3.06 | 4.32 | 6.55 | 9.48 | 13.74 | 18.97 |
| H | 20 | 0.228 | 0.356 | 0.531 | 0.864 | 1.33 | 2.03 | 2.93 | 4.10 | 6.18 | 8.95 | 13.01 | 18.07 |
| I | 25 | 0.250 | 0.378 | 0.551 | 0.874 | 1.32 | 2.00 | 2.86 | 3.97 | 5.98 | 8.65 | 12.60 | 17.55 |
| J | 35 | 0.253 | 0.373 | 0.534 | 0.833 | 1.24 | 1.87 | 2.66 | 3.70 | 5.58 | 8.11 | 11.89 | 16.67 |
| K | 50 | 0.243 | 0.355 | 0.503 | 0.778 | 1.16 | 1.73 | 2.47 | 3.44 | 5.21 | 7.61 | 11.23 | 15.87 |
| L | 75 | 0.225 | 0.326 | 0.461 | 0.711 | 1.06 | 1.59 | 2.27 | 3.17 | 4.83 | 7.10 | 10.58 | 15.07 |
| M | 100 | 0.218 | 0.315 | 0.444 | 0.684 | 1.02 | 1.52 | 2.18 | 3.06 | 4.67 | 6.88 | 10.29 | 14.71 |
| N | 150 | 0.202 | 0.292 | 0.412 | 0.636 | 0.946 | 1.42 | 2.05 | 2.88 | 4.42 | 6.56 | 9.86 | 14.18 |
| P | 200 | 0.204 | 0.294 | 0.414 | 0.637 | 0.945 | 1.42 | 2.04 | 2.86 | 4.39 | 6.52 | 9.80 | 14.11 |
| | | .10 | .15 | .25 | .40 | .65 | 1.00 | 1.50 | 2.50 | 4.00 | 6.50 | 10.00 | |

AQLs (tightened inspection)

All AQL values are in percent nonconforming. *T* denotes plan used exclusively on tightened inspection and provides symbol for identification of appropriate OC curve.

↓ = Use first sampling plan below arrow; that is, both sample size and *M* value. When sample size equals or exceeds lot size, every item in the lot must be inspected.

**Table 9.12** ANSI/ASQ Z1.9-2003 Table B-5: Table for estimating the lot percent nonconforming using the standard deviation method. Values tabulated are read as percentages. *Source:* ANSI/ASQ (2003b).

| $Q_U$ or $Q_L$ | Sample size | | | | | | | | | | | | | | |
|---|---|---|---|---|---|---|---|---|---|---|---|---|---|---|---|
| | 3 | 4 | 5 | 7 | 10 | 15 | 20 | 25 | 30 | 35 | 50 | 75 | 100 | 150 | 200 |
| 0 | 50.00 | 50.00 | 50.00 | 50.00 | 50.00 | 50.00 | 50.00 | 50.00 | 50.00 | 50.00 | 50.00 | 50.00 | 50.00 | 50.00 | 50.00 |
| .1 | 47.24 | 46.67 | 46.44 | 46.26 | 46.16 | 46.10 | 46.08 | 46.06 | 46.05 | 46.05 | 46.04 | 46.03 | 46.03 | 46.02 | 46.02 |
| .2 | 44.46 | 43.33 | 42.90 | 42.54 | 42.35 | 42.24 | 42.19 | 42.16 | 42.15 | 42.13 | 42.11 | 42.10 | 42.09 | 42.09 | 42.08 |
| .3 | 41.63 | 40.00 | 39.37 | 38.87 | 38.60 | 38.44 | 38.37 | 38.33 | 38.31 | 38.29 | 38.27 | 38.25 | 38.24 | 38.23 | 38.22 |
| .31 | 41.35 | 39.67 | 39.02 | 38.50 | 38.23 | 38.06 | 37.99 | 37.95 | 37.93 | 37.91 | 37.89 | 37.87 | 37.86 | 37.85 | 37.84 |
| .32 | 41.06 | 39.33 | 38.67 | 38.14 | 37.86 | 37.69 | 37.62 | 37.58 | 37.55 | 37.54 | 37.51 | 37.49 | 37.48 | 37.47 | 37.46 |
| .33 | 40.77 | 39.00 | 38.32 | 37.78 | 37.49 | 37.31 | 37.24 | 37.20 | 37.18 | 37.16 | 37.13 | 37.11 | 37.10 | 37.09 | 37.08 |
| .34 | 40.49 | 38.67 | 37.97 | 37.42 | 37.12 | 36.94 | 36.87 | 36.83 | 36.80 | 36.78 | 36.75 | 36.73 | 36.72 | 36.71 | 36.71 |
| .35 | 40.20 | 38.33 | 37.62 | 37.06 | 36.75 | 36.57 | 36.49 | 36.45 | 36.43 | 36.41 | 36.38 | 36.36 | 36.35 | 36.34 | 36.33 |
| .36 | 39.91 | 38.00 | 37.28 | 36.69 | 36.38 | 36.20 | 36.12 | 36.08 | 36.05 | 36.04 | 36.01 | 35.98 | 35.97 | 35.96 | 35.96 |
| .37 | 39.62 | 37.67 | 36.93 | 36.33 | 36.02 | 35.83 | 35.75 | 35.71 | 35.68 | 35.66 | 35.63 | 35.61 | 35.60 | 35.59 | 35.58 |
| .38 | 39.33 | 37.33 | 36.58 | 35.98 | 35.65 | 35.46 | 35.38 | 35.34 | 35.31 | 35.29 | 35.26 | 35.24 | 35.23 | 35.22 | 35.21 |
| .39 | 39.03 | 37.00 | 36.23 | 35.62 | 35.29 | 35.10 | 35.02 | 34.97 | 34.94 | 34.93 | 34.89 | 34.87 | 34.86 | 34.85 | 34.84 |
| .40 | 38.74 | 36.67 | 35.88 | 35.26 | 34.93 | 34.73 | 34.65 | 34.60 | 34.58 | 34.56 | 34.53 | 34.50 | 34.49 | 34.48 | 34.47 |
| .41 | 38.45 | 36.33 | 35.54 | 34.90 | 34.57 | 34.37 | 34.28 | 34.24 | 34.21 | 34.19 | 34.16 | 34.13 | 34.12 | 34.11 | 34.11 |
| .42 | 38.15 | 36.00 | 35.19 | 34.55 | 34.21 | 34.00 | 33.92 | 33.87 | 33.85 | 33.83 | 33.79 | 33.77 | 33.76 | 33.75 | 33.74 |
| .43 | 37.85 | 35.67 | 34.85 | 34.19 | 33.85 | 33.64 | 33.56 | 33.51 | 33.48 | 33.46 | 33.43 | 33.40 | 33.39 | 33.38 | 33.38 |
| .44 | 37.56 | 35.33 | 34.50 | 33.84 | 33.49 | 33.28 | 33.20 | 33.15 | 33.12 | 33.10 | 33.07 | 33.04 | 33.03 | 33.02 | 33.01 |
| .45 | 37.26 | 35.00 | 34.16 | 33.49 | 33.13 | 32.92 | 32.84 | 32.79 | 32.76 | 32.74 | 32.71 | 32.68 | 32.67 | 32.66 | 32.65 |
| .46 | 36.96 | 34.67 | 33.81 | 33.13 | 32.78 | 32.57 | 32.48 | 32.43 | 32.40 | 32.38 | 32.35 | 32.32 | 32.31 | 32.30 | 32.29 |
| .47 | 36.66 | 34.33 | 33.47 | 32.78 | 32.42 | 32.21 | 32.12 | 32.07 | 32.04 | 32.02 | 31.99 | 31.96 | 31.95 | 31.94 | 31.93 |
| .48 | 36.35 | 34.00 | 33.12 | 32.43 | 32.07 | 31.85 | 31.77 | 31.72 | 31.69 | 31.67 | 31.63 | 31.61 | 31.60 | 31.58 | 31.58 |
| .49 | 36.05 | 33.67 | 32.78 | 32.08 | 31.72 | 31.50 | 31.41 | 31.36 | 31.33 | 31.31 | 31.28 | 31.25 | 31.24 | 31.23 | 31.22 |
| .50 | 35.75 | 33.33 | 32.44 | 31.74 | 31.37 | 31.15 | 31.06 | 31.01 | 30.98 | 30.96 | 30.93 | 30.90 | 30.89 | 30.88 | 30.87 |
| .51 | 35.44 | 33.00 | 32.10 | 31.39 | 31.02 | 30.80 | 30.71 | 30.66 | 30.63 | 30.61 | 30.57 | 30.55 | 30.54 | 30.53 | 30.52 |
| .52 | 35.13 | 32.67 | 31.76 | 31.04 | 30.67 | 30.45 | 30.36 | 30.31 | 30.28 | 30.26 | 30.23 | 30.20 | 30.19 | 30.18 | 30.17 |
| .53 | 34.82 | 32.33 | 31.42 | 30.70 | 30.32 | 30.10 | 30.01 | 29.96 | 29.93 | 29.91 | 29.88 | 29.85 | 29.84 | 29.83 | 29.82 |
| .54 | 34.51 | 32.00 | 31.08 | 30.36 | 29.98 | 29.76 | 29.67 | 29.62 | 29.59 | 29.57 | 29.53 | 29.51 | 29.49 | 29.48 | 29.48 |
| .55 | 34.20 | 31.67 | 30.74 | 30.01 | 29.64 | 29.41 | 29.32 | 29.27 | 29.24 | 29.22 | 29.19 | 29.16 | 29.15 | 29.14 | 29.13 |
| .56 | 33.88 | 31.33 | 30.40 | 29.67 | 29.29 | 29.07 | 28.98 | 28.93 | 28.90 | 28.88 | 28.85 | 28.82 | 28.81 | 28.80 | 28.79 |
| .57 | 33.57 | 31.00 | 30.06 | 29.33 | 28.95 | 28.73 | 28.64 | 28.59 | 28.56 | 28.54 | 28.51 | 28.48 | 28.47 | 28.46 | 28.45 |
| .58 | 33.25 | 30.67 | 29.73 | 28.99 | 28.61 | 28.39 | 28.30 | 28.25 | 28.22 | 28.20 | 28.17 | 28.14 | 28.13 | 28.12 | 28.11 |
| .59 | 32.93 | 30.33 | 29.39 | 28.66 | 28.28 | 28.05 | 27.96 | 27.92 | 27.89 | 27.87 | 27.83 | 27.81 | 27.79 | 27.78 | 27.78 |
| .60 | 32.61 | 30.00 | 29.05 | 28.32 | 27.94 | 27.72 | 27.63 | 27.58 | 27.55 | 27.53 | 27.50 | 27.47 | 27.46 | 27.45 | 27.44 |
| .61 | 32.28 | 29.67 | 28.72 | 27.98 | 27.60 | 27.39 | 27.30 | 27.25 | 27.22 | 27.20 | 27.16 | 27.14 | 27.13 | 27.11 | 27.11 |

**Table 9.12** (Continued)

| $Q_U$ or $Q_L$ | 3 | 4 | 5 | 7 | 10 | 15 | 20 | 25 | 30 | 35 | 50 | 75 | 100 | 150 | 200 |
|---|---|---|---|---|---|---|---|---|---|---|---|---|---|---|---|
| | | | | | | | **Sample size** | | | | | | | | |
| .62 | 31.96 | 29.33 | 28.39 | 27.65 | 27.27 | 27.05 | 26.96 | 26.92 | 26.89 | 26.87 | 26.83 | 26.81 | 26.80 | 26.78 | 26.78 |
| .63 | 31.63 | 29.00 | 28.05 | 27.32 | 26.94 | 26.72 | 26.63 | 26.59 | 26.56 | 26.54 | 26.50 | 26.48 | 26.47 | 26.46 | 26.45 |
| .64 | 31.30 | 28.67 | 27.72 | 26.99 | 26.61 | 26.39 | 26.31 | 26.26 | 26.23 | 26.21 | 26.18 | 26.15 | 26.14 | 26.13 | 26.12 |
| .65 | 30.97 | 28.33 | 27.39 | 26.66 | 26.28 | 26.07 | 25.98 | 25.93 | 25.90 | 25.88 | 25.85 | 25.83 | 25.82 | 25.81 | 25.80 |
| .66 | 30.63 | 28.00 | 27.06 | 26.33 | 25.96 | 25.74 | 25.66 | 25.61 | 25.58 | 25.56 | 25.53 | 25.51 | 25.49 | 25.48 | 25.48 |
| .67 | 30.30 | 27.67 | 26.73 | 26.00 | 25.63 | 25.42 | 25.33 | 25.29 | 25.26 | 25.24 | 25.21 | 25.19 | 25.17 | 25.16 | 25.16 |
| .68 | 29.96 | 27.33 | 26.40 | 25.68 | 25.31 | 25.10 | 25.01 | 24.97 | 24.94 | 24.92 | 24.89 | 24.87 | 24.86 | 24.85 | 24.84 |
| .69 | 29.61 | 27.00 | 26.07 | 25.35 | 24.99 | 24.78 | 24.70 | 24.65 | 24.62 | 24.60 | 24.57 | 24.55 | 24.54 | 24.53 | 24.52 |
| .70 | 29.27 | 26.67 | 25.74 | 25.03 | 24.67 | 24.46 | 24.38 | 24.33 | 24.31 | 24.29 | 24.26 | 24.24 | 24.23 | 24.22 | 24.21 |
| .71 | 28.92 | 26.33 | 25.41 | 24.71 | 24.35 | 24.15 | 24.06 | 24.02 | 23.99 | 23.98 | 23.95 | 23.92 | 23.91 | 23.90 | 23.90 |
| .72 | 28.57 | 26.00 | 25.09 | 24.39 | 24.03 | 23.83 | 23.75 | 23.71 | 23.68 | 23.67 | 23.64 | 23.61 | 23.60 | 23.59 | 23.59 |
| .73 | 28.22 | 25.67 | 24.76 | 24.07 | 23.72 | 23.52 | 23.44 | 23.40 | 23.37 | 23.36 | 23.33 | 23.31 | 23.30 | 23.29 | 23.28 |
| .74 | 27.86 | 25.33 | 24.44 | 23.75 | 23.41 | 23.21 | 23.13 | 23.09 | 23.07 | 23.05 | 23.02 | 23.00 | 22.99 | 22.98 | 22.98 |
| .75 | 27.50 | 25.00 | 24.11 | 23.44 | 23.10 | 22.90 | 22.83 | 22.79 | 22.76 | 22.75 | 22.72 | 22.70 | 22.69 | 22.68 | 22.68 |
| .76 | 27.13 | 24.67 | 23.79 | 23.12 | 22.79 | 22.60 | 22.52 | 22.48 | 22.46 | 22.44 | 22.42 | 22.40 | 22.39 | 22.38 | 22.38 |
| .77 | 26.76 | 24.33 | 23.47 | 22.81 | 22.48 | 22.30 | 22.22 | 22.18 | 22.16 | 22.14 | 22.12 | 22.10 | 22.09 | 22.08 | 22.08 |
| .78 | 26.39 | 24.00 | 23.15 | 22.50 | 22.18 | 21.99 | 21.92 | 21.89 | 21.86 | 21.85 | 21.82 | 21.80 | 21.78 | 21.79 | 21.78 |
| .79 | 26.02 | 23.67 | 22.83 | 22.19 | 21.87 | 21.70 | 21.63 | 21.59 | 21.57 | 21.55 | 21.53 | 21.51 | 21.50 | 21.49 | 21.49 |
| .80 | 25.64 | 23.33 | 22.51 | 21.88 | 21.57 | 21.40 | 21.33 | 21.29 | 21.27 | 21.26 | 21.23 | 21.22 | 21.21 | 21.20 | 21.20 |
| .81 | 25.25 | 23.00 | 22.19 | 21.58 | 21.27 | 21.10 | 21.04 | 21.00 | 20.98 | 20.97 | 20.94 | 20.93 | 20.92 | 20.91 | 20.91 |
| .82 | 24.86 | 22.67 | 21.87 | 21.27 | 20.98 | 20.81 | 20.75 | 20.71 | 20.69 | 20.68 | 20.65 | 20.64 | 20.63 | 20.62 | 20.62 |
| .83 | 24.47 | 22.33 | 21.56 | 29.97 | 29.68 | 20.52 | 20.46 | 20.42 | 20.40 | 20.39 | 20.37 | 20.35 | 20.35 | 20.34 | 20.34 |
| .84 | 24.07 | 22.00 | 21.24 | 20.67 | 20.39 | 20.23 | 20.17 | 20.14 | 20.12 | 20.11 | 20.09 | 20.07 | 20.06 | 20.06 | 20.05 |
| .85 | 23.67 | 21.67 | 20.93 | 20.37 | 20.10 | 19.94 | 19.89 | 19.86 | 19.84 | 19.82 | 19.80 | 19.79 | 19.78 | 19.78 | 19.77 |
| .86 | 23.26 | 21.33 | 20.62 | 20.07 | 19.81 | 19.66 | 19.60 | 19.57 | 19.56 | 19.54 | 19.53 | 19.51 | 19.51 | 19.50 | 19.50 |
| .87 | 22.84 | 21.00 | 20.31 | 19.78 | 19.52 | 19.38 | 19.32 | 19.30 | 19.28 | 19.27 | 19.25 | 19.24 | 19.23 | 19.23 | 19.22 |
| .88 | 22.42 | 20.67 | 20.00 | 19.48 | 19.23 | 19.10 | 19.05 | 19.02 | 19.00 | 18.99 | 18.98 | 18.96 | 18.96 | 18.95 | 18.95 |
| .89 | 21.99 | 20.33 | 19.69 | 19.19 | 18.95 | 18.82 | 18.77 | 18.74 | 18.73 | 18.72 | 18.70 | 18.69 | 18.69 | 18.68 | 18.68 |
| .90 | 21.55 | 20.00 | 19.38 | 18.90 | 18.67 | 18.54 | 18.50 | 18.47 | 18.46 | 18.45 | 18.43 | 18.42 | 18.42 | 18.41 | 18.41 |
| .91 | 21.11 | 19.67 | 19.07 | 18.61 | 18.39 | 18.27 | 18.23 | 18.20 | 18.19 | 18.18 | 18.17 | 18.16 | 18.15 | 18.15 | 18.15 |
| .92 | 20.66 | 19.33 | 18.77 | 18.33 | 18.11 | 18.00 | 17.96 | 17.94 | 17.92 | 17.92 | 17.90 | 17.89 | 17.89 | 17.89 | 17.88 |
| .93 | 20.19 | 19.00 | 18.46 | 18.04 | 17.84 | 17.73 | 17.69 | 17.67 | 17.66 | 17.65 | 17.64 | 17.63 | 17.63 | 17.62 | 17.62 |
| .94 | 19.73 | 18.67 | 18.16 | 17.76 | 17.56 | 17.46 | 17.43 | 17.41 | 17.40 | 17.39 | 17.38 | 17.37 | 17.37 | 17.37 | 17.36 |
| .95 | 19.25 | 18.33 | 17.86 | 17.48 | 17.29 | 17.20 | 17.17 | 17.16 | 17.14 | 17.13 | 17.12 | 17.12 | 17.11 | 17.11 | 17.11 |
| .96 | 18.75 | 18.00 | 17.55 | 17.20 | 17.03 | 16.94 | 16.90 | 16.89 | 16.88 | 16.88 | 16.87 | 16.86 | 16.86 | 16.86 | 16.86 |

*(Continued)*

**Table 9.12** (Continued)

| $Q_U$ or $Q_L$ | 3 | 4 | 5 | 7 | 10 | 15 | 20 | 25 | 30 | 35 | 50 | 75 | 100 | 150 | 200 |
|---|---|---|---|---|---|---|---|---|---|---|---|---|---|---|---|
| | | | | | | | Sample size | | | | | | | | |
| .97 | 18.25 | 17.67 | 17.25 | 16.92 | 16.76 | 16.68 | 16.65 | 16.63 | 16.62 | 16.62 | 16.61 | 16.61 | 16.61 | 16.61 | 16.60 |
| .98 | 17.74 | 17.33 | 16.96 | 16.65 | 16.49 | 16.42 | 16.39 | 16.38 | 16.37 | 16.37 | 16.36 | 16.36 | 16.36 | 16.36 | 16.36 |
| .99 | 17.21 | 17.00 | 16.66 | 16.37 | 16.23 | 16.16 | 16.14 | 16.13 | 16.12 | 16.12 | 16.12 | 16.11 | 16.11 | 16.11 | 16.11 |
| 1.00 | 16.67 | 16.67 | 16.36 | 16.10 | 15.97 | 15.91 | 15.89 | 15.88 | 15.88 | 15.87 | 15.87 | 15.87 | 15.87 | 15.87 | 15.87 |
| 1.01 | 16.11 | 16.33 | 16.07 | 15.83 | 15.72 | 15.66 | 15.64 | 15.63 | 15.63 | 15.63 | 15.63 | 15.62 | 15.62 | 15.62 | 15.62 |
| 1.02 | 15.53 | 16.00 | 15.78 | 15.56 | 15.46 | 15.41 | 15.40 | 15.39 | 15.39 | 15.38 | 15.38 | 15.38 | 15.38 | 15.39 | 15.39 |
| 1.03 | 14.93 | 15.67 | 15.48 | 15.30 | 15.21 | 15.17 | 15.15 | 15.15 | 15.15 | 15.15 | 15.15 | 15.15 | 15.15 | 15.15 | 15.15 |
| 1.04 | 14.31 | 15.33 | 15.19 | 15.03 | 14.96 | 14.92 | 14.91 | 14.91 | 14.91 | 14.91 | 14.91 | 14.91 | 14.91 | 14.91 | 14.91 |
| 1.05 | 13.66 | 15.00 | 14.91 | 14.77 | 14.71 | 14.68 | 14.67 | 14.67 | 14.67 | 14.67 | 14.68 | 14.68 | 14.68 | 14.68 | 14.68 |
| 1.06 | 12.98 | 14.67 | 14.62 | 14.51 | 14.46 | 14.44 | 14.44 | 14.44 | 14.44 | 14.44 | 14.45 | 14.45 | 14.45 | 14.45 | 14.45 |
| 1.07 | 12.27 | 14.33 | 14.33 | 14.26 | 14.22 | 14.20 | 14.20 | 14.21 | 14.21 | 14.21 | 14.22 | 14.22 | 14.22 | 14.23 | 14.22 |
| 1.08 | 11.51 | 14.00 | 14.05 | 14.00 | 13.97 | 13.97 | 13.97 | 13.98 | 13.98 | 13.98 | 13.99 | 13.99 | 14.00 | 14.00 | 14.00 |
| 1.09 | 10.71 | 13.67 | 13.76 | 13.75 | 13.73 | 13.74 | 13.74 | 13.75 | 13.75 | 13.76 | 13.77 | 13.77 | 13.77 | 13.78 | 13.78 |
| 1.10 | 9.84 | 13.33 | 13.48 | 13.49 | 13.50 | 13.51 | 13.52 | 13.52 | 13.53 | 13.54 | 13.54 | 13.55 | 13.55 | 13.56 | 13.56 |
| 1.11 | 8.89 | 13.00 | 13.20 | 13.25 | 13.26 | 13.28 | 13.29 | 13.30 | 13.31 | 13.31 | 13.32 | 13.33 | 13.34 | 13.34 | 13.34 |
| 1.12 | 7.82 | 12.67 | 12.93 | 13.00 | 13.03 | 13.05 | 13.07 | 13.08 | 13.09 | 13.10 | 13.11 | 13.12 | 13.12 | 13.13 | 13.13 |
| 1.13 | 6.60 | 12.33 | 12.65 | 12.75 | 12.80 | 12.83 | 12.85 | 12.86 | 12.87 | 12.88 | 12.89 | 12.90 | 12.91 | 12.91 | 12.92 |
| 1.14 | 5.08 | 12.00 | 12.37 | 12.51 | 12.57 | 12.61 | 12.63 | 12.65 | 12.66 | 12.67 | 12.68 | 12.69 | 12.70 | 12.70 | 12.71 |
| 1.15 | 2.87 | 11.67 | 12.10 | 12.27 | 12.34 | 12.39 | 12.42 | 12.44 | 12.45 | 12.46 | 12.47 | 12.48 | 12.49 | 12.49 | 12.50 |
| 1.16 | 0.00 | 11.33 | 11.83 | 12.03 | 12.12 | 12.18 | 12.21 | 12.22 | 12.24 | 12.25 | 12.26 | 12.28 | 12.28 | 12.29 | 12.29 |
| 1.17 | 0.00 | 11.00 | 11.56 | 11.79 | 11.90 | 11.96 | 12.00 | 12.02 | 12.03 | 12.04 | 12.06 | 12.07 | 12.08 | 12.09 | 12.09 |
| 1.18 | 0.00 | 10.67 | 11.29 | 11.56 | 11.68 | 11.75 | 11.79 | 11.81 | 11.82 | 11.84 | 11.85 | 11.87 | 11.88 | 11.88 | 11.89 |
| 1.19 | 0.00 | 10.33 | 11.02 | 11.33 | 11.46 | 11.54 | 11.58 | 11.61 | 11.62 | 11.63 | 11.65 | 11.67 | 11.68 | 11.69 | 11.69 |
| 1.20 | 0.00 | 10.00 | 10.76 | 11.10 | 11.24 | 11.34 | 11.38 | 11.41 | 11.42 | 11.43 | 11.46 | 11.47 | 11.48 | 11.49 | 11.49 |
| 1.21 | 0.00 | 9.67 | 10.50 | 10.87 | 11.03 | 11.13 | 11.18 | 11.21 | 11.22 | 11.24 | 11.26 | 11.28 | 11.29 | 11.30 | 11.30 |
| 1.22 | 0.00 | 9.33 | 10.23 | 10.65 | 10.82 | 10.93 | 10.98 | 11.01 | 11.03 | 11.04 | 11.07 | 11.09 | 11.09 | 11.10 | 11.11 |
| 1.23 | 0.00 | 9.00 | 9.97 | 10.42 | 10.61 | 10.73 | 10.78 | 10.81 | 10.84 | 10.85 | 10.88 | 10.90 | 10.91 | 10.92 | 10.92 |
| 1.24 | 0.00 | 8.67 | 9.72 | 10.20 | 10.41 | 10.53 | 10.59 | 10.62 | 10.64 | 10.66 | 10.69 | 10.71 | 10.72 | 10.73 | 10.73 |
| 1.25 | 0.00 | 8.33 | 9.46 | 9.98 | 10.21 | 10.34 | 10.40 | 10.43 | 10.46 | 10.47 | 10.50 | 10.52 | 10.53 | 10.54 | 10.55 |
| 1.26 | 0.00 | 8.00 | 9.21 | 9.77 | 10.00 | 10.15 | 10.21 | 10.25 | 10.27 | 10.29 | 10.32 | 10.34 | 10.35 | 10.36 | 10.37 |
| 1.27 | 0.00 | 7.67 | 8.96 | 9.55 | 9.81 | 9.96 | 10.02 | 10.06 | 10.09 | 10.10 | 10.13 | 10.16 | 10.17 | 10.18 | 10.19 |
| 1.28 | 0.00 | 7.33 | 8.71 | 9.34 | 9.61 | 9.77 | 9.84 | 9.88 | 9.90 | 9.92 | 9.95 | 9.98 | 9.99 | 10.00 | 10.01 |
| 1.29 | 0.00 | 7.00 | 8.46 | 9.13 | 9.42 | 9.58 | 9.66 | 9.70 | 9.72 | 9.74 | 9.78 | 9.80 | 9.82 | 9.83 | 9.83 |
| 1.30 | 0.00 | 6.67 | 8.21 | 8.93 | 9.22 | 9.40 | 9.48 | 9.52 | 9.55 | 9.57 | 9.60 | 9.63 | 9.64 | 9.65 | 9.66 |
| 1.31 | 0.00 | 6.33 | 7.97 | 8.72 | 9.03 | 9.22 | 9.30 | 9.34 | 9.37 | 9.39 | 9.43 | 9.46 | 9.47 | 9.48 | 9.49 |
| 1.32 | 0.00 | 6.00 | 7.73 | 8.52 | 8.85 | 9.04 | 9.12 | 9.17 | 9.20 | 9.22 | 9.26 | 9.29 | 9.30 | 9.31 | 9.32 |

**Table 9.12** (Continued)

| $Q_U$ or $Q_L$ | Sample size 3 | 4 | 5 | 7 | 10 | 15 | 20 | 25 | 30 | 35 | 50 | 75 | 100 | 150 | 200 |
|------|------|------|------|------|------|------|------|------|------|------|------|------|------|------|------|
| 1.33 | 0.00 | 5.67 | 7.49 | 8.32 | 8.66 | 8.86 | 8.95 | 9.00 | 9.03 | 9.05 | 9.09 | 9.12 | 9.13 | 9.15 | 9.15 |
| 1.34 | 0.00 | 5.33 | 7.25 | 8.12 | 8.48 | 8.69 | 8.78 | 8.83 | 8.86 | 8.88 | 8.92 | 8.95 | 8.97 | 8.98 | 8.99 |
| 1.35 | 0.00 | 5.00 | 7.02 | 7.92 | 8.30 | 8.52 | 8.61 | 8.66 | 8.69 | 8.72 | 8.76 | 8.79 | 8.81 | 8.82 | 8.83 |
| 1.36 | 0.00 | 4.67 | 6.79 | 7.73 | 8.12 | 8.35 | 8.44 | 8.50 | 8.53 | 8.55 | 8.60 | 8.63 | 8.65 | 8.66 | 8.67 |
| 1.37 | 0.00 | 4.33 | 6.56 | 7.54 | 7.95 | 8.18 | 8.28 | 8.33 | 8.37 | 8.39 | 8.44 | 8.47 | 8.49 | 8.50 | 8.51 |
| 1.38 | 0.00 | 4.00 | 6.33 | 7.35 | 7.77 | 8.01 | 8.12 | 8.17 | 8.21 | 8.24 | 8.28 | 8.31 | 8.33 | 8.35 | 8.36 |
| 1.39 | 0.00 | 3.67 | 6.10 | 7.17 | 7.60 | 7.85 | 7.96 | 8.01 | 8.05 | 8.08 | 8.12 | 8.16 | 8.18 | 8.19 | 8.20 |
| 1.40 | 0.00 | 3.33 | 5.88 | 6.98 | 7.44 | 7.69 | 7.80 | 7.86 | 7.90 | 7.92 | 7.97 | 8.01 | 8.02 | 8.04 | 8.05 |
| 1.41 | 0.00 | 3.00 | 5.66 | 6.80 | 7.27 | 7.53 | 7.64 | 7.70 | 7.74 | 7.77 | 7.82 | 7.86 | 7.87 | 7.89 | 7.90 |
| 1.42 | 0.00 | 2.67 | 5.44 | 6.62 | 7.10 | 7.37 | 7.49 | 7.55 | 7.59 | 7.62 | 7.67 | 7.71 | 7.73 | 7.74 | 7.75 |
| 1.43 | 0.00 | 2.33 | 5.23 | 6.45 | 6.94 | 7.22 | 7.34 | 7.40 | 7.44 | 7.47 | 7.52 | 7.56 | 7.58 | 7.60 | 7.61 |
| 1.44 | 0.00 | 2.00 | 5.02 | 6.27 | 6.78 | 7.07 | 7.19 | 7.26 | 7.30 | 7.33 | 7.38 | 7.42 | 7.44 | 7.46 | 7.47 |
| 1.45 | 0.00 | 1.67 | 4.81 | 6.10 | 6.63 | 6.92 | 7.04 | 7.11 | 7.15 | 7.18 | 7.24 | 7.28 | 7.30 | 7.32 | 7.32 |
| 1.46 | 0.00 | 1.33 | 4.60 | 5.93 | 6.47 | 6.77 | 6.90 | 6.97 | 7.01 | 7.04 | 7.10 | 7.14 | 7.16 | 7.18 | 7.19 |
| 1.47 | 0.00 | 1.00 | 4,39 | 5.77 | 6.32 | 6.63 | 6.75 | 6.83 | 6.87 | 6.90 | 6.96 | 7.00 | 7.02 | 7.04 | 7.05 |
| 1.48 | 0.00 | .67 | 4.19 | 5.60 | 6.17 | 6.48 | 6.61 | 6.69 | 6.73 | 6.77 | 6.82 | 6.86 | 6.88 | 6.90 | 6.91 |
| 1.49 | 0.00 | .33 | 3.99 | 5.44 | 6.02 | 6.34 | 6.48 | 6.55 | 6.60 | 6.63 | 6.69 | 6.73 | 6.75 | 6.77 | 6.78 |
| 1.50 | 0.00 | 0.00 | 3.80 | 5.28 | 5.87 | 6.20 | 6.34 | 6.41 | 6.46 | 6.50 | 6.55 | 6.60 | 6.62 | 6.64 | 6.65 |
| 1.51 | 0.00 | 0.00 | 3.61 | 5.13 | 5.73 | 6.06 | 6.20 | 6.28 | 6.33 | 6.36 | 6.42 | 6.47 | 6.49 | 6.51 | 6.52 |
| 1.52 | 0.00 | 0.00 | 3.42 | 4.97 | 5.59 | 5.93 | 6.07 | 6.15 | 6.20 | 6.23 | 6.29 | 6.34 | 6.36 | 6.38 | 6.39 |
| 1.53 | 0.00 | 0.00 | 3.23 | 4.82 | 5.45 | 5.80 | 5.94 | 6.02 | 6.07 | 6.11 | 6.17 | 6.21 | 6.24 | 6.26 | 6.27 |
| 1.54 | 0.00 | 0.00 | 3.05 | 4.67 | 5.31 | 5.67 | 5.81 | 5.89 | 5.95 | 5.98 | 6.04 | 6.09 | 6.11 | 6.13 | 6.15 |
| 1.55 | 0.00 | 0.00 | 2.87 | 4.52 | 5.18 | 5.54 | 5.69 | 5.77 | 5.82 | 5.86 | 5.92 | 5.97 | 5.99 | 6.01 | 6.02 |
| 1.56 | 0.00 | 0.00 | 2.69 | 4.38 | 5.05 | 5.41 | 5.56 | 5.65 | 5.70 | 5.74 | 5.80 | 5.85 | 5.87 | 5.89 | 5.90 |
| 1.57 | 0.00 | 0.00 | 2.52 | 4.24 | 4.92 | 5.29 | 5.44 | 5.53 | 5.58 | 5.62 | 5.68 | 5.73 | 5.75 | 5.78 | 5.79 |
| 1.58 | 0.00 | 0.00 | 2.35 | 4.10 | 4.79 | 5.16 | 5.32 | 5.41 | 5.46 | 5.50 | 5.56 | 5.61 | 5.64 | 5.66 | 5.67 |
| 1.59 | 0.00 | 0.00 | 2.19 | 3.96 | 4.66 | 5.04 | 5.20 | 5.29 | 5.34 | 5.38 | 5.45 | 5.50 | 5.52 | 5.55 | 5.56 |
| 1.60 | 0.00 | 0.00 | 2.03 | 3.83 | 4.54 | 4.92 | 5.08 | 5.17 | 5.23 | 5.27 | 5.33 | 5.38 | 5.41 | 5.43 | 5.44 |
| 1.61 | 0.00 | 0.00 | 1.87 | 3.69 | 4.41 | 4.81 | 4.97 | 5.06 | 5.12 | 5.16 | 5.22 | 5.27 | 5.30 | 5.32 | 5.33 |
| 1.62 | 0.00 | 0.00 | 1.72 | 3.57 | 4.30 | 4.69 | 4.86 | 4.95 | 5.01 | 5.04 | 5.11 | 5.16 | 5.19 | 5.21 | 5.23 |
| 1.63 | 0.00 | 0.00 | 1.57 | 3.44 | 4.18 | 4.58 | 4.75 | 4.84 | 4.90 | 4.94 | 5.01 | 5.06 | 5.08 | 5.11 | 5.12 |
| 1.64 | 0.00 | 0.00 | 1.42 | 3.31 | 4.06 | 4.47 | 4.64 | 4.73 | 4.79 | 4.83 | 4.90 | 4.95 | 4.98 | 5.00 | 5.01 |
| 1.65 | 0.00 | 0.00 | 1.28 | 3.19 | 3.95 | 4.36 | 4.53 | 4.62 | 4.68 | 4.72 | 4.79 | 4.85 | 4.87 | 4.90 | 4.91 |
| 1.66 | 0.00 | 0.00 | 1.15 | 3.07 | 3.84 | 4.25 | 4.43 | 4.52 | 4.58 | 4.62 | 4.69 | 4.74 | 4.77 | 4.80 | 4.81 |
| 1.67 | 0.00 | 0.00 | 1.02 | 2.95 | 3.73 | 4.15 | 4.32 | 4.42 | 4.48 | 4.52 | 4.59 | 4.64 | 4.67 | 4.70 | 4.71 |

*(Continued)*

**Table 9.12** (Continued)

| $Q_U$ or $Q_L$ | \ Sample size 3 | 4 | 5 | 7 | 10 | 15 | 20 | 25 | 30 | 35 | 50 | 75 | 100 | 150 | 200 |
|---|---|---|---|---|---|---|---|---|---|---|---|---|---|---|---|
| 1.68 | 0.00 | 0.00 | 0.89 | 2.84 | 3.62 | 4.05 | 4.22 | 4.32 | 4.38 | 4.42 | 4.49 | 4.55 | 4.57 | 4.60 | 4.61 |
| 1.69 | 0.00 | 0.00 | 0.77 | 2.73 | 3.52 | 3.94 | 4.12 | 4.22 | 4.28 | 4.32 | 4.39 | 4.45 | 4.47 | 4.50 | 4.51 |
| 1.70 | 0.00 | 0.00 | 0.66 | 2.62 | 3.41 | 3.84 | 4.02 | 4.12 | 4.18 | 4.22 | 4.30 | 4.35 | 4.38 | 4.41 | 4.42 |
| 1.71 | 0.00 | 0.00 | 0.55 | 2.51 | 3.31 | 3.75 | 3.93 | 4.02 | 4.09 | 4.13 | 4.20 | 4.26 | 4.29 | 4.31 | 4.32 |
| 1.72 | 0.00 | 0.00 | 0.45 | 2.41 | 3.21 | 3.65 | 3.83 | 3.93 | 3.99 | 4.04 | 4.11 | 4.17 | 4.19 | 4.22 | 4.23 |
| 1.73 | 0.00 | 0.00 | 0.36 | 2.30 | 3.11 | 3.56 | 3.74 | 3.84 | 3.90 | 3.94 | 4.02 | 4.08 | 4.10 | 4.13 | 4.14 |
| 1.74 | 0.00 | 0.00 | 0.27 | 2.20 | 3.02 | 3.46 | 3.65 | 3.75 | 3.81 | 3.85 | 3.93 | 3.99 | 4.01 | 4.04 | 4.05 |
| 1.75 | 0.00 | 0.00 | 0.19 | 2.11 | 2.93 | 3.37 | 3.56 | 3.66 | 3.72 | 3.77 | 3.84 | 3.90 | 3.93 | 3.95 | 3.97 |
| 1.76 | 0.00 | 0.00 | 0.12 | 2.01 | 2.83 | 3.28 | 3.47 | 3.57 | 3.63 | 3.68 | 3.76 | 3.81 | 3.84 | 3.87 | 3.88 |
| 1.77 | 0.00 | 0.00 | 0.06 | 1.92 | 2.74 | 3.20 | 3.38 | 3.48 | 3.55 | 3.59 | 3.67 | 3.73 | 3.76 | 3.78 | 3.80 |
| 1.78 | 0.00 | 0.00 | 0.02 | 1.83 | 2.66 | 3.11 | 3.30 | 3.40 | 3.47 | 3.51 | 3.59 | 3.64 | 3.67 | 3.70 | 3.71 |
| 1.79 | 0.00 | 0.00 | 0.00 | 1.74 | 2.57 | 3.03 | 3.21 | 3.32 | 3.38 | 3.43 | 3.51 | 3.56 | 3.59 | 3.62 | 3.63 |
| 1.80 | 0.00 | 0.00 | 0.00 | 1.65 | 2.49 | 2.94 | 3.13 | 3.24 | 3.30 | 3.35 | 3.43 | 3.48 | 3.51 | 3.54 | 3.55 |
| 1.81 | 0.00 | 0.00 | 0.00 | 1.57 | 2.40 | 2.86 | 3.05 | 31.6 | 3.22 | 3.27 | 3.35 | 3.40 | 3.43 | 3.46 | 3.47 |
| 1.82 | 0.00 | 0.00 | 0.00 | 1.49 | 2.32 | 2.79 | 2.98 | 3.08 | 3.15 | 3.19 | 3.27 | 3.33 | 3.36 | 3.38 | 3.40 |
| 1.83 | 0.00 | 0.00 | 0.00 | 1.41 | 2.25 | 2.71 | 2.90 | 3.00 | 3.07 | 3.11 | 3.19 | 3.25 | 3.28 | 3.31 | 3.32 |
| 1.84 | 0.00 | 0.00 | 0.00 | 1.34 | 2.17 | 2.63 | 2.82 | 2.93 | 2.99 | 3.04 | 3.12 | 3.18 | 3.21 | 3.23 | 3.25 |
| 1.85 | 0.00 | 0.00 | 0.00 | 1.26 | 2.09 | 2.56 | 2.75 | 2.85 | 2.92 | 2.97 | 3.05 | 3.10 | 3.13 | 3.16 | 3.17 |
| 1.86 | 0.00 | 0.00 | 0.00 | 1.19 | 2.02 | 2.48 | 2.68 | 2.78 | 2.85 | 2.89 | 2.97 | 3.03 | 3.06 | 3.09 | 3.10 |
| 1.87 | 0.00 | 0.00 | 0.00 | 1.12 | 1.95 | 2.41 | 2.61 | 2.71 | 2.78 | 2.82 | 2.90 | 2.96 | 2.99 | 3.02 | 3.03 |
| 1.88 | 0.00 | 0.00 | 0.00 | 1.06 | 1.88 | 2.34 | 2.54 | 2.64 | 2.71 | 2.75 | 2.83 | 2.89 | 2.92 | 2.95 | 2.96 |
| 1.89 | 0.00 | 0.00 | 0.00 | 0.99 | 1.81 | 2.28 | 2.47 | 2.57 | 2.64 | 2.69 | 2.77 | 2.83 | 2.85 | 2.88 | 2.90 |
| 1.90 | 0.00 | 0.00 | 0.00 | 0.93 | 1.75 | 2.21 | 2.40 | 2.51 | 2.57 | 2.62 | 2.70 | 2.76 | 2.79 | 2.82 | 2.83 |
| 1.91 | 0.00 | 0.00 | 0.00 | 0.87 | 1.68 | 2.14 | 2.34 | 2.44 | 2.51 | 2.56 | 2.63 | 2.69 | 2.72 | 2.75 | 2.77 |
| 1.92 | 0.00 | 0.00 | 0.00 | 0.81 | 1.62 | 2.08 | 2.27 | 2.38 | 2.45 | 2.49 | 2.57 | 2.63 | 2.66 | 2.69 | 2.70 |
| 1.93 | 0.00 | 0.00 | 0.00 | 0.76 | 1.56 | 2.02 | 2.21 | 2.32 | 2.38 | 2.43 | 2.51 | 2.57 | 2.60 | 2.63 | 2.64 |
| 1.94 | 0.00 | 0.00 | 0.00 | 0.70 | 1.50 | 1.96 | 2.15 | 2.25 | 2.32 | 2.37 | 2.45 | 2.51 | 2.54 | 2.56 | 2.58 |
| 1.95 | 0.00 | 0.00 | 0.00 | 0.65 | 1.44 | 1.90 | 2.09 | 2.19 | 2.26 | 2.31 | 2.39 | 2.45 | 2.48 | 2.50 | 2.52 |
| 1.96 | 0.00 | 0.00 | 0.00 | 0.60 | 1.38 | 1.84 | 2.03 | 2.14 | 2.20 | 2.25 | 2.33 | 2.39 | 2.42 | 2.44 | 2.46 |
| 1.97 | 0.00 | 0.00 | 0.00 | 0.56 | 1.33 | 1.78 | 1.97 | 2.08 | 2.14 | 2.19 | 2.27 | 2.33 | 2.36 | 2.39 | 2.40 |
| 1.98 | 0.00 | 0.00 | 0.00 | 0.51 | 1.27 | 1.73 | 1.92 | 2.02 | 2.09 | 2.13 | 2.21 | 2.27 | 2.30 | 2.33 | 2.34 |
| 1.99 | 0.00 | 0.00 | 0.00 | 0.47 | 1.22 | 1.67 | 1.86 | 1.97 | 2.03 | 2.08 | 2.16 | 2.22 | 2.25 | 2.27 | 2.29 |
| 2.00 | 0.00 | 0.00 | 0.00 | 0.43 | 1.17 | 1.62 | 1.81 | 1.91 | 1.98 | 2.03 | 2.10 | 2.16 | 2.19 | 2.22 | 2.23 |
| 2.01 | 0.00 | 0.00 | 0.00 | 0.39 | 1.12 | 1.57 | 1.76 | 1.86 | 1.93 | 1.97 | 2.05 | 2.11 | 2.14 | 2.17 | 2.18 |
| 2.02 | 0.00 | 0.00 | 0.00 | 0.36 | 1.07 | 1.52 | 1.71 | 1.81 | 1.87 | 1.92 | 2.00 | 2.06 | 2.09 | 2.11 | 2.13 |
| 2.03 | 0.00 | 0.00 | 0.00 | 0.32 | 1.03 | 1.47 | 1.66 | 1.76 | 1.82 | 1.87 | 1.95 | 2.01 | 2.04 | 2.06 | 2.08 |

**Table 9.12** (Continued)

| $Q_U$ or $Q_L$ | 3 | 4 | 5 | 7 | 10 | 15 | 20 | 25 | 30 | 35 | 50 | 75 | 100 | 150 | 200 |
|---|---|---|---|---|---|---|---|---|---|---|---|---|---|---|---|
| 2.04 | 0.00 | 0.00 | 0.00 | 0.29 | 0.98 | 1.42 | 1.61 | 1.71 | 1.77 | 1.82 | 1.90 | 1.96 | 1.99 | 2.01 | 2.03 |
| 2.05 | 0.00 | 0.00 | 0.00 | 0.26 | 0.94 | 1.37 | 1.56 | 1.66 | 1.73 | 1.77 | 1.85 | 1.91 | 1.94 | 1.96 | 1.98 |
| 2.06 | 0.00 | 0.00 | 0.00 | 0.23 | 0.90 | 1.33 | 1.51 | 1.61 | 1.68 | 1.72 | 1.80 | 1.86 | 1.89 | 1.92 | 1.93 |
| 2.07 | 0.00 | 0.00 | 0.00 | 0.21 | 0.86 | 1.28 | 1.47 | 1.57 | 1.63 | 1.68 | 1.76 | 1.81 | 1.84 | 1.87 | 1.88 |
| 2.08 | 0.00 | 0.00 | 0.00 | 0.18 | 0.82 | 1.24 | 1.42 | 1.52 | 1.59 | 1.63 | 1.71 | 1.77 | 1.79 | 1.82 | 1.84 |
| 2.09 | 0.00 | 0.00 | 0.00 | 0.16 | 0.78 | 1.20 | 1.38 | 1.48 | 1.54 | 1.59 | 1.66 | 1.72 | 1.75 | 1.78 | 1.79 |
| 2.10 | 0.00 | 0.00 | 0.00 | 0.14 | 0.74 | 1.16 | 1.34 | 1.44 | 1.50 | 1.54 | 1.62 | 1.68 | 1.71 | 1.73 | 1.75 |
| 2.11 | 0.00 | 0.00 | 0.00 | 0.12 | 0.71 | 1.12 | 1.30 | 1.39 | 1.46 | 1.50 | 1.58 | 1.63 | 1.66 | 1.69 | 1.70 |
| 2.12 | 0.00 | 0.00 | 0.00 | 0.10 | 0.67 | 1.08 | 1.26 | 1.35 | 1.42 | 1.46 | 1.54 | 1.59 | 1.62 | 1.65 | 1.66 |
| 2.13 | 0.00 | 0.00 | 0.00 | 0.08 | 0.64 | 1.04 | 1.22 | 1.31 | 1.38 | 1.42 | 1.50 | 1.55 | 1.58 | 1.61 | 1.62 |
| 2.14 | 0.00 | 0.00 | 0.00 | 0.07 | 0.61 | 1.00 | 1.18 | 1.28 | 1.34 | 1.38 | 1.46 | 1.51 | 1.54 | 1.57 | 1.58 |
| 2.15 | 0.00 | 0.00 | 0.00 | 0.06 | 0.58 | 0.97 | 1.14 | 1.24 | 1.30 | 1.34 | 1.42 | 1.47 | 1.50 | 1.53 | 1.54 |
| 2.16 | 0.00 | 0.00 | 0.00 | 0.05 | 0.55 | 0.93 | 1.10 | 1.20 | 1.26 | 1.30 | 1.38 | 1.43 | 1.46 | 1.49 | 1.50 |
| 2.17 | 0.00 | 0.00 | 0.00 | 0.04 | 0.52 | 0.90 | 1.07 | 1.16 | 1.22 | 1.27 | 1.34 | 1.40 | 1.42 | 1.45 | 1.46 |
| 2.18 | 0.00 | 0.00 | 0.00 | 0.03 | 0.49 | 0.87 | 1.03 | 1.13 | 1.19 | 1.23 | 1.30 | 1.36 | 1.39 | 1.41 | 1.42 |
| 2.19 | 0.00 | 0.00 | 0.00 | 0.02 | 0.46 | 0.83 | 1.00 | 1.09 | 1.15 | 1.20 | 1.27 | 1.32 | 1.35 | 1.38 | 1.39 |
| 2.20 | 0.000 | 0.000 | 0.000 | 0.015 | 0.437 | 0.803 | 0.968 | 1.160 | 1.120 | 1.160 | 1.233 | 1.287 | 1.314 | 1.340 | 1.352 |
| 2.21 | 0.000 | 0.000 | 0.000 | 0.010 | 0.413 | 0.772 | 0.936 | 1.028 | 1.087 | 1.128 | 1.199 | 1.253 | 1.279 | 1.305 | 1.318 |
| 2.22 | 0.000 | 0.000 | 0.000 | 0.006 | 0.389 | 0.734 | 0.905 | 0.996 | 1.054 | 1.095 | 1.166 | 1.219 | 1.245 | 1.271 | 1.284 |
| 2.23 | 0.000 | 0.000 | 0.000 | 0.003 | 0.366 | 0.715 | 0.874 | 0.965 | 1.023 | 1.063 | 1.134 | 1.186 | 1.212 | 1.238 | 1.250 |
| 2.24 | 0.000 | 0.000 | 0.000 | 0.002 | 0.345 | 0.687 | 0.845 | 0.935 | 0.992 | 1.032 | 1.102 | 1.154 | 1.180 | 1.205 | 1.218 |
| 2.25 | 0.000 | 0.000 | 0.000 | 0.001 | 0.324 | 0.660 | 0.816 | 0.905 | 0.962 | 1.002 | 1.071 | 1.123 | 1.148 | 1.173 | 1.186 |
| 2.26 | 0.000 | 0.000 | 0.000 | 0.000 | 0.304 | 0.634 | 0.789 | 0.876 | 0.933 | 0.972 | 1.041 | 1.092 | 1.117 | 1.142 | 1.155 |
| 2.27 | 0.000 | 0.000 | 0.000 | 0.000 | 0.285 | 0.609 | 0.762 | 0.848 | 0.904 | 0.943 | 1.011 | 1.062 | 1.087 | 1.112 | 1.124 |
| 2.28 | 0.000 | 0.000 | 0.000 | 0.000 | 0.267 | 0.585 | 0.735 | 0.821 | 0.876 | 0.915 | 0.982 | 1.033 | 1.058 | 1.082 | 1.095 |
| 2.29 | 0.000 | 0.000 | 0.000 | 0.000 | 0.250 | 0.561 | 0.710 | 0.794 | 0.849 | 0.887 | 0.954 | 1.004 | 1.029 | 1.053 | 1.065 |
| 2.30 | 0.000 | 0.000 | 0.000 | 0.000 | 0.233 | 0.538 | 0.685 | 0.769 | 0.823 | 0.861 | 0.927 | 0.977 | 1.001 | 1.025 | 1.037 |
| 2.31 | 0.000 | 0.000 | 0.000 | 0.000 | 0.218 | 0.516 | 0.661 | 0.743 | 0.797 | 0.834 | 0.900 | 0.949 | 0.974 | 0.998 | 1.009 |
| 2.32 | 0.000 | 0.000 | 0.000 | 0.000 | 0.203 | 0.495 | 0.637 | 0.719 | 0.772 | 0.809 | 0.874 | 0.923 | 0.947 | 0.971 | 0.982 |
| 2.33 | 0.000 | 0.000 | 0.000 | 0.000 | 0.189 | 0.474 | 0.614 | 0.695 | 0.748 | 0.784 | 0.848 | 0.897 | 0.921 | 0.944 | 0.956 |
| 2.34 | 0.000 | 0.000 | 0.000 | 0.000 | 0.175 | 0.454 | 0.592 | 0.672 | 0.724 | 0.760 | 0.824 | 0.872 | 0.895 | 0.919 | 0.930 |
| 2.35 | 0.000 | 0.000 | 0.000 | 0.000 | 0.163 | 0.435 | 0.571 | 0.650 | 0.701 | 0.736 | 0.799 | 0.847 | 0.870 | 0.893 | 0.905 |
| 2.36 | 0.000 | 0.000 | 0.000 | 0.000 | 0.151 | 0.416 | 0.550 | 0.628 | 0.678 | 0.714 | 0.776 | 0.823 | 0.846 | 0.869 | 0.880 |
| 2.37 | 0.000 | 0.000 | 0.000 | 0.000 | 0.139 | 0.398 | 0.530 | 0.606 | 0.656 | 0.691 | 0.753 | 0.799 | 0.822 | 0.845 | 0.856 |
| 2.38 | 0.000 | 0.000 | 0.000 | 0.000 | 0.128 | 0.381 | 0.510 | 0.586 | 0.635 | 0.670 | 0.730 | 0.777 | 0.799 | 0.822 | 0.833 |
| 2.39 | 0.000 | 0.000 | 0.000 | 0.000 | 0.118 | 0.364 | 0.491 | 0.566 | 0.614 | 0.648 | 0.709 | 0.754 | 0.777 | 0.799 | 0.810 |

*(Continued)*

**Table 9.12** (Continued)

| $Q_U$ or $Q_L$ | Sample size | | | | | | | | | | | | | | |
|---|---|---|---|---|---|---|---|---|---|---|---|---|---|---|---|
| | 3 | 4 | 5 | 7 | 10 | 15 | 20 | 25 | 30 | 35 | 50 | 75 | 100 | 150 | 200 |
| 2.40 | 0.000 | 0.000 | 0.000 | 0.000 | 0.109 | 0.348 | 0.473 | 0.546 | 0.594 | 0.628 | 0.687 | 0.732 | 0.755 | 0.777 | 0.787 |
| 2.41 | 0.000 | 0.000 | 0.000 | 0.000 | 0.100 | 0.332 | 0.455 | 0.527 | 0.575 | 0.608 | 0.667 | 0.711 | 0.733 | 0.755 | 0.766 |
| 2.42 | 0.000 | 0.000 | 0.000 | 0.000 | 0.091 | 0.317 | 0.437 | 0.509 | 0.555 | 0.588 | 0.646 | 0.691 | 0.712 | 0.734 | 0.744 |
| 2.43 | 0.000 | 0.000 | 0.000 | 0.000 | 0.083 | 0.302 | 0.421 | 0.491 | 0.537 | 0.569 | 0.627 | 0.670 | 0.692 | 0.713 | 0.724 |
| 2.44 | 0.000 | 0.000 | 0.000 | 0.000 | 0.076 | 0.288 | 0.404 | 0.474 | 0.519 | 0.551 | 0.608 | 0.651 | 0.672 | 0.693 | 0.703 |
| 2.45 | 0.000 | 0.000 | 0.000 | 0.000 | 0.069 | 0.275 | 0.389 | 0.457 | 0.501 | 0.533 | 0.589 | 0.632 | 0.653 | 0.673 | 0.684 |
| 2.46 | 0.000 | 0.000 | 0.000 | 0.000 | 0.063 | 0.262 | 0.373 | 0.440 | 0.484 | 0.516 | 0.571 | 0.613 | 0.634 | 0.654 | 0.664 |
| 2.47 | 0.000 | 0.000 | 0.000 | 0.000 | 0.057 | 0.249 | 0.359 | 0.425 | 0.468 | 0.499 | 0.553 | 0.595 | 0.615 | 0.636 | 0.646 |
| 2.48 | 0.000 | 0.000 | 0.000 | 0.000 | 0.051 | 0.237 | 0.345 | 0.409 | 0.452 | 0.482 | 0.536 | 0.577 | 0.597 | 0.617 | 0.627 |
| 2.49 | 0.000 | 0.000 | 0.000 | 0.000 | 0.046 | 0.226 | 0.331 | 0.394 | 0.436 | 0.466 | 0.519 | 0.560 | 0.580 | 0.600 | 0.609 |
| 2.50 | 0.000 | 0.000 | 0.000 | 0.000 | 0.041 | 0.214 | 0.317 | 0.380 | 0.421 | 0.451 | 0.503 | 0.543 | 0.563 | 0.582 | 0.592 |
| 2.51 | 0.000 | 0.000 | 0.000 | 0.000 | 0.037 | 0.204 | 0.305 | 0.366 | 0.407 | 0.436 | 0.487 | 0.527 | 0.546 | 0.565 | 0.575 |
| 2.52 | 0.000 | 0.000 | 0.000 | 0.000 | 0.033 | 0.193 | 0.292 | 0.352 | 0.392 | 0.421 | 0.472 | 0.511 | 0.530 | 0.549 | 0.559 |
| 2.53 | 0.000 | 0.000 | 0.000 | 0.000 | 0.029 | 0.184 | 0.280 | 0.339 | 0.379 | 0.407 | 0.457 | 0.495 | 0.514 | 0.533 | 0.542 |
| 2.54 | 0.000 | 0.000 | 0.000 | 0.000 | 0.026 | 0.174 | 0.268 | 0.326 | 0.365 | 0.393 | 0.442 | 0.480 | 0.499 | 0.517 | 0.527 |
| 2.55 | 0.000 | 0.000 | 0.000 | 0.000 | 0.023 | 0.165 | 0.257 | 0.314 | 0.352 | 0.379 | 0.428 | 0.465 | 0.484 | 0.502 | 0.511 |
| 2.56 | 0.000 | 0.000 | 0.000 | 0.000 | 0.020 | 0.156 | 0.246 | 0.302 | 0.340 | 0.366 | 0.414 | 0.451 | 0.469 | 0.487 | 0.496 |
| 2.57 | 0.000 | 0.000 | 0.000 | 0.000 | 0.017 | 0.148 | 0.236 | 0.291 | 0.327 | 0.354 | 0.401 | 0.437 | 0.455 | 0.473 | 0.482 |
| 2.58 | 0.000 | 0.000 | 0.000 | 0.000 | 0.015 | 0.140 | 0.226 | 0.279 | 0.316 | 0.341 | 0.388 | 0.424 | 0.441 | 0.459 | 0.468 |
| 2.59 | 0.000 | 0.000 | 0.000 | 0.000 | 0.013 | 0.133 | 0.216 | 0.269 | 0.304 | 0.330 | 0.375 | 0.410 | 0.428 | 0.445 | 0.454 |
| 2.60 | 0.000 | 0.000 | 0.000 | 0.000 | 0.011 | 0.125 | 0.207 | 0.258 | 0.293 | 0.318 | 0.363 | 0.398 | 0.415 | 0.432 | 0.441 |
| 2.61 | 0.000 | 0.000 | 0.000 | 0.000 | 0.009 | 0.118 | 0.198 | 0.248 | 0.282 | 0.307 | 0.351 | 0.385 | 0.402 | 0.419 | 0.428 |
| 2.62 | 0.000 | 0.000 | 0.000 | 0.000 | 0.008 | 0.112 | 0.189 | 0.238 | 0.272 | 0.296 | 0.339 | 0.373 | 0.390 | 0.406 | 0.415 |
| 2.63 | 0.000 | 0.000 | 0.000 | 0.000 | 0.007 | 0.105 | 0.181 | 0.229 | 0.262 | 0.285 | 0.328 | 0.361 | 0.378 | 0.394 | 0.402 |
| 2.64 | 0.000 | 0.000 | 0.000 | 0.000 | 0.006 | 0.099 | 0.172 | 0.220 | 0.252 | 0.275 | 0.317 | 0.350 | 0.366 | 0.382 | 0.390 |
| 2.65 | 0.000 | 0.000 | 0.000 | 0.000 | 0.005 | 0.094 | 0.165 | 0.211 | 0.242 | 0.265 | 0.307 | 0.339 | 0.355 | 0.371 | 0.379 |
| 2.66 | 0.000 | 0.000 | 0.000 | 0.000 | 0.004 | 0.088 | 0.157 | 0.202 | 0.233 | 0.256 | 0.296 | 0.328 | 0.344 | 0.359 | 0.367 |
| 2.67 | 0.000 | 0.000 | 0.000 | 0.000 | 0.003 | 0.083 | 0.150 | 0.194 | 0.224 | 0.246 | 0.286 | 0.317 | 0.333 | 0.348 | 0.356 |
| 2.68 | 0.000 | 0.000 | 0.000 | 0.000 | 0.002 | 0.078 | 0.143 | 0.186 | 0.216 | 0.237 | 0.277 | 0.307 | 0.322 | 0.338 | 0.345 |
| 2.69 | 0.000 | 0.000 | 0.000 | 0.000 | 0.002 | 0.073 | 0.136 | 0.179 | 0.208 | 0.229 | 0.267 | 0.297 | 0.312 | 0.327 | 0.335 |
| 2.70 | 0.000 | 0.000 | 0.000 | 0.000 | 0.001 | 0.069 | 0.130 | 0.171 | 0.200 | 0.220 | 0.258 | 0.288 | 0.302 | 0.317 | 0.325 |
| 2.71 | 0.000 | 0.000 | 0.000 | 0.000 | 0.001 | 0.064 | 0.124 | 0.164 | 0.192 | 0.212 | 0.249 | 0.278 | 0.293 | 0.307 | 0.315 |
| 2.72 | 0.000 | 0.000 | 0.000 | 0.000 | 0.001 | 0.060 | 0.118 | 0.157 | 0.184 | 0.204 | 0.241 | 0.269 | 0.283 | 0.298 | 0.305 |
| 2.73 | 0.000 | 0.000 | 0.000 | 0.000 | 0.001 | 0.057 | 0.112 | 0.151 | 0.177 | 0.197 | 0.232 | 0.260 | 0.274 | 0.288 | 0.296 |
| 2.74 | 0.000 | 0.000 | 0.000 | 0.000 | 0.000 | 0.053 | 0.107 | 0.144 | 0.170 | 0.189 | 0.224 | 0.252 | 0.266 | 0.279 | 0.286 |
| 2.75 | 0.000 | 0.000 | 0.000 | 0.000 | 0.000 | 0.049 | 0.102 | 0.138 | 0.163 | 0.182 | 0.216 | 0.243 | 0.257 | 0.271 | 0.277 |
| 2.76 | 0.000 | 0.000 | 0.000 | 0.000 | 0.000 | 0.046 | 0.097 | 0.132 | 0.157 | 0.175 | 0.209 | 0.235 | 0.249 | 0.262 | 0.269 |

**Table 9.12** (Continued)

| $Q_U$ or $Q_L$ | Sample size | | | | | | | | | | | | | | |
|---|---|---|---|---|---|---|---|---|---|---|---|---|---|---|---|
| | 3 | 4 | 5 | 7 | 10 | 15 | 20 | 25 | 30 | 35 | 50 | 75 | 100 | 150 | 200 |
| 2.77 | 0.000 | 0.000 | 0.000 | 0.000 | 0.000 | 0.043 | 0.092 | 0.126 | 0.151 | 0.168 | 0.201 | 0.227 | 0.241 | 0.254 | 0.260 |
| 2.78 | 0.000 | 0.000 | 0.000 | 0.000 | 0.000 | 0.040 | 0.087 | 0.121 | 0.145 | 0.162 | 0.194 | 0.220 | 0.223 | 0.246 | 0.252 |
| 2.79 | 0.000 | 0.000 | 0.000 | 0.000 | 0.000 | 0.037 | 0.083 | 0.115 | 0.139 | 0.156 | 0.187 | 0.212 | 0.220 | 0.238 | 0.244 |
| 2.80 | 0.000 | 0.000 | 0.000 | 0.000 | 0.000 | 0.035 | 0.079 | 0.110 | 0.133 | 0.150 | 0.181 | 0.205 | 0.218 | 0.230 | 0.237 |
| 2.81 | 0.000 | 0.000 | 0.000 | 0.000 | 0.000 | 0.032 | 0.075 | 0.105 | 0.128 | 0.144 | 0.174 | 0.198 | 0.211 | 0.223 | 0.229 |
| 2.82 | 0.000 | 0.000 | 0.000 | 0.000 | 0.000 | 0.030 | 0.071 | 0.101 | 0.122 | 0.138 | 0.168 | 0.192 | 0.204 | 0.216 | 0.222 |
| 2.83 | 0.000 | 0.000 | 0.000 | 0.000 | 0.000 | 0.028 | 0.067 | 0.096 | 0.117 | 0.133 | 0.162 | 0.185 | 0.197 | 0.209 | 0.215 |
| 2.84 | 0.000 | 0.000 | 0.000 | 0.000 | 0.000 | 0.026 | 0.064 | 0.092 | 0.112 | 0.128 | 0.156 | 0.179 | 0.190 | 0.202 | 0.208 |
| 2.85 | 0.000 | 0.000 | 0.000 | 0.000 | 0.000 | 0.024 | 0.060 | 0.088 | 0.108 | 0.122 | 0.150 | 0.173 | 0.184 | 0.195 | 0.201 |
| 2.86 | 0.000 | 0.000 | 0.000 | 0.000 | 0.000 | 0.022 | 0.057 | 0.084 | 0.103 | 0.118 | 0.145 | 0.167 | 0.178 | 0.189 | 0.195 |
| 2.87 | 0.000 | 0.000 | 0.000 | 0.000 | 0.000 | 0.020 | 0.054 | 0.080 | 0.099 | 0.113 | 0.139 | 0.161 | 0.172 | 0.183 | 0.188 |
| 2.88 | 0.000 | 0.000 | 0.000 | 0.000 | 0.000 | 0.019 | 0.051 | 0.076 | 0.094 | 0.108 | 0.134 | 0.155 | 0.166 | 0.177 | 0.182 |
| 2.89 | 0.000 | 0.000 | 0.000 | 0.000 | 0.000 | 0.017 | 0.048 | 0.073 | 0.090 | 0.104 | 0.129 | 0.150 | 0.160 | 0.171 | 0.176 |
| 2.90 | 0.000 | 0.000 | 0.000 | 0.000 | 0.000 | 0.016 | 0.046 | 0.069 | 0.087 | 0.100 | 0.125 | 0.145 | 0.155 | 0.165 | 0.171 |
| 2.91 | 0.000 | 0.000 | 0.000 | 0.000 | 0.000 | 0.015 | 0.043 | 0.066 | 0.083 | 0.096 | 0.120 | 0.140 | 0.150 | 0.160 | 0.165 |
| 2.92 | 0.000 | 0.000 | 0.000 | 0.000 | 0.000 | 0.013 | 0.041 | 0.063 | 0.079 | 0.092 | 0.115 | 0.135 | 0.145 | 0.155 | 0.160 |
| 2.93 | 0.000 | 0.000 | 0.000 | 0.000 | 0.000 | 0.012 | 0.038 | 0.060 | 0.076 | 0.088 | 0.111 | 0.130 | 0.140 | 0.149 | 0.154 |
| 2.94 | 0.000 | 0.000 | 0.000 | 0.000 | 0.000 | 0.011 | 0.036 | 0.057 | 0.072 | 0.084 | 0.107 | 0.125 | 0.135 | 0.144 | 0.149 |
| 2.95 | 0.000 | 0.000 | 0.000 | 0.000 | 0.000 | 0.010 | 0.034 | 0.054 | 0.069 | 0.081 | 0.103 | 0.121 | 0.130 | 0.140 | 0.144 |
| 2.96 | 0.000 | 0.000 | 0.000 | 0.000 | 0.000 | 0.009 | 0.032 | 0.051 | 0.066 | 0.077 | 0.099 | 0.117 | 0.126 | 0.135 | 0.140 |
| 2.97 | 0.000 | 0.000 | 0.000 | 0.000 | 0.000 | 0.009 | 0.030 | 0.049 | 0.063 | 0.074 | 0.095 | 0.112 | 0.121 | 0.130 | 0.135 |
| 2.98 | 0.000 | 0.000 | 0.000 | 0.000 | 0.000 | 0.008 | 0.028 | 0.046 | 0.060 | 0.071 | 0.091 | 0.108 | 0.117 | 0.126 | 0.130 |
| 2.99 | 0.000 | 0.000 | 0.000 | 0.000 | 0.000 | 0.007 | 0.027 | 0.044 | 0.057 | 0.068 | 0.088 | 0.104 | 0.113 | 0.122 | 0.126 |
| 3.00 | 0.000 | 0.000 | 0.000 | 0.000 | 0.000 | 0.006 | 0.025 | 0.042 | 0.055 | 0.065 | 0.084 | 0.101 | 0.109 | 0.118 | 0.122 |
| 3.01 | 0.000 | 0.000 | 0.000 | 0.000 | 0.000 | 0.006 | 0.024 | 0.040 | 0.052 | 0.062 | 0.081 | 0.097 | 0.105 | 0.113 | 0.118 |
| 3.02 | 0.000 | 0.000 | 0.000 | 0.000 | 0.000 | 0.005 | 0.022 | 0.038 | 0.050 | 0.059 | 0.078 | 0.093 | 0.101 | 0.110 | 0.114 |
| 3.03 | 0.000 | 0.000 | 0.000 | 0.000 | 0.000 | 0.005 | 0.021 | 0.036 | 0.048 | 0.057 | 0.075 | 0.090 | 0.098 | 0.106 | 0.110 |
| 3.04 | 0.000 | 0.000 | 0.000 | 0.000 | 0.000 | 0.004 | 0.019 | 0.034 | 0.045 | 0.054 | 0.072 | 0.087 | 0.094 | 0.102 | 0.106 |
| 3.05 | 0.000 | 0.000 | 0.000 | 0.000 | 0.000 | 0.004 | 0.018 | 0.032 | 0.043 | 0.052 | 0.069 | 0.083 | 0.091 | 0.099 | 0.103 |
| 3.06 | 0.000 | 0.000 | 0.000 | 0.000 | 0.000 | 0.003 | 0.017 | 0.030 | 0.041 | 0.050 | 0.066 | 0.080 | 0.088 | 0.095 | 0.099 |
| 3.07 | 0.000 | 0.000 | 0.000 | 0.000 | 0.000 | 0.003 | 0.016 | 0.029 | 0.039 | 0.047 | 0.064 | 0.077 | 0.085 | 0.092 | 0.096 |
| 3.08 | 0.000 | 0.000 | 0.000 | 0.000 | 0.000 | 0.003 | 0.015 | 0.027 | 0.037 | 0.045 | 0.061 | 0.074 | 0.081 | 0.089 | 0.092 |
| 3.09 | 0.000 | 0.000 | 0.000 | 0.000 | 0.000 | 0.002 | 0.014 | 0.026 | 0.036 | 0.043 | 0.059 | 0.072 | 0.079 | 0.086 | 0.089 |
| 3.10 | 0.000 | 0.000 | 0.000 | 0.000 | 0.000 | 0.002 | 0.013 | 0.024 | 0.034 | 0.041 | 0.056 | 0.069 | 0.076 | 0.083 | 0.086 |
| 3.11 | 0.000 | 0.000 | 0.000 | 0.000 | 0.000 | 0.002 | 0.012 | 0.023 | 0.032 | 0.039 | 0.054 | 0.066 | 0.073 | 0.080 | 0.083 |

*(Continued)*

**Table 9.12** (Continued)

| $Q_U$ or $Q_L$ | \multicolumn{15}{c}{Sample size} |
|------|-------|-------|-------|-------|-------|-------|-------|-------|-------|-------|-------|-------|-------|-------|-------|
| | 3 | 4 | 5 | 7 | 10 | 15 | 20 | 25 | 30 | 35 | 50 | 75 | 100 | 150 | 200 |
| 3.12 | 0.000 | 0.000 | 0.000 | 0.000 | 0.000 | 0.002 | 0.011 | 0.022 | 0.031 | 0.038 | 0.052 | 0.064 | 0.070 | 0.077 | 0.080 |
| 3.13 | 0.000 | 0.000 | 0.000 | 0.000 | 0.000 | 0.002 | 0.011 | 0.021 | 0.029 | 0.036 | 0.050 | 0.061 | 0.068 | 0.074 | 0.077 |
| 3.14 | 0.000 | 0.000 | 0.000 | 0.000 | 0.000 | 0.001 | 0.010 | 0.019 | 0.028 | 0.034 | 0.048 | 0.059 | 0.065 | 0.071 | 0.075 |
| 3.15 | 0.000 | 0.000 | 0.000 | 0.000 | 0.000 | 0.001 | 0.009 | 0.018 | 0.026 | 0.033 | 0.046 | 0.057 | 0.063 | 0.069 | 0.072 |
| 3.16 | 0.000 | 0.000 | 0.000 | 0.000 | 0.000 | 0.001 | 0.009 | 0.017 | 0.025 | 0.031 | 0.044 | 0.055 | 0.060 | 0.066 | 0.069 |
| 3.17 | 0.000 | 0.000 | 0.000 | 0.000 | 0.000 | 0.001 | 0.008 | 0.016 | 0.024 | 0.030 | 0.042 | 0.053 | 0.058 | 0.064 | 0.067 |
| 3.18 | 0.000 | 0.000 | 0.000 | 0.000 | 0.000 | 0.001 | 0.007 | 0.015 | 0.022 | 0.028 | 0.040 | 0.050 | 0.056 | 0.062 | 0.065 |
| 3.19 | 0.000 | 0.000 | 0.000 | 0.000 | 0.000 | 0.001 | 0.007 | 0.015 | 0.021 | 0.027 | 0.038 | 0.049 | 0.054 | 0.059 | 0.062 |
| 3.20 | 0.000 | 0.000 | 0.000 | 0.000 | 0.000 | 0.001 | 0.006 | 0.014 | 0.020 | 0.026 | 0.037 | 0.047 | 0.052 | 0.057 | 0.060 |
| 3.21 | 0.000 | 0.000 | 0.000 | 0.000 | 0.000 | 0.000 | 0.006 | 0.013 | 0.019 | 0.024 | 0.035 | 0.045 | 0.050 | 0.055 | 0.058 |
| 3.22 | 0.000 | 0.000 | 0.000 | 0.000 | 0.000 | 0.000 | 0.005 | 0.012 | 0.018 | 0.023 | 0.034 | 0.043 | 0.048 | 0.053 | 0.056 |
| 3.23 | 0.000 | 0.000 | 0.000 | 0.000 | 0.000 | 0.000 | 0.005 | 0.011 | 0.017 | 0.022 | 0.032 | 0.041 | 0.046 | 0.051 | 0.054 |
| 3.24 | 0.000 | 0.000 | 0.000 | 0.000 | 0.000 | 0.000 | 0.005 | 0.011 | 0.016 | 0.021 | 0.031 | 0.040 | 0.044 | 0.049 | 0.052 |
| 3.25 | 0.000 | 0.000 | 0.000 | 0.000 | 0.000 | 0.000 | 0.004 | 0.010 | 0.015 | 0.020 | 0.030 | 0.038 | 0.043 | 0.048 | 0.050 |
| 3.26 | 0.000 | 0.000 | 0.000 | 0.000 | 0.000 | 0.000 | 0.004 | 0.009 | 0.015 | 0.019 | 0.028 | 0.037 | 0.042 | 0.046 | 0.048 |
| 3.27 | 0.000 | 0.000 | 0.000 | 0.000 | 0.000 | 0.000 | 0.004 | 0.009 | 0.014 | 0.018 | 0.027 | 0.035 | 0.040 | 0.044 | 0.046 |
| 3.28 | 0.000 | 0.000 | 0.000 | 0.000 | 0.000 | 0.000 | 0.003 | 0.008 | 0.013 | 0.017 | 0.026 | 0.034 | 0.038 | 0.042 | 0.045 |
| 3.29 | 0.000 | 0.000 | 0.000 | 0.000 | 0.000 | 0.000 | 0.003 | 0.008 | 0.012 | 0.016 | 0.025 | 0.032 | 0.037 | 0.041 | 0.043 |
| 3.30 | 0.000 | 0.000 | 0.000 | 0.000 | 0.000 | 0.000 | 0.003 | 0.007 | 0.012 | 0.015 | 0.024 | 0.031 | 0.035 | 0.039 | 0.042 |
| 3.31 | 0.000 | 0.000 | 0.000 | 0.000 | 0.000 | 0.000 | 0.003 | 0.007 | 0.011 | 0.015 | 0.023 | 0.030 | 0.034 | 0.038 | 0.040 |
| 3.32 | 0.000 | 0.000 | 0.000 | 0.000 | 0.000 | 0.000 | 0.002 | 0.006 | 0.010 | 0.014 | 0.022 | 0.029 | 0.032 | 0.036 | 0.038 |
| 3.33 | 0.000 | 0.000 | 0.000 | 0.000 | 0.000 | 0.000 | 0.002 | 0.006 | 0.010 | 0.013 | 0.021 | 0.027 | 0.031 | 0.035 | 0.037 |
| 3.34 | 0.000 | 0.000 | 0.000 | 0.000 | 0.000 | 0.000 | 0.002 | 0.006 | 0.009 | 0.013 | 0.020 | 0.026 | 0.030 | 0.034 | 0.036 |
| 3.35 | 0.000 | 0.000 | 0.000 | 0.000 | 0.000 | 0.000 | 0.002 | 0.005 | 0.009 | 0.012 | 0.019 | 0.025 | 0.029 | 0.032 | 0.034 |
| 3.36 | 0.000 | 0.000 | 0.000 | 0.000 | 0.000 | 0.000 | 0.002 | 0.005 | 0.008 | 0.011 | 0.018 | 0.024 | 0.028 | 0.031 | 0.033 |
| 3.37 | 0.000 | 0.000 | 0.000 | 0.000 | 0.000 | 0.000 | 0.002 | 0.005 | 0.008 | 0.011 | 0.017 | 0.023 | 0.026 | 0.030 | 0.032 |
| 3.38 | 0.000 | 0.000 | 0.000 | 0.000 | 0.000 | 0.000 | 0.001 | 0.004 | 0.007 | 0.010 | 0.016 | 0.022 | 0.025 | 0.029 | 0.031 |
| 3.39 | 0.000 | 0.000 | 0.000 | 0.000 | 0.000 | 0.000 | 0.001 | 0.004 | 0.007 | 0.010 | 0.016 | 0.021 | 0.024 | 0.028 | 0.029 |
| 3.40 | 0.000 | 0.000 | 0.000 | 0.000 | 0.000 | 0.000 | 0.001 | 0.004 | 0.007 | 0.009 | 0.015 | 0.020 | 0.023 | 0.027 | 0.028 |
| 3.41 | 0.000 | 0.000 | 0.000 | 0.000 | 0.000 | 0.000 | 0.001 | 0.003 | 0.006 | 0.009 | 0.014 | 0.020 | 0.022 | 0.026 | 0.027 |
| 3.42 | 0.000 | 0.000 | 0.000 | 0.000 | 0.000 | 0.000 | 0.001 | 0.003 | 0.006 | 0.008 | 0.014 | 0.019 | 0.022 | 0.025 | 0.026 |
| 3.43 | 0.000 | 0.000 | 0.000 | 0.000 | 0.000 | 0.000 | 0.001 | 0.003 | 0.005 | 0.008 | 0.013 | 0.018 | 0.021 | 0.024 | 0.025 |
| 3.44 | 0.000 | 0.000 | 0.000 | 0.000 | 0.000 | 0.000 | 0.001 | 0.003 | 0.005 | 0.007 | 0.012 | 0.017 | 0.020 | 0.023 | 0.024 |
| 3.45 | 0.000 | 0.000 | 0.000 | 0.000 | 0.000 | 0.000 | 0.001 | 0.003 | 0.005 | 0.007 | 0.012 | 0.016 | 0.019 | 0.022 | 0.023 |
| 3.46 | 0.000 | 0.000 | 0.000 | 0.000 | 0.000 | 0.000 | 0.001 | 0.002 | 0.005 | 0.007 | 0.011 | 0.016 | 0.018 | 0.021 | 0.022 |
| 3.47 | 0.000 | 0.000 | 0.000 | 0.000 | 0.000 | 0.000 | 0.001 | 0.002 | 0.004 | 0.006 | 0.011 | 0.015 | 0.018 | 0.020 | 0.022 |
| 3.48 | 0.000 | 0.000 | 0.000 | 0.000 | 0.000 | 0.000 | 0.001 | 0.002 | 0.004 | 0.006 | 0.010 | 0.014 | 0.017 | 0.019 | 0.021 |
| 3.49 | 0.000 | 0.000 | 0.000 | 0.000 | 0.000 | 0.000 | 0.000 | 0.002 | 0.004 | 0.005 | 0.010 | 0.014 | 0.016 | 0.019 | 0.020 |

**Table 9.12** (Continued)

| $Q_U$ or $Q_L$ | 3 | 4 | 5 | 7 | 10 | 15 | 20 | 25 | 30 | 35 | 50 | 75 | 100 | 150 | 200 |
|---|---|---|---|---|---|---|---|---|---|---|---|---|---|---|---|
| | | | | | | | Sample size | | | | | | | | |
| 3.50 | 0.000 | 0.000 | 0.000 | 0.000 | 0.000 | 0.000 | 0.000 | 0.002 | 0.003 | 0.005 | 0.009 | 0.013 | 0.015 | 0.018 | 0.019 |
| 3.51 | 0.000 | 0.000 | 0.000 | 0.000 | 0.000 | 0.000 | 0.000 | 0.002 | 0.003 | 0.005 | 0.009 | 0.013 | 0.015 | 0.017 | 0.018 |
| 3.52 | 0.000 | 0.000 | 0.000 | 0.000 | 0.000 | 0.000 | 0.000 | 0.002 | 0.003 | 0.005 | 0.008 | 0.012 | 0.014 | 0.016 | 0.018 |
| 3.53 | 0.000 | 0.000 | 0.000 | 0.000 | 0.000 | 0.000 | 0.000 | 0.001 | 0.003 | 0.004 | 0.008 | 0.011 | 0.014 | 0.016 | 0.017 |
| 3.54 | 0.000 | 0.000 | 0.000 | 0.000 | 0.000 | 0.000 | 0.000 | 0.001 | 0.003 | 0.004 | 0.008 | 0.011 | 0.013 | 0.015 | 0.016 |
| 3.55 | 0.000 | 0.000 | 0.000 | 0.000 | 0.000 | 0.000 | 0.000 | 0.001 | 0.003 | 0.004 | 0.007 | 0.011 | 0.012 | 0.015 | 0.016 |
| 3.56 | 0.000 | 0.000 | 0.000 | 0.000 | 0.000 | 0.000 | 0.000 | 0.001 | 0.002 | 0.004 | 0.007 | 0.010 | 0.012 | 0.014 | 0.015 |
| 3.57 | 0.000 | 0.000 | 0.000 | 0.000 | 0.000 | 0.000 | 0.000 | 0.001 | 0.002 | 0.003 | 0.006 | 0.010 | 0.011 | 0.013 | 0.014 |
| 3.58 | 0.000 | 0.000 | 0.000 | 0.000 | 0.000 | 0.000 | 0.000 | 0.001 | 0.002 | 0.003 | 0.006 | 0.009 | 0.011 | 0.013 | 0.014 |
| 3.59 | 0.000 | 0.000 | 0.000 | 0.000 | 0.000 | 0.000 | 0.000 | 0.001 | 0.002 | 0.003 | 0.006 | 0.009 | 0.010 | 0.012 | 0.013 |
| 3.60 | 0.000 | 0.000 | 0.000 | 0.000 | 0.000 | 0.000 | 0.000 | 0.001 | 0.002 | 0.003 | 0.006 | 0.008 | 0.010 | 0.012 | 0.013 |
| 3.61 | 0.000 | 0.000 | 0.000 | 0.000 | 0.000 | 0.000 | 0.000 | 0.001 | 0.002 | 0.003 | 0.005 | 0.008 | 0.010 | 0.011 | 0.012 |
| 3.62 | 0.000 | 0.000 | 0.000 | 0.000 | 0.000 | 0.000 | 0.000 | 0.001 | 0.002 | 0.003 | 0.005 | 0.008 | 0.009 | 0.011 | 0.012 |
| 3.63 | 0.000 | 0.000 | 0.000 | 0.000 | 0.000 | 0.000 | 0.000 | 0.001 | 0.001 | 0.002 | 0.005 | 0.007 | 0.009 | 0.010 | 0.011 |
| 3.64 | 0.000 | 0.000 | 0.000 | 0.000 | 0.000 | 0.000 | 0.000 | 0.001 | 0.001 | 0.002 | 0.004 | 0.007 | 0.008 | 0.010 | 0.011 |
| 3.65 | 0.000 | 0.000 | 0.000 | 0.000 | 0.000 | 0.000 | 0.000 | 0.001 | 0.001 | 0.002 | 0.004 | 0.007 | 0.008 | 0.010 | 0.010 |
| 3.66 | 0.000 | 0.000 | 0.000 | 0.000 | 0.000 | 0.000 | 0.000 | 0.000 | 0.001 | 0.002 | 0.004 | 0.006 | 0.008 | 0.009 | 0.010 |
| 3.67 | 0.000 | 0.000 | 0.000 | 0.000 | 0.000 | 0.000 | 0.000 | 0.000 | 0.001 | 0.002 | 0.004 | 0.006 | 0.007 | 0.009 | 0.010 |
| 3.68 | 0.000 | 0.000 | 0.000 | 0.000 | 0.000 | 0.000 | 0.000 | 0.000 | 0.001 | 0.002 | 0.004 | 0.006 | 0.007 | 0.008 | 0.009 |
| 3.69 | 0.000 | 0.000 | 0.000 | 0.000 | 0.000 | 0.000 | 0.000 | 0.000 | 0.001 | 0.002 | 0.003 | 0.005 | 0.007 | 0.008 | 0.009 |
| 3.70 | 0.000 | 0.000 | 0.000 | 0.000 | 0.000 | 0.000 | 0.000 | 0.000 | 0.001 | 0.002 | 0.003 | 0.005 | 0.006 | 0.008 | 0.008 |
| 3.71 | 0.000 | 0.000 | 0.000 | 0.000 | 0.000 | 0.000 | 0.000 | 0.000 | 0.001 | 0.001 | 0.003 | 0.005 | 0.006 | 0.007 | 0.008 |
| 3.72 | 0.000 | 0.000 | 0.000 | 0.000 | 0.000 | 0.000 | 0.000 | 0.000 | 0.001 | 0.001 | 0.003 | 0.005 | 0.006 | 0.007 | 0.008 |
| 3.73 | 0.000 | 0.000 | 0.000 | 0.000 | 0.000 | 0.000 | 0.000 | 0.000 | 0.001 | 0.001 | 0.003 | 0.005 | 0.006 | 0.007 | 0.007 |
| 3.74 | 0.000 | 0.000 | 0.000 | 0.000 | 0.000 | 0.000 | 0.000 | 0.000 | 0.001 | 0.001 | 0.003 | 0.004 | 0.005 | 0.006 | 0.007 |
| 3.75 | 0.000 | 0.000 | 0.000 | 0.000 | 0.000 | 0.000 | 0.000 | 0.000 | 0.001 | 0.001 | 0.002 | 0.004 | 0.005 | 0.006 | 0.007 |
| 3.76 | 0.000 | 0.000 | 0.000 | 0.000 | 0.000 | 0.000 | 0.000 | 0.000 | 0.001 | 0.001 | 0.002 | 0.004 | 0.005 | 0.006 | 0.007 |
| 3.77 | 0.000 | 0.000 | 0.000 | 0.000 | 0.000 | 0.000 | 0.000 | 0.000 | 0.001 | 0.001 | 0.002 | 0.004 | 0.005 | 0.006 | 0.006 |
| 3.78 | 0.000 | 0.000 | 0.000 | 0.000 | 0.000 | 0.000 | 0.000 | 0.000 | 0.000 | 0.001 | 0.002 | 0.004 | 0.004 | 0.005 | 0.006 |
| 3.79 | 0.000 | 0.000 | 0.000 | 0.000 | 0.000 | 0.000 | 0.000 | 0.000 | 0.000 | 0.001 | 0.002 | 0.003 | 0.004 | 0.005 | 0.006 |
| 3.80 | 0.000 | 0.000 | 0.000 | 0.000 | 0.000 | 0.000 | 0.000 | 0.000 | 0.000 | 0.001 | 0.002 | 0.003 | 0.004 | 0.005 | 0.006 |
| 3.81 | 0.000 | 0.000 | 0.000 | 0.000 | 0.000 | 0.000 | 0.000 | 0.000 | 0.000 | 0.001 | 0.002 | 0.003 | 0.004 | 0.005 | 0.005 |
| 3.82 | 0.000 | 0.000 | 0.000 | 0.000 | 0.000 | 0.000 | 0.000 | 0.000 | 0.000 | 0.001 | 0.002 | 0.003 | 0.004 | 0.005 | 0.005 |
| 3.83 | 0.000 | 0.000 | 0.000 | 0.000 | 0.000 | 0.000 | 0.000 | 0.000 | 0.000 | 0.001 | 0.002 | 0.003 | 0.004 | 0.004 | 0.005 |
| 3.84 | 0.000 | 0.000 | 0.000 | 0.000 | 0.000 | 0.000 | 0.000 | 0.000 | 0.000 | 0.001 | 0.001 | 0.003 | 0.003 | 0.004 | 0.005 |

*(Continued)*

**Table 9.12** (Continued)

| $Q_U$ or $Q_L$ | Sample size | | | | | | | | | | | | | | |
|---|---|---|---|---|---|---|---|---|---|---|---|---|---|---|---|
| | 3 | 4 | 5 | 7 | 10 | 15 | 20 | 25 | 30 | 35 | 50 | 75 | 100 | 150 | 200 |
| 3.85 | 0.000 | 0.000 | 0.000 | 0.000 | 0.000 | 0.000 | 0.000 | 0.000 | 0.000 | 0.001 | 0.001 | 0.002 | 0.003 | 0.004 | 0.004 |
| 3.86 | 0.000 | 0.000 | 0.000 | 0.000 | 0.000 | 0.000 | 0.000 | 0.000 | 0.000 | 0.000 | 0.001 | 0.002 | 0.003 | 0.004 | 0.004 |
| 3.87 | 0.000 | 0.000 | 0.000 | 0.000 | 0.000 | 0.000 | 0.000 | 0.000 | 0.000 | 0.000 | 0.001 | 0.002 | 0.003 | 0.004 | 0.004 |
| 3.88 | 0.000 | 0.000 | 0.000 | 0.000 | 0.000 | 0.000 | 0.000 | 0.000 | 0.000 | 0.000 | 0.001 | 0.002 | 0.003 | 0.003 | 0.004 |
| 3.89 | 0.000 | 0.000 | 0.000 | 0.000 | 0.000 | 0.000 | 0.000 | 0.000 | 0.000 | 0.000 | 0.001 | 0.002 | 0.003 | 0.003 | 0.004 |
| 3.90 | 0.000 | 0.000 | 0.000 | 0.000 | 0.000 | 0.000 | 0.000 | 0.000 | 0.000 | 0.000 | 0.001 | 0.002 | 0.003 | 0.003 | 0.004 |

# 10

# Computer Resources to Support SQC: Minitab, R, JMP, and Python

## Introduction

In the past three decades, the use of technology to analyze complicated data has increased substantially, which has not only made the analysis much easier but also reduced the time required to complete the analysis. To facilitate statistical analysis, many companies use personal computer (PC)-based statistical application software. Several software packages are available, including Minitab, R, Python, JMP, SAS, SPSS, and SYSTAT, to name a few. A great deal of effort has been expended in the development of these software packages to create graphical user interfaces that allow software users to complete statistical analysis activities without having to know a computer programming or scripting language. I believe that publishing a book on statistical quality control (SQC) without acknowledging and addressing the importance and usefulness of statistical software is not in the best interests of the reader. Accordingly, this chapter briefly discusses four of the statistical packages that are commonly used by academicians and practitioners: Minitab, R, Python, and JMP. It is my explicit intent not to endorse a software package, as each has different strengths and features/capabilities. In order to gain the most benefit from these statistical software packages, you should have some basic computing literacy.

**Note:**

Chapter 10 in its entirety is not available in the printed version of this text, but can be downloaded from the book's website: www.wiley.com/college/gupta/SQC.

*Statistical Quality Control: Using Minitab, R, JMP, and Python*, First Edition. Bhisham C. Gupta.
© 2021 John Wiley & Sons, Inc. Published 2021 by John Wiley & Sons, Inc.
Companion website: www.wiley.com/go/college/gupta/SQC

## The Topics covered in this Chapter are the following:

# Appendix A

# Statistical Tables

**Table A.1**  Random numbers.

| | | | | | | | | | |
|---|---|---|---|---|---|---|---|---|---|
| 051407 | 989018 | 492019 | 104768 | 575186 | 245627 | 286990 | 734378 | 453966 | 057822 |
| 269053 | 276922 | 626639 | 672429 | 195157 | 261315 | 654257 | 422375 | 431234 | 118589 |
| 367431 | 749277 | 842766 | 168999 | 210133 | 816278 | 847625 | 664969 | 065701 | 024018 |
| 124630 | 013237 | 179229 | 435437 | 550763 | 752891 | 089084 | 292255 | 199266 | 557418 |
| 235703 | 291002 | 385271 | 207907 | 360800 | 276579 | 676139 | 769805 | 783328 | 436849 |
| 690077 | 456559 | 436334 | 395621 | 700837 | 781531 | 186054 | 821361 | 983046 | 055051 |
| 064522 | 297716 | 600883 | 381178 | 169364 | 100801 | 596694 | 928310 | 703015 | 277547 |
| 764938 | 805569 | 604184 | 977595 | 363240 | 078850 | 996467 | 690208 | 334904 | 842078 |
| 875941 | 644067 | 510442 | 811601 | 829395 | 040948 | 746376 | 609475 | 676581 | 258998 |
| 758163 | 303864 | 360595 | 406956 | 613170 | 659663 | 165049 | 285017 | 508337 | 823585 |
| 805127 | 590014 | 144389 | 672585 | 094987 | 111625 | 331838 | 818612 | 481421 | 401552 |
| 525959 | 799809 | 141968 | 625825 | 297508 | 334761 | 860898 | 960450 | 785312 | 746866 |
| 351524 | 456015 | 143766 | 420487 | 368857 | 730553 | 815900 | 317512 | 047606 | 283084 |
| 940666 | 599608 | 558502 | 853032 | 057656 | 056246 | 479494 | 975590 | 713502 | 116101 |
| 557125 | 106544 | 069601 | 752609 | 897074 | 240681 | 209045 | 145960 | 683943 | 437854 |
| 190980 | 359006 | 623535 | 763922 | 122217 | 220988 | 416186 | 312541 | 738818 | 490698 |
| 992339 | 518042 | 207523 | 781965 | 792693 | 594357 | 758633 | 193427 | 143471 | 502953 |
| 915848 | 881688 | 291695 | 447687 | 462282 | 802405 | 706686 | 055756 | 580658 | 814693 |
| 197116 | 180139 | 716829 | 291097 | 056602 | 424613 | 236547 | 415732 | 423617 | 644397 |
| 118122 | 936037 | 305685 | 509440 | 108748 | 215414 | 684961 | 684762 | 362416 | 133571 |
| 283321 | 359369 | 900968 | 211269 | 865878 | 952056 | 233151 | 978019 | 775520 | 968944 |
| 474018 | 149319 | 582300 | 362831 | 346320 | 692174 | 547654 | 948322 | 384851 | 801187 |
| 809947 | 466717 | 020564 | 975865 | 223465 | 112251 | 403475 | 222629 | 379671 | 270475 |
| 224102 | 479858 | 549809 | 622585 | 751051 | 468493 | 018852 | 493268 | 146506 | 178368 |
| 366694 | 104709 | 967459 | 251556 | 079166 | 652152 | 505645 | 639175 | 028598 | 404765 |
| 734814 | 311724 | 026072 | 962867 | 814804 | 190999 | 740559 | 023023 | 327014 | 811488 |
| 319937 | 808873 | 539157 | 307523 | 098627 | 909137 | 770359 | 114529 | 881772 | 145209 |
| 036430 | 039847 | 167620 | 072545 | 428240 | 600695 | 003392 | 565195 | 332140 | 503965 |
| 894345 | 168655 | 706409 | 748967 | 876037 | 365212 | 660673 | 571480 | 558421 | 426590 |
| 227929 | 567801 | 552407 | 365578 | 580152 | 897712 | 858336 | 400702 | 406915 | 830437 |
| 720918 | 315830 | 269847 | 043686 | 006433 | 277134 | 378624 | 907969 | 762816 | 959970 |
| 291797 | 701820 | 728789 | 699785 | 715058 | 750720 | 536696 | 611293 | 544362 | 402326 |
| 564482 | 758563 | 645279 | 943094 | 588786 | 125794 | 749337 | 615120 | 568039 | 899783 |
| 236422 | 473016 | 993530 | 507143 | 335475 | 436568 | 798873 | 027549 | 940155 | 530141 |
| 689701 | 926465 | 003731 | 242454 | 058491 | 385395 | 519231 | 042314 | 955428 | 238312 |
| 857239 | 581295 | 661440 | 496859 | 529204 | 410573 | 528164 | 003660 | 587030 | 270332 |
| 209684 | 798568 | 429214 | 353484 | 193667 | 287780 | 342053 | 706113 | 193544 | 818766 |
| 780527 | 198360 | 307604 | 179501 | 891015 | 513358 | 300694 | 204837 | 681840 | 231955 |
| 753734 | 619631 | 790026 | 637123 | 101453 | 454308 | 147441 | 686401 | 027541 | 945805 |
| 823316 | 720549 | 567136 | 213060 | 266102 | 621525 | 708377 | 251598 | 278505 | 802855 |
| 967448 | 479578 | 890643 | 687587 | 046236 | 580267 | 798545 | 062865 | 752600 | 335860 |

**Table A.1**  (Continued)

| | | | | | | | | | |
|---|---|---|---|---|---|---|---|---|---|
| 582204 | 247423 | 235450 | 566691 | 086168 | 455891 | 197764 | 140909 | 747406 | 253775 |
| 801682 | 781300 | 754834 | 224141 | 068082 | 893656 | 002893 | 039025 | 414661 | 882745 |
| 386489 | 999069 | 053767 | 557623 | 688263 | 306146 | 836909 | 609168 | 823938 | 499821 |
| 242456 | 974476 | 979505 | 641408 | 240580 | 428127 | 532147 | 666926 | 018437 | 291907 |
| 935535 | 398184 | 874762 | 563669 | 548471 | 998446 | 436267 | 489528 | 430501 | 311211 |
| 838423 | 749391 | 911628 | 800272 | 143947 | 918833 | 130208 | 783122 | 827365 | 308491 |
| 821829 | 694139 | 038590 | 889019 | 212883 | 739878 | 121333 | 242205 | 312241 | 777086 |
| 589642 | 722828 | 677276 | 169636 | 465933 | 525376 | 836387 | 969518 | 231291 | 330460 |
| 634530 | 779956 | 167305 | 517950 | 851658 | 764485 | 341043 | 689067 | 402153 | 061227 |

**Table A.2** Factors helpful in constructing control charts for variables.

| $n$ | $A$ | $A_2$ | $A_3$ | $C_4$ | $C_5$ | $B_3$ | $B_4$ | $B_5$ | $B_6$ | $d_2$ | $1/d_2$ | $d_3$ | $D_1$ | $D_2$ | $D_3$ | $D_4$ |
|---|---|---|---|---|---|---|---|---|---|---|---|---|---|---|---|---|
| 2 | 2.12130 | 1.88060 | 2.65870 | 0.79788 | 0.60282 | 0.00000 | 3.26657 | 0.00000 | 2.60633 | 1.128 | 0.88652 | 0.853 | 0.000 | 3.687 | 0.0000 | 3.26862 |
| 3 | 1.73205 | 1.02307 | 1.95440 | 0.88623 | 0.46325 | 0.00000 | 2.56814 | 0.00000 | 2.27597 | 1.693 | 0.59067 | 0.888 | 0.000 | 4.357 | 0.0000 | 2.57354 |
| 4 | 1.50000 | 0.72851 | 1.62810 | 0.92132 | 0.38881 | 0.00000 | 2.26603 | 0.00000 | 2.08774 | 2.059 | 0.48567 | 0.880 | 0.000 | 4.699 | 0.0000 | 2.28218 |
| 5 | 1.34164 | 0.57680 | 1.42729 | 0.93999 | 0.34120 | 0.00000 | 2.08895 | 0.00000 | 1.96360 | 2.326 | 0.42992 | 0.864 | 0.000 | 4.918 | 0.0000 | 2.11436 |
| 6 | 1.22474 | 0.48332 | 1.28713 | 0.95153 | 0.30756 | 0.03033 | 1.96967 | 0.02886 | 1.87420 | 2.534 | 0.39463 | 0.848 | 0.000 | 5.078 | 0.0000 | 2.00395 |
| 7 | 1.13389 | 0.41934 | 1.18191 | 0.95937 | 0.28215 | 0.11770 | 1.88230 | 0.11292 | 1.80582 | 2.704 | 0.36982 | 0.833 | 0.205 | 5.203 | 0.0758 | 1.92419 |
| 8 | 1.06066 | 0.37255 | 1.09910 | 0.96503 | 0.26214 | 0.18508 | 1.81492 | 0.17861 | 1.75145 | 2.847 | 0.35125 | 0.820 | 0.387 | 5.307 | 0.1359 | 1.86407 |
| 9 | 1.00000 | 0.33670 | 1.03166 | 0.96931 | 0.24584 | 0.23912 | 1.76088 | 0.23179 | 1.70683 | 2.970 | 0.33670 | 0.808 | 0.546 | 5.394 | 0.1838 | 1.81616 |
| 10 | 0.94868 | 0.30821 | 0.97535 | 0.97266 | 0.23223 | 0.28372 | 1.71628 | 0.27596 | 1.66936 | 3.078 | 0.32489 | 0.797 | 0.687 | 5.469 | 0.2232 | 1.77680 |
| 11 | 0.90453 | 0.28507 | 0.92739 | 0.97535 | 0.22066 | 0.32128 | 1.67872 | 0.31336 | 1.63734 | 3.173 | 0.31516 | 0.787 | 0.812 | 5.534 | 0.2559 | 1.74409 |
| 12 | 0.86603 | 0.26582 | 0.88591 | 0.97756 | 0.21066 | 0.35352 | 1.64648 | 0.34559 | 1.60953 | 3.258 | 0.30694 | 0.778 | 0.924 | 5.592 | 0.2836 | 1.71639 |
| 13 | 0.83205 | 0.24942 | 0.84954 | 0.97941 | 0.20188 | 0.38162 | 1.61838 | 0.37377 | 1.58505 | 3.336 | 0.29976 | 0.770 | 1.026 | 5.646 | 0.3076 | 1.69245 |
| 14 | 0.80178 | 0.23533 | 0.81734 | 0.98097 | 0.19416 | 0.40622 | 1.59378 | 0.39849 | 1.56345 | 3.407 | 0.29351 | 0.763 | 1.118 | 5.696 | 0.3282 | 1.67185 |
| 15 | 0.77460 | 0.22310 | 0.78854 | 0.98232 | 0.18721 | 0.42826 | 1.57174 | 0.42069 | 1.54395 | 3.472 | 0.28802 | 0.756 | 1.204 | 5.740 | 0.3468 | 1.65323 |
| 16 | 0.75000 | 0.21234 | 0.76260 | 0.98348 | 0.18102 | 0.44783 | 1.55217 | 0.44043 | 1.52653 | 3.532 | 0.28313 | 0.750 | 1.282 | 5.782 | 0.3630 | 1.63703 |
| 17 | 0.72761 | 0.20279 | 0.73905 | 0.98451 | 0.17533 | 0.46574 | 1.53426 | 0.45852 | 1.51050 | 3.588 | 0.27871 | 0.744 | 1.356 | 5.820 | 0.3779 | 1.62207 |
| 18 | 0.70711 | 0.19426 | 0.71758 | 0.98541 | 0.17020 | 0.48185 | 1.51815 | 0.47482 | 1.49600 | 3.640 | 0.27473 | 0.739 | 1.423 | 5.857 | 0.3909 | 1.60907 |
| 19 | 0.68825 | 0.18657 | 0.69787 | 0.98621 | 0.16550 | 0.49656 | 1.50344 | 0.48971 | 1.48271 | 3.689 | 0.27108 | 0.734 | 1.487 | 5.891 | 0.4031 | 1.59691 |
| 20 | 0.67082 | 0.17960 | 0.67970 | 0.98693 | 0.16115 | 0.51015 | 1.48985 | 0.50348 | 1.47038 | 3.735 | 0.26774 | 0.729 | 1.548 | 5.922 | 0.4145 | 1.58554 |
| 21 | 0.65465 | 0.17328 | 0.66289 | 0.98758 | 0.15712 | 0.52272 | 1.47728 | 0.51623 | 1.45893 | 3.778 | 0.26469 | 0.724 | 1.606 | 5.950 | 0.4251 | 1.57491 |
| 22 | 0.63960 | 0.16748 | 0.64726 | 0.98817 | 0.15336 | 0.53440 | 1.46560 | 0.52808 | 1.44826 | 3.819 | 0.26185 | 0.720 | 1.659 | 5.979 | 0.4344 | 1.56559 |
| 23 | 0.62554 | 0.16214 | 0.63269 | 0.98870 | 0.14991 | 0.54514 | 1.45486 | 0.53898 | 1.43842 | 3.858 | 0.25920 | 0.716 | 1.710 | 6.006 | 0.4432 | 1.55677 |
| 24 | 0.61237 | 0.15869 | 0.61906 | 0.98919 | 0.14664 | 0.55527 | 1.44473 | 0.54927 | 1.42911 | 3.859 | 0.25913 | 0.712 | 1.723 | 5.995 | 0.4465 | 1.55351 |
| 25 | 0.60000 | 0.15263 | 0.60628 | 0.98964 | 0.14357 | 0.56478 | 1.43522 | 0.55893 | 1.42035 | 3.931 | 0.25439 | 0.708 | 1.807 | 6.055 | 0.4597 | 1.54032 |

**Table A.3** Values of $K_1$ for computing repeatability using the range method.

| | Number of trials | | | | | | | | | | | | | |
|---|---|---|---|---|---|---|---|---|---|---|---|---|---|---|
| $n$ | 2 | 3 | 4 | 5 | 6 | 7 | 8 | 9 | 10 | 11 | 12 | 13 | 14 | 15 |
| 1 | 3.65 | 2.70 | 2.30 | 2.08 | 1.93 | 1.82 | 1.74 | 1.67 | 1.62 | 1.57 | 1.54 | 1.51 | 1.48 | 1.47 |
| 2 | 4.02 | 2.85 | 2.40 | 2.15 | 1.98 | 1.86 | 1.77 | 1.71 | 1.65 | 1.60 | 1.56 | 1.52 | 1.49 | 1.47 |
| 3 | 4.19 | 2.91 | 2.43 | 2.16 | 2.00 | 1.87 | 1.78 | 1.71 | 1.66 | 1.60 | 1.57 | 1.53 | 1.50 | 1.47 |
| 4 | 4.26 | 2.94 | 2.44 | 2.17 | 2.00 | 1.88 | 1.79 | 1.72 | 1.66 | 1.61 | 1.57 | 1.53 | 1.50 | 1.48 |
| 5 | 4.33 | 2.96 | 2.45 | 2.18 | 2.01 | 1.89 | 1.79 | 1.72 | 1.66 | 1.61 | 1.57 | 1.54 | 1.51 | 1.48 |
| 6 | 4.36 | 2.98 | 2.46 | 2.19 | 2.01 | 1.89 | 1.79 | 1.72 | 1.66 | 1.61 | 1.57 | 1.54 | 1.51 | 1.48 |
| 7 | 4.40 | 2.98 | 2.46 | 2.19 | 2.02 | 1.89 | 1.79 | 1.72 | 1.66 | 1.61 | 1.57 | 1.54 | 1.51 | 1.48 |
| 8 | 4.44 | 2.99 | 2.48 | 2.19 | 2.02 | 1.89 | 1.79 | 1.73 | 1.67 | 1.61 | 1.57 | 1.54 | 1.51 | 1.48 |
| 9 | 4.44 | 2.99 | 2.48 | 2.20 | 2.02 | 1.89 | 1.80 | 1.73 | 1.67 | 1.62 | 1.57 | 1.54 | 1.51 | 1.48 |
| 10 | 4.44 | 2.99 | 2.48 | 2.20 | 2.02 | 1.89 | 1.80 | 1.73 | 1.67 | 1.62 | 1.57 | 1.54 | 1.51 | 1.48 |
| 11 | 4.44 | 3.01 | 2.48 | 2.20 | 2.02 | 1.89 | 1.80 | 1.73 | 1.67 | 1.62 | 1.57 | 1.54 | 1.51 | 1.48 |
| 12 | 4.48 | 3.01 | 2.49 | 2.20 | 2.02 | 1.89 | 1.81 | 1.73 | 1.67 | 1.62 | 1.57 | 1.54 | 1.51 | 1.48 |
| 13 | 4.48 | 3.01 | 2.49 | 2.20 | 2.02 | 1.90 | 1.81 | 1.73 | 1.67 | 1.62 | 1.57 | 1.54 | 1.51 | 1.48 |
| 14 | 4.48 | 3.01 | 2.49 | 2.20 | 2.03 | 1.90 | 1.81 | 1.73 | 1.67 | 1.62 | 1.57 | 1.54 | 1.51 | 1.48 |
| 15 | 4.48 | 3.01 | 2.49 | 2.20 | 2.03 | 1.90 | 1.81 | 1.73 | 1.67 | 1.62 | 1.58 | 1.54 | 1.51 | 1.48 |
| $n \geq 16$ | 4.56 | 3.05 | 2.50 | 2.21 | 2.04 | 1.91 | 1.81 | 1.73 | 1.67 | 1.62 | 1.58 | 1.54 | 1.51 | 1.48 |

$n = (\# \text{ of parts (samples)}) \cdot (\# \text{ of operators})$

**Table A.4** Values of $K_2$ for computing reproducibility using the range method.

| Number of operators | | | | | | | | | | | | |
|---|---|---|---|---|---|---|---|---|---|---|---|---|
| 3 | 4 | 5 | 6 | 7 | 8 | 9 | 10 | 11 | 12 | 13 | 14 | 15 |
| 2.70 | 2.30 | 2.08 | 1.93 | 1.82 | 1.74 | 1.67 | 1.62 | 1.57 | 1.54 | 1.51 | 1.48 | 1.47 |

**Table A.5**  Binomial probabilities $b(x) = \binom{n}{x} p^x (1-p)^{n-x}$.

| | | | | | | $p$ | | | | | | |
|---|---|---|---|---|---|---|---|---|---|---|---|---|
| $n$ | $x$ | .05 | .10 | .20 | .30 | .40 | .50 | .60 | .70 | .80 | .90 | .95 |
| 1 | 0 | .950 | .900 | .800 | .700 | .600 | .500 | .400 | .300 | .200 | .100 | .050 |
| | 1 | .050 | .100 | .200 | .300 | .400 | .500 | .600 | .700 | .800 | .900 | .950 |
| 2 | 0 | .902 | .810 | .640 | .490 | .360 | .250 | .160 | .090 | .040 | .010 | .003 |
| | 1 | .095 | .180 | .320 | .420 | .480 | .500 | .480 | .420 | .320 | .180 | .095 |
| | 2 | .003 | .010 | .040 | .090 | .160 | .250 | .360 | .490 | .640 | .810 | .902 |
| 3 | 0 | .857 | .729 | .512 | .343 | .216 | .125 | .064 | .027 | .008 | .001 | .000 |
| | 1 | .136 | .243 | .384 | .441 | .432 | .375 | .288 | .189 | .096 | .027 | .007 |
| | 2 | .007 | .027 | .096 | .189 | .288 | .375 | .432 | .441 | .384 | .243 | .135 |
| | 3 | .000 | .001 | .008 | .027 | .064 | .125 | .216 | .343 | .512 | .729 | .857 |
| 4 | 0 | .815 | .656 | .410 | .240 | .130 | .062 | .025 | .008 | .002 | .000 | .000 |
| | 1 | .171 | .292 | .410 | .412 | .346 | .250 | .154 | .076 | .026 | .004 | .001 |
| | 2 | .014 | .048 | .154 | .265 | .345 | .375 | .346 | .264 | .154 | .048 | .014 |
| | 3 | .000 | .004 | .025 | .075 | .154 | .250 | .346 | .412 | .409 | .292 | .171 |
| | 4 | .000 | .000 | .001 | .008 | .025 | .063 | .129 | .240 | .409 | .656 | .815 |
| 5 | 0 | .774 | .591 | .328 | .168 | .078 | .031 | .010 | .002 | .000 | .000 | .000 |
| | 1 | .204 | .328 | .410 | .360 | .259 | .156 | .077 | .028 | .006 | .001 | .000 |
| | 2 | .021 | .073 | .205 | .309 | .346 | .312 | .230 | .132 | .051 | .008 | .001 |
| | 3 | .001 | .008 | .051 | .132 | .230 | .312 | .346 | .308 | .205 | .073 | .021 |
| | 4 | .000 | .000 | .006 | .028 | .077 | .156 | .259 | .360 | .410 | .328 | .204 |
| | 5 | .000 | .000 | .000 | .003 | .010 | .031 | .078 | .168 | .328 | .590 | .774 |
| 6 | 0 | .735 | .531 | .262 | .118 | .047 | .016 | .004 | .001 | .000 | .000 | .000 |
| | 1 | .232 | .354 | .393 | .302 | .187 | .094 | .037 | .010 | .002 | .000 | .000 |
| | 2 | .031 | .098 | .246 | .324 | .311 | .234 | .138 | .059 | .015 | .001 | .000 |
| | 3 | .002 | .015 | .082 | .185 | .276 | .313 | .277 | .185 | .082 | .015 | .002 |
| | 4 | .000 | .001 | .015 | .059 | .138 | .234 | .311 | .324 | .246 | .098 | .031 |
| | 5 | .000 | .000 | .002 | .010 | .037 | .094 | .186 | .302 | .393 | .354 | .232 |
| | 6 | .000 | .000 | .000 | .001 | .004 | .015 | .047 | .118 | .262 | .531 | .735 |
| 7 | 0 | .698 | .478 | .210 | .082 | .028 | .008 | .002 | .000 | .000 | .000 | .000 |
| | 1 | .257 | .372 | .367 | .247 | .131 | .055 | .017 | .004 | .000 | .000 | .000 |
| | 2 | .041 | .124 | .275 | .318 | .261 | .164 | .077 | .025 | .004 | .000 | .000 |
| | 3 | .004 | .023 | .115 | .227 | .290 | .273 | .194 | .097 | .029 | .003 | .000 |
| | 4 | .000 | .003 | .029 | .097 | .194 | .273 | .290 | .227 | .115 | .023 | .004 |
| | 5 | .000 | .000 | .004 | .025 | .077 | .164 | .261 | .318 | .275 | .124 | .041 |
| | 6 | .000 | .000 | .000 | .004 | .017 | .055 | .131 | .247 | .367 | .372 | .257 |
| | 7 | .000 | .000 | .000 | .000 | .002 | .008 | .028 | .082 | .210 | .478 | .698 |
| 8 | 0 | .663 | .430 | .168 | .058 | .017 | .004 | .001 | .000 | .000 | .000 | .000 |
| | 1 | .279 | .383 | .335 | .198 | .089 | .031 | .008 | .001 | .000 | .000 | .000 |
| | 2 | .052 | .149 | .294 | .296 | .209 | .109 | .041 | .010 | .001 | .000 | .000 |
| | 3 | .005 | .033 | .147 | .254 | .279 | .219 | .124 | .048 | .009 | .000 | .000 |
| | 4 | .000 | .005 | .046 | .136 | .232 | .273 | .232 | .136 | .046 | .005 | .000 |
| | 5 | .000 | .000 | .009 | .047 | .124 | .219 | .279 | .254 | .147 | .033 | .005 |
| | 6 | .000 | .000 | .001 | .010 | .041 | .110 | .209 | .296 | .294 | .149 | .052 |
| | 7 | .000 | .000 | .000 | .001 | .008 | .031 | .089 | .198 | .335 | .383 | .279 |
| | 8 | .000 | .000 | .000 | .000 | .001 | .004 | .017 | .057 | .168 | .430 | .664 |
| 9 | 0 | .630 | .387 | .134 | .040 | .010 | .002 | .000 | .000 | .000 | .000 | .000 |
| | 1 | .298 | .387 | .302 | .156 | .061 | .018 | .004 | .000 | .000 | .000 | .000 |
| | 2 | .063 | .172 | .302 | .267 | .161 | .070 | .021 | .004 | .000 | .000 | .000 |

**Table A.5** (Continued)

| | | | | | | | $p$ | | | | | |
|---|---|---|---|---|---|---|---|---|---|---|---|---|
| $n$ | $x$ | .05 | .10 | .20 | .30 | .40 | .50 | .60 | .70 | .80 | .90 | .95 |
| | 3 | .008 | .045 | .176 | .267 | .251 | .164 | .074 | .021 | .003 | .000 | .000 |
| | 4 | .001 | .007 | .066 | .172 | .251 | .246 | .167 | .073 | .017 | .001 | .000 |
| | 5 | .000 | .001 | .017 | .073 | .167 | .246 | .251 | .172 | .066 | .007 | .001 |
| | 6 | .000 | .000 | .003 | .021 | .074 | .164 | .251 | .267 | .176 | .045 | .008 |
| | 7 | .000 | .000 | .000 | .004 | .021 | .070 | .161 | .267 | .302 | .172 | .063 |
| | 8 | .000 | .000 | .000 | .000 | .004 | .018 | .060 | .156 | .302 | .387 | .298 |
| | 9 | .000 | .000 | .000 | .000 | .000 | .002 | .010 | .040 | .134 | .387 | .630 |
| 10 | 0 | .599 | .349 | .107 | .028 | .006 | .001 | .000 | .000 | .000 | .000 | .000 |
| | 1 | .315 | .387 | .268 | .121 | .040 | .010 | .002 | .000 | .000 | .000 | .000 |
| | 2 | .075 | .194 | .302 | .234 | .121 | .044 | .011 | .001 | .000 | .000 | .000 |
| | 3 | .010 | .057 | .201 | .267 | .215 | .117 | .042 | .009 | .001 | .000 | .000 |
| | 4 | .001 | .011 | .088 | .200 | .251 | .205 | .111 | .037 | .006 | .000 | .000 |
| | 5 | .000 | .002 | .026 | .103 | .201 | .246 | .201 | .103 | .026 | .002 | .000 |
| | 6 | .000 | .000 | .006 | .037 | .111 | .205 | .251 | .200 | .088 | .011 | .001 |
| | 7 | .000 | .000 | .001 | .009 | .042 | .117 | .215 | .267 | .201 | .057 | .011 |
| | 8 | .000 | .000 | .000 | .001 | .011 | .044 | .121 | .234 | .302 | .194 | .075 |
| | 9 | .000 | .000 | .000 | .000 | .002 | .010 | .040 | .121 | .268 | .387 | .315 |
| | 10 | .000 | .000 | .000 | .000 | .000 | .001 | .006 | .028 | .107 | .349 | .599 |
| 11 | 0 | .569 | .314 | .086 | .020 | .004 | .001 | .000 | .000 | .000 | .000 | .000 |
| | 1 | .329 | .384 | .236 | .093 | .027 | .005 | .001 | .000 | .000 | .000 | .000 |
| | 2 | .087 | .213 | .295 | .200 | .089 | .027 | .005 | .001 | .000 | .000 | .000 |
| | 3 | .014 | .071 | .222 | .257 | .177 | .081 | .023 | .004 | .000 | .000 | .000 |
| | 4 | .001 | .016 | .111 | .220 | .237 | .161 | .070 | .017 | .002 | .000 | .000 |
| | 5 | .000 | .003 | .039 | .132 | .221 | .226 | .147 | .057 | .010 | .000 | .000 |
| | 6 | .000 | .000 | .010 | .057 | .147 | .226 | .221 | .132 | .039 | .003 | .000 |
| | 7 | .000 | .000 | .002 | .017 | .070 | .161 | .237 | .220 | .111 | .016 | .001 |
| | 8 | .000 | .000 | .000 | .004 | .023 | .081 | .177 | .257 | .222 | .071 | .014 |
| | 9 | .000 | .000 | .000 | .001 | .005 | .027 | .089 | .200 | .295 | .213 | .087 |
| | 10 | .000 | .000 | .000 | .000 | .001 | .005 | .027 | .093 | .236 | .384 | .329 |
| | 11 | .000 | .000 | .000 | .000 | .000 | .001 | .004 | .020 | .086 | .314 | .569 |
| 12 | 0 | .540 | .282 | .069 | .014 | .002 | .000 | .000 | .000 | .000 | .000 | .000 |
| | 1 | .341 | .377 | .206 | .071 | .017 | .003 | .000 | .000 | .000 | .000 | .000 |
| | 2 | .099 | .230 | .283 | .168 | .064 | .016 | .003 | .000 | .000 | .000 | .000 |
| | 3 | :017 | .085 | .236 | .240 | .142 | .054 | .012 | .002 | .000 | .000 | .000 |
| | 4 | .002 | .021 | .133 | .231 | .213 | .121 | .042 | .008 | .001 | .000 | .000 |
| | 5 | .000 | .004 | .053 | .159 | .227 | .193 | .101 | .030 | .003 | .000 | .000 |
| | 6 | .000 | .001 | .016 | .079 | .177 | .226 | .177 | .079 | .016 | .001 | .000 |
| | 7 | .000 | .000 | .003 | .029 | .101 | .193 | .227 | .159 | .053 | .004 | .000 |
| | 8 | .000 | .000 | .001 | .008 | .042 | .121 | .213 | .231 | .133 | .021 | .002 |
| | 9 | .000 | .000 | .000 | .001 | .013 | .054 | .142 | .240 | .236 | .085 | .017 |
| | 10 | .000 | .000 | .000 | .000 | .003 | .016 | .064 | .168 | .283 | .230 | .099 |
| | 11 | .000 | .000 | .000 | .000 | .000 | .003 | .017 | .071 | .206 | .377 | .341 |
| | 12 | .000 | .000 | .000 | .000 | .000 | .000 | .002 | .014 | .069 | .282 | .540 |
| 13 | 0 | .513 | .254 | .055 | .010 | .001 | .000 | .000 | .000 | .000 | .000 | .000 |
| | 1 | .351 | .367 | .179 | .054 | .011 | .002 | .000 | .000 | .000 | .000 | .000 |
| | 2 | .111 | .245 | .268 | .139 | .045 | .010 | .001 | .000 | .000 | .000 | .000 |
| | 3 | .021 | .010 | .246 | .218 | .111 | .035 | .007 | .001 | .000 | .000 | .000 |
| | 4 | .003 | .028 | .154 | .234 | .185 | .087 | .024 | .003 | .000 | .000 | .000 |
| | 5 | .000 | .006 | .069 | .180 | .221 | .157 | .066 | .014 | .001 | .000 | .000 |

(Continued)

**Table A.5** (Continued)

| n | x | .05 | .10 | .20 | .30 | .40 | .50 | .60 | .70 | .80 | .90 | .95 |
|---|---|-----|-----|-----|-----|-----|-----|-----|-----|-----|-----|-----|
|  | 6 | .000 | .001 | .023 | .103 | .197 | .210 | .131 | .044 | .006 | .000 | .000 |
|  | 7 | .000 | .000 | .006 | .044 | .131 | .210 | .197 | .103 | .023 | .001 | .000 |
|  | 8 | .000 | .000 | .001 | .014 | .066 | .157 | .221 | .180 | .069 | .006 | .000 |
|  | 9 | .000 | .000 | .000 | .003 | .024 | .087 | .184 | .234 | .154 | .028 | .003 |
|  | 10 | .000 | .000 | .000 | .001 | .007 | .035 | .111 | .218 | .246 | .100 | .021 |
|  | 11 | .000 | .000 | .000 | .000 | .001 | .010 | .045 | .139 | .268 | .245 | .111 |
|  | 12 | .000 | .000 | .000 | .000 | .000 | .000 | .011 | .054 | .179 | .367 | .351 |
|  | 13 | .000 | .000 | .000 | .000 | .000 | .000 | .001 | .0100 | .055 | .254 | .513 |
| 14 | 0 | .488 | .229 | .044 | .007 | .001 | .000 | .000 | .000 | .000 | .000 | .000 |
|  | 1 | .359 | .356 | .154 | .041 | .007 | .001 | .000 | .000 | .000 | .000 | .000 |
|  | 2 | .123 | .257 | .250 | .113 | .032 | .006 | .001 | .000 | .000 | .000 | .000 |
|  | 3 | .026 | .114 | .250 | .194 | .085 | .022 | .003 | .000 | .000 | .000 | .000 |
|  | 4 | .004 | .035 | .172 | .229 | .155 | .061 | .014 | .001 | .000 | .000 | .000 |
|  | 5 | .000 | .008 | .086 | .196 | .207 | .122 | .041 | .007 | .000 | .000 | .000 |
|  | 6 | .000 | .001 | .032 | .126 | .207 | .183 | .092 | .023 | .002 | .000 | .000 |
|  | 7 | .000 | .000 | .009 | .062 | .157 | .210 | .157 | .062 | .010 | .000 | .000 |
|  | 8 | .000 | .000 | .002 | .023 | .092 | .183 | .207 | .126 | .032 | .001 | .000 |
|  | 9 | .000 | .000 | .0003 | .0066 | .0408 | .1222 | .2066 | .1963 | .0860 | .0078 | .000 |
|  | 10 | .000 | .000 | .000 | .001 | .014 | .061 | .155 | .229 | .172 | .114 | .004 |
|  | 11 | .000 | .000 | .000 | .000 | .003 | .022 | .085 | .194 | .250 | .114 | .026 |
|  | 12 | .000 | .000 | .000 | .000 | .001 | .006 | .032 | .113 | .250 | .257 | .123 |
|  | 13 | .000 | .000 | .000 | .000 | .000 | .001 | .007 | .041 | .154 | .356 | .359 |
|  | 14 | .000 | .000 | .000 | .000 | .000 | .000 | .001 | .007 | .044 | .229 | .488 |
| 15 | 0 | .463 | .206 | .035 | .005 | .001 | .000 | .000 | .000 | .000 | .000 | .000 |
|  | 1 | .366 | .343 | .132 | .031 | .005 | .001 | .000 | .000 | .000 | .000 | .000 |
|  | 2 | .135 | .267 | .231 | .092 | .022 | .003 | .000 | .000 | .000 | .000 | .000 |
|  | 3 | .031 | .129 | .250 | .170 | .063 | .014 | .002 | .000 | .000 | .000 | .000 |
|  | 4 | .005 | .043 | .188 | .219 | .127 | .042 | .007 | .001 | .000 | .000 | .000 |
|  | 5 | .001 | .011 | .103 | .206 | .186 | .092 | .025 | .003 | .000 | .000 | .000 |
|  | 6 | .000 | .002 | .043 | .147 | .207 | .153 | .061 | .012 | .001 | .000 | .000 |
|  | 7 | .000 | .000 | .014 | .081 | .177 | .196 | .118 | .035 | .004 | .000 | .000 |
|  | 8 | .000 | .000 | .004 | .035 | .118 | .196 | .177 | .081 | .014 | .000 | .000 |
|  | 9 | .000 | .000 | .001 | .012 | .061 | .153 | .207 | .147 | .043 | .002 | .000 |
|  | 10 | .000 | .000 | .000 | .003 | .025 | .092 | .186 | .206 | .103 | .011 | .001 |
|  | 11 | .000 | .000 | .000 | .001 | .007 | .042 | .127 | .219 | .188 | .043 | .005 |
|  | 12 | .000 | .000 | .000 | .000 | .002 | .014 | .063 | .170 | .250 | .129 | .031 |
|  | 13 | .000 | .000 | .000 | .000 | .000 | .003 | .022 | .092 | .231 | .267 | .135 |
|  | 14 | .000 | .000 | .000 | .000 | .000 | .001 | .005 | .031 | .132 | .343 | .366 |
|  | 15 | .000 | .000 | .000 | .000 | .000 | .000 | .001 | .005 | .035 | .206 | .463 |

**Table A.6** Poisson probabilities $p(x) = \dfrac{e^{-k}k^x}{x!}, \quad x = 0, 1, 2, \cdots$.

| x | k | | | | | | | | | |
|---|-----|-----|-----|-----|-----|-----|-----|-----|-----|-----|
|   | 0.1 | 0.2 | 0.3 | 0.4 | 0.5 | 0.6 | 0.7 | 0.8 | 0.9 | 1.0 |
| 0 | .905 | .819 | .741 | .670 | .607 | .549 | .497 | .449 | .407 | .368 |
| 1 | .091 | .164 | .222 | .268 | .303 | .329 | .348 | .360 | .366 | .368 |
| 2 | .005 | .016 | .033 | .054 | .076 | .099 | .122 | .144 | .165 | .184 |
| 3 | .000 | .001 | .003 | .007 | .013 | .020 | .028 | .038 | .049 | .061 |
| 4 | .000 | .000 | .000 | .000 | .002 | .003 | .005 | .008 | .011 | .015 |
| 5 | .000 | .000 | .000 | .000 | .000 | .000 | .001 | .001 | .002 | .003 |
| 6 | .000 | .000 | .000 | .000 | .000 | .000 | .000 | .000 | .000 | .001 |
| 7 | .000 | .000 | .000 | .000 | .000 | .000 | .000 | .000 | .000 | .000 |

| x | k | | | | | | | | | |
|---|-----|-----|-----|-----|-----|-----|-----|-----|-----|-----|
|   | 1.1 | 1.2 | 1.3 | 1.4 | 1.5 | 1.6 | 1.7 | 1.8 | 1.9 | 2.0 |
| 0 | .333 | .301 | .273 | .247 | .223 | .202 | .183 | .165 | .150 | .135 |
| 1 | .366 | .361 | .354 | .345 | .335 | .323 | .311 | .298 | .284 | .271 |
| 2 | .201 | .217 | .230 | .242 | .251 | .258 | .264 | .268 | .270 | .271 |
| 3 | .074 | .087 | .100 | .113 | .126 | .138 | .150 | .161 | .171 | .180 |
| 4 | .020 | .026 | .032 | .040 | .047 | .055 | .064 | .072 | .081 | .090 |
| 5 | .005 | .006 | .008 | .011 | .014 | .018 | .022 | .026 | .031 | .036 |
| 6 | .001 | .001 | .002 | .003 | .004 | .005 | .006 | .008 | .010 | .012 |
| 7 | .000 | .000 | .000 | .001 | .001 | .001 | .002 | .002 | .003 | .003 |
| 8 | .000 | .000 | .000 | .000 | .000 | .000 | .000 | .001 | .001 | .001 |
| 9 | .000 | .000 | .000 | .000 | .000 | .000 | .000 | .000 | .000 | .000 |

| x | k | | | | | | | | | |
|---|-----|-----|-----|-----|-----|-----|-----|-----|-----|-----|
|   | 2.1 | 2.2 | 2.3 | 2.4 | 2.5 | 2.6 | 2.7 | 2.8 | 2.9 | 3.0 |
| 0 | .123 | .111 | .100 | .091 | .082 | .074 | .067 | .061 | .055 | .050 |
| 1 | .257 | .244 | .231 | .218 | .205 | .193 | .182 | .170 | .160 | .149 |
| 2 | .270 | .268 | .265 | .261 | .257 | .251 | .245 | .238 | .231 | .224 |
| 3 | .189 | .197 | .203 | .209 | .214 | .218 | .221 | .223 | .224 | .224 |
| 4 | .099 | .108 | .117 | .125 | .134 | .141 | .149 | .156 | .162 | .168 |
| 5 | .042 | .048 | .054 | .060 | .067 | .074 | .080 | .087 | .094 | .101 |
| 6 | .015 | .017 | .021 | .024 | .028 | .032 | .036 | .041 | .046 | .050 |
| 7 | .004 | .006 | .007 | .008 | .010 | .012 | .014 | .016 | .019 | .022 |

*(Continued)*

**Table A.6** (Continued)

| | | | | | *k* | | | | | |
|---|---|---|---|---|---|---|---|---|---|---|
| *x* | 2.1 | 2.2 | 2.3 | 2.4 | 2.5 | 2.6 | 2.7 | 2.8 | 2.9 | 3.0 |
| 8 | .001 | .002 | .002 | .003 | .003 | .004 | .005 | .006 | .007 | .008 |
| 9 | .000 | .000 | .001 | .001 | .001 | .001 | .001 | .002 | .002 | .003 |
| 10 | .000 | .000 | .000 | .000 | .000 | .000 | .000 | .001 | .001 | .001 |
| 11 | .000 | .000 | .000 | .000 | .000 | .000 | .000 | .000 | .000 | .000 |
| 12 | .000 | .000 | .000 | .000 | .000 | .000 | .000 | .000 | .000 | .000 |

| | | | | | *k* | | | | | |
|---|---|---|---|---|---|---|---|---|---|---|
| *x* | 3.1 | 3.2 | 3.3 | 3.4 | 3.5 | 3.6 | 3.7 | 3.8 | 3.9 | 4.0 |
| 0 | .045 | .041 | .037 | .033 | .030 | .027 | .025 | .022 | .020 | .018 |
| 1 | .140 | .130 | .122 | .114 | .106 | .098 | .092 | .085 | .079 | .073 |
| 2 | .217 | .209 | .201 | .193 | .185 | .177 | .169 | .162 | .154 | .147 |
| 3 | .224 | .223 | .221 | .219 | .213 | .209 | .205 | .200 | .195 | .195 |
| 4 | .173 | .178 | .182 | .186 | .191 | .193 | .194 | .195 | .195 | .195 |
| 5 | .107 | .114 | .120 | .132 | .138 | .143 | .148 | .152 | .156 | .156 |
| 6 | .056 | .061 | .066 | .077 | .083 | .088 | .094 | .099 | .104 | .104 |
| 7 | .025 | .028 | .031 | .039 | .043 | .047 | .051 | .055 | .060 | .060 |
| 8 | .010 | .011 | .013 | .017 | .019 | .022 | .024 | .027 | .030 | .030 |
| 9 | .003 | .004 | .005 | .007 | .008 | .009 | .010 | .012 | .013 | .013 |
| 10 | .001 | .001 | .002 | .002 | .002 | .003 | .003 | .004 | .005 | .005 |
| 11 | .000 | .000 | .001 | .001 | .001 | .001 | .001 | .001 | .002 | .002 |
| 12 | .000 | .000 | .000 | .000 | .000 | .000 | .000 | .000 | .001 | .001 |
| 13 | .000 | .000 | .0000 | .000 | .000 | .000 | .000 | .000 | .000 | .000 |
| 14 | .000 | .000 | .000 | .000 | .000 | .000 | .000 | .000 | .000 | .000 |

| | | | | | *k* | | | | | |
|---|---|---|---|---|---|---|---|---|---|---|
| *x* | 4.1 | 4.2 | 4.3 | 4.4 | 4.5 | 4.6 | 4.7 | 4.8 | 4.9 | 5.0 |
| 0 | .017 | .015 | .014 | .012 | .011 | .010 | .009 | .008 | .007 | .007 |
| 1 | .068 | .063 | .058 | .054 | .050 | .046 | .043 | .040 | .037 | .034 |
| 2 | .139 | .132 | .125 | .119 | .113 | .106 | .101 | .095 | .089 | .084 |
| 3 | .190 | .185 | .180 | .174 | .169 | .163 | .157 | .152 | .146 | .140 |
| 4 | .195 | .194 | .193 | .192 | .190 | .188 | .185 | .182 | .179 | .176 |
| 5 | .160 | .163 | .166 | .169 | .171 | .173 | .174 | .175 | .175 | .176 |
| 6 | .109 | .114 | .119 | .124 | .128 | .132 | .136 | .140 | .143 | .146 |
| 7 | .064 | .069 | .073 | .078 | .082 | .087 | .091 | .096 | .100 | .104 |
| 8 | .033 | .036 | .039 | .043 | .046 | .050 | .054 | .058 | .061 | .065 |

**Table A.6** (Continued)

| | | | | | k | | | | | |
|---|---|---|---|---|---|---|---|---|---|---|
| *x* | 4.1 | 4.2 | 4.3 | 4.4 | 4.5 | 4.6 | 4.7 | 4.8 | 4.9 | 5.0 |
| 9 | .015 | .017 | .019 | .021 | .023 | .026 | .028 | .031 | .033 | .036 |
| 10 | .006 | .007 | .008 | .009 | .010 | .012 | .013 | .015 | .016 | .018 |
| 11 | .002 | .003 | .003 | .004 | .004 | .005 | .006 | .006 | .007 | .008 |
| 12 | .009 | .001 | .001 | .001 | .002 | .002 | .002 | .003 | .003 | .003 |
| 13 | .000 | .000 | .000 | .001 | .001 | .001 | .001 | .001 | .001 | .001 |
| 14 | .000 | .000 | .000 | .000 | .000 | .000 | .000 | .000 | .000 | .000 |
| 15 | .000 | .000 | .000 | .000 | .000 | .000 | .000 | .000 | .000 | .000 |

| | | | | | k | | | | | |
|---|---|---|---|---|---|---|---|---|---|---|
| *x* | 5.1 | 5.2 | 5.3 | 5.4 | 5.5 | 5.6 | 5.7 | 5.8 | 5.9 | 6.0 |
| 0 | .006 | .006 | .005 | .005 | .004 | .004 | .003 | .003 | .003 | .002 |
| 1 | .031 | .029 | .027 | .024 | .022 | .021 | .019 | .018 | .016 | .015 |
| 2 | .079 | .075 | .070 | .066 | .062 | .058 | .054 | .051 | .048 | .045 |
| 3 | .135 | .129 | .124 | .119 | .113 | .108 | .103 | .099 | .094 | .089 |
| 4 | .172 | .168 | .164 | .160 | .156 | .152 | .147 | .143 | .138 | .134 |
| 5 | .175 | .175 | .174 | .173 | .171 | .170 | .168 | .166 | .163 | .161 |
| 6 | .149 | .151 | .154 | .156 | .157 | .158 | .159 | .160 | .161 | .161 |
| 7 | .109 | .113 | .116 | .120 | .123 | .127 | .130 | .133 | .135 | .138 |
| 8 | .069 | .073 | .077 | .081 | .085 | .089 | .093 | .096 | .100 | .103 |
| 9 | .039 | .042 | .045 | .049 | .052 | .055 | .059 | .062 | .065 | .069 |
| 10 | .020 | .022 | .024 | .026 | .029 | .031 | .033 | .036 | .039 | .041 |
| 11 | .009 | .010 | .012 | .013 | .014 | .016 | .017 | .019 | .021 | .023 |
| 12 | .004 | .005 | .005 | .006 | .007 | .007 | .008 | .009 | .010 | .011 |
| 13 | .002 | .002 | .002 | .002 | .003 | .003 | .004 | .004 | .005 | .005 |
| 14 | .001 | .001 | .001 | .001 | .001 | .001 | .002 | .002 | .002 | .002 |
| 15 | .000 | .000 | .000 | .000 | .000 | .000 | .001 | .001 | .001 | .001 |
| 16 | .000 | .000 | .000 | .000 | .000 | .000 | .000 | .000 | .000 | .000 |
| 17 | .000 | .000 | .000 | .000 | .000 | .000 | .000 | .000 | .000 | .000 |

| | | | | | k | | | | | |
|---|---|---|---|---|---|---|---|---|---|---|
| *x* | 6.1 | 6.2 | 6.3 | 6.4 | 6.5 | 6.6 | 6.7 | 6.8 | 6.9 | 7.0 |
| 0 | .002 | .002 | .002 | .002 | .002 | .001 | .001 | .001 | .001 | .001 |
| 1 | .014 | .013 | .012 | .011 | .010 | .010 | .008 | .008 | .007 | .007 |

(*Continued*)

**Table A.6** (Continued)

| x | 6.1 | 6.2 | 6.3 | 6.4 | 6.5 | 6.6 | 6.7 | 6.8 | 6.9 | 7.0 |
|---|------|------|------|------|------|------|------|------|------|------|
| | | | | | $k$ | | | | | |
| 2 | .042 | .040 | .036 | .034 | .032 | .029 | .028 | .026 | .024 | .022 |
| 3 | .085 | .081 | .077 | .073 | .069 | .065 | .062 | .058 | .055 | .052 |
| 4 | .129 | .125 | .121 | .116 | .112 | .108 | .103 | .099 | .095 | .091 |
| 5 | .158 | .155 | .152 | .149 | .145 | .142 | .139 | .135 | .131 | .128 |
| 6 | .160 | .160 | .159 | .159 | .158 | .156 | .155 | .153 | .151 | .149 |
| 7 | .140 | .142 | .144 | .145 | .146 | .147 | .148 | .149 | .149 | .149 |
| 8 | .107 | .110 | .113 | .116 | .119 | .122 | .124 | .126 | .128 | .130 |
| 9 | .072 | .076 | .079 | .083 | .086 | .089 | .092 | .095 | .098 | .101 |
| 10 | .044 | .047 | .050 | .053 | .056 | .059 | .062 | .065 | .068 | .071 |
| 11 | .024 | .026 | .029 | .031 | .033 | .035 | .038 | .040 | .043 | .045 |
| 12 | .012 | .014 | .015 | .016 | .018 | .019 | .021 | .023 | .025 | .026 |
| 13 | .006 | .007 | .007 | .008 | .009 | .010 | .011 | .012 | .013 | .014 |
| 14 | .003 | .003 | .003 | .004 | .004 | .005 | .005 | .006 | .006 | .007 |
| 15 | .001 | .001 | .001 | .002 | .002 | .002 | .002 | .003 | .003 | .003 |
| 16 | .000 | .001 | .001 | .001 | .001 | .001 | .001 | .001 | .001 | .001 |
| 17 | .000 | .000 | .000 | .000 | .000 | .000 | .000 | .000 | .001 | .001 |
| 18 | .000 | .000 | .000 | .000 | .000 | .000 | .000 | .000 | .000 | .000 |
| 19 | .000 | .000 | .000 | .000 | .000 | .000 | .000 | .000 | .000 | .000 |

**Table A.7**  Standard normal distribution.

**Tabulated values are $P(0 \le Z \le z)$ = shaded area under the standard normal curve.**

| z | .00 | .01 | .02 | .03 | .04 | .05 | .06 | .07 | .08 | .09 |
|---|---|---|---|---|---|---|---|---|---|---|
| 0.0 | .0000 | .0040 | .0080 | .0120 | .0160 | .0199 | .0239 | .0279 | .0319 | .0359 |
| 0.1 | .0398 | .0438 | .0478 | .0517 | .0557 | .0596 | .0636 | .0675 | .0714 | .0753 |
| 0.2 | .0793 | .0832 | .0871 | .0910 | .0948 | .0987 | .1026 | .1064 | .1103 | .1141 |
| 0.3 | .1179 | .1217 | .1255 | .1293 | .1331 | .1368 | .1406 | .1443 | .1480 | .1517 |
| 0.4 | .1554 | .1591 | .1628 | .1664 | .1700 | .1736 | .1772 | .1808 | .1844 | .1879 |
| 0.5 | .1915 | .1950 | .1985 | .2019 | .2054 | .2088 | .2123 | .2157 | .2190 | .2224 |
| 0.6 | .2257 | .2291 | .2324 | .2357 | .2389 | .2422 | .2454 | .2486 | .2517 | .2549 |
| 0.7 | .2580 | .2611 | .2642 | .2673 | .2704 | .2734 | .2764 | .2794 | .2823 | .2852 |
| 0.8 | .2881 | .2910 | .2939 | .2967 | .2995 | .3023 | .3051 | .3078 | .3106 | .3133 |
| 0.9 | .3159 | .3186 | .3212 | .3238 | .3264 | .3289 | .3315 | .3340 | .3365 | .3389 |
| 1.0 | .3413 | .3438 | .3461 | .3485 | .3508 | .3531 | .3554 | .3577 | .3599 | .3621 |
| 1.1 | .3643 | .3665 | .3686 | .3708 | .3729 | .3749 | .3770 | .3790 | .3810 | .3830 |
| 1.2 | .3849 | .3869 | .3888 | .3907 | .3925 | .3944 | .3962 | .3980 | .3997 | .4015 |
| 1.3 | .4032 | .4049 | .4066 | .4082 | .4099 | .4115 | .4131 | .4147 | .4162 | .4177 |
| 1.4 | .4192 | .4207 | .4222 | .4236 | .4251 | .4265 | .4279 | .4292 | .4306 | .4319 |
| 1.5 | .4332 | .4345 | .4357 | .4370 | .4382 | .4394 | .4406 | .4418 | .4429 | .4441 |
| 1.6 | .4452 | .4463 | .4474 | .4484 | .4495 | .4505 | .4515 | .4525 | .4535 | .4545 |
| 1.7 | .4554 | .4564 | .4573 | .4582 | .4591 | .4599 | .4608 | .4616 | .4625 | .4633 |
| 1.8 | .4641 | .4649 | .4656 | .4664 | .4671 | .4678 | .4686 | .4693 | .4699 | .4706 |
| 1.9 | .4713 | .4719 | .4726 | .4732 | .4738 | .4744 | .4750 | .4756 | .4761 | .4767 |
| 2.0 | .4772 | .4778 | .4783 | .4788 | .4793 | .4798 | .4803 | .4808 | .4812 | .4817 |
| 2.1 | .4821 | .4826 | .4830 | .4834 | .4838 | .4842 | .4846 | .4850 | .4854 | .4857 |
| 2.2 | .4861 | .4864 | .4868 | .4871 | .4875 | .4878 | .4881 | .4884 | .4887 | .4890 |
| 2.3 | .4893 | .4896 | .4898 | .4901 | .4904 | .4906 | .4909 | .4911 | .4913 | .4916 |
| 2.4 | .4918 | .4920 | .4922 | .4925 | .4927 | .4929 | .4931 | .4932 | .4934 | .4936 |
| 2.5 | .4938 | .4940 | .4941 | .4943 | .4945 | .4946 | .4948 | .4949 | .4951 | .4952 |
| 2.6 | .4953 | .4955 | .4956 | .4957 | .4959 | .4960 | .4961 | .4962 | .4963 | .4964 |
| 2.7 | .4965 | .4966 | .4967 | .4968 | .4969 | .4970 | .4971 | .4972 | .4973 | .4974 |
| 2.8 | .4974 | .4975 | .4976 | .4977 | .4977 | .4978 | .4979 | .4979 | .4980 | .4981 |
| 2.9 | .4981 | .4982 | .4982 | .4983 | .4984 | .4984 | .4985 | .4985 | .4986 | .4986 |
| 3.0 | .4987 | .4987 | .4987 | .4988 | .4988 | .4989 | .4989 | .4989 | .4990 | .4990 |

*For negative values of z, the probabilities are found by using the symmetric property.*

**Table A.8**  Critical values of $\chi^2$ with $\nu$ degrees of freedom.

| $\nu$ | $\chi^2_{0.995}$ | $\chi^2_{0.990}$ | $\chi^2_{0.975}$ | $\chi^2_{0.950}$ | $\chi^2_{0.900}$ |
|---|---|---|---|---|---|
| 1 | 0.00004 | 0.00016 | 0.00098 | 0.00393 | 0.01589 |
| 2 | 0.0100 | 0.0201 | 0.0506 | 0.1026 | 0.2107 |
| 3 | 0.0717 | 0.1148 | 0.2158 | 0.3518 | 0.5844 |
| 4 | 0.2070 | 0.2971 | 0.4844 | 0.7107 | 1.0636 |
| 5 | 0.4117 | 0.5543 | 0.8312 | 1.1455 | 1.6103 |
| 6 | 0.6757 | 0.8720 | 1.2373 | 1.6354 | 2.2041 |
| 7 | 0.9893 | 1.2390 | 1.6899 | 2.1674 | 2.8331 |
| 8 | 1.3444 | 1.6465 | 2.1797 | 2.7326 | 3.4895 |
| 9 | 1.7349 | 2.0879 | 2.7004 | 3.3251 | 4.1682 |
| 10 | 2.1559 | 2.5582 | 3.2470 | 3.9403 | 4.8652 |
| 11 | 2.6032 | 3.0535 | 3.8158 | 4.5748 | 5.5778 |
| 12 | 3.0738 | 3.5706 | 4.4038 | 5.2260 | 6.3038 |
| 13 | 3.5650 | 4.1069 | 5.0087 | 5.8919 | 7.0415 |
| 14 | 4.0747 | 4.6604 | 5.6287 | 6.5706 | 7.7895 |
| 15 | 4.6009 | 5.2294 | 6.2621 | 7.2609 | 8.5468 |
| 16 | 5.1422 | 5.8122 | 6.9077 | 7.9616 | 9.3122 |
| 17 | 5.6972 | 6.4078 | 7.5642 | 8.6718 | 10.085 |
| 18 | 6.2648 | 7.0149 | 8.2308 | 9.3905 | 10.865 |
| 19 | 6.8440 | 7.6327 | 8.9066 | 10.1170 | 11.6509 |
| 20 | 7.4339 | 8.2604 | 9.5908 | 10.8508 | 12.4426 |
| 21 | 8.0337 | 8.8972 | 10.2829 | 11.5913 | 13.2396 |
| 22 | 8.6427 | 9.5425 | 10.9823 | 12.3380 | 14.0415 |
| 23 | 9.2604 | 10.1957 | 11.6885 | 13.0905 | 14.8479 |
| 24 | 9.8862 | 10.8564 | 12.4011 | 13.8484 | 15.6587 |
| 25 | 10.5197 | 11.5240 | 13.1197 | 14.6114 | 16.4734 |
| 26 | 11.1603 | 12.1981 | 13.8439 | 15.3791 | 17.2919 |
| 27 | 11.8076 | 12.8786 | 14.5733 | 16.1513 | 18.1138 |
| 28 | 12.4613 | 13.5648 | 15.3079 | 16.9279 | 18.9392 |
| 29 | 13.1211 | 14.2565 | 16.0471 | 17.7083 | 19.7677 |
| 30 | 13.7867 | 14.9535 | 16.7908 | 18.4926 | 20.5992 |
| 40 | 20.7065 | 22.1643 | 24.4331 | 26.5093 | 29.0505 |
| 50 | 27.9907 | 29.7067 | 32.3574 | 34.7642 | 37.6886 |
| 60 | 35.5346 | 37.4848 | 40.4817 | 43.1879 | 46.4589 |
| 70 | 43.2752 | 45.4418 | 48.7576 | 51.7393 | 55.3290 |
| 80 | 51.1720 | 53.5400 | 57.1532 | 60.3915 | 64.2778 |
| 90 | 59.1963 | 61.7541 | 65.6466 | 69.1260 | 73.2912 |
| 100 | 67.3276 | 70.0648 | 74.2219 | 77.9295 | 82.3581 |

**Table A.8** (Continued)

| $v$ | $x^2_{0.100}$ | $x^2_{0.050}$ | $x^2_{0.025}$ | $x^2_{0.010}$ | $x^2_{0.005}$ |
|---|---|---|---|---|---|
| 1 | 2.7055 | 3.8415 | 5.0239 | 6.6349 | 7.8794 |
| 2 | 4.6052 | 5.9915 | 7.3778 | 9.2103 | 10.5966 |
| 3 | 6.2514 | 7.8147 | 9.3484 | 11.3449 | 12.8381 |
| 4 | 7.7794 | 9.4877 | 11.1433 | 13.2767 | 14.8602 |
| 5 | 9.2364 | 11.0705 | 12.8325 | 15.0863 | 16.7496 |
| 6 | 10.6446 | 12.5916 | 14.4494 | 16.8119 | 18.5476 |
| 7 | 12.0170 | 14.0671 | 16.0128 | 18.4753 | 20.2777 |
| 8 | 13.3616 | 15.5073 | 17.5346 | 20.0902 | 21.9550 |
| 9 | 14.6837 | 16.9190 | 19.0228 | 21.6660 | 23.5893 |
| 10 | 15.9871 | 18.3070 | 20.4831 | 23.2093 | 25.1882 |
| 11 | 17.2750 | 19.6751 | 21.9200 | 24.7250 | 26.7569 |
| 12 | 18.5494 | 21.0261 | 23.3367 | 26.2170 | 28.2995 |
| 13 | 19.8119 | 22.3621 | 24.7356 | 27.6883 | 29.8194 |
| 14 | 21.0642 | 23.6848 | 26.1190 | 29.1413 | 31.3193 |
| 15 | 22.3072 | 24.9958 | 27.4884 | 30.5779 | 32.8013 |
| 16 | 23.5418 | 26.2962 | 28.8454 | 31.9999 | 34.2672 |
| 17 | 24.7690 | 27.5871 | 30.1910 | 33.4087 | 35.7185 |
| 18 | 25.9894 | 28.8693 | 31.5264 | 34.8053 | 37.1564 |
| 19 | 27.2036 | 30.1435 | 32.8523 | 36.1908 | 38.5822 |
| 20 | 28.4120 | 31.4104 | 34.1696 | 37.5662 | 39.9968 |
| 21 | 29.6151 | 32.6705 | 35.4789 | 38.9321 | 41.4010 |
| 22 | 30.8133 | 33.9244 | 36.7807 | 40.2894 | 42.7956 |
| 23 | 32.0069 | 35.1725 | 38.0757 | 41.6384 | 44.1813 |
| 24 | 33.1963 | 36.4151 | 39.3641 | 42.9798 | 45.5585 |
| 25 | 34.3816 | 37.6525 | 40.6465 | 44.3141 | 46.9278 |
| 26 | 35.5631 | 38.8852 | 41.9232 | 45.6417 | 48.2899 |
| 27 | 36.7412 | 40.1133 | 43.1944 | 46.9630 | 49.6449 |
| 28 | 37.9159 | 41.3372 | 44.4607 | 48.2782 | 50.9933 |
| 29 | 39.0875 | 42.5569 | 45.7222 | 49.5879 | 52.3356 |
| 30 | 40.2560 | 43.7729 | 46.9792 | 50.8922 | 53.6720 |
| 40 | 51.8050 | 55.7585 | 59.3417 | 63.6907 | 66.7659 |
| 50 | 63.1671 | 67.5048 | 71.4202 | 76.1539 | 79.4900 |
| 60 | 74.3970 | 79.0819 | 83.2976 | 88.3794 | 91.9517 |
| 70 | 85.5271 | 90.5312 | 95.0231 | 100.4251 | 104.2148 |
| 80 | 96.5782 | 101.8795 | 106.6285 | 112.3288 | 116.3210 |
| 90 | 107.5650 | 113.1452 | 118.1360 | 124.1162 | 128.2290 |
| 100 | 118.4980 | 124.3421 | 129.5613 | 135.8070 | 140.1697 |

**Table A.9**  Critical values of t with $\nu$ degrees of freedom.

| $\nu$ | $t_{.100}$ | $t_{.050}$ | $t_{.025}$ | $t_{.010}$ | $t_{.005}$ | $t_{.0005}$ |
|---|---|---|---|---|---|---|
| 1 | 3.078 | 6.314 | 12.706 | 31.821 | 63.657 | 636.619 |
| 2 | 1.886 | 2.920 | 4.303 | 6.965 | 9.925 | 31.599 |
| 3 | 1.638 | 2.353 | 3.182 | 4.541 | 5.841 | 12.924 |
| 4 | 1.533 | 2.132 | 2.776 | 3.747 | 4.604 | 8.610 |
| 5 | 1.476 | 2.015 | 2.571 | 3.365 | 4.032 | 6.869 |
| 6 | 1.440 | 1.943 | 2.447 | 3.143 | 3.707 | 5.959 |
| 7 | 1.415 | 1.895 | 2.365 | 2.998 | 3.499 | 5.408 |
| 8 | 1.397 | 1.860 | 2.306 | 2.896 | 3.355 | 5.041 |
| 9 | 1.383 | 1.833 | 2.262 | 2.821 | 3.250 | 4.781 |
| 10 | 1.372 | 1.812 | 2.228 | 2.764 | 3.169 | 4.587 |
| 11 | 1.363 | 1.796 | 2.201 | 2.718 | 3.106 | 4.437 |
| 12 | 1.356 | 1.782 | 2.179 | 2.681 | 3.055 | 4.318 |
| 13 | 1.350 | 1.771 | 2.160 | 2.650 | 3.012 | 4.221 |
| 14 | 1.345 | 1.761 | 2.145 | 2.624 | 2.977 | 4.140 |
| 15 | 1.341 | 1.753 | 2.131 | 2.602 | 2.947 | 4.073 |
| 16 | 1.337 | 1.746 | 2.120 | 2.583 | 2.921 | 4.015 |
| 17 | 1.333 | 1.740 | 2.110 | 2.567 | 2.898 | 3.965 |
| 18 | 1.330 | 1.734 | 2.101 | 2.552 | 2.878 | 3.922 |
| 19 | 1.328 | 1.729 | 2.093 | 2.539 | 2.861 | 3.883 |
| 20 | 1.325 | 1.725 | 2.086 | 2.528 | 2.845 | 3.850 |
| 21 | 1.323 | 1.721 | 2.080 | 2.518 | 2.831 | 3.819 |
| 22 | 1.321 | 1.717 | 2.074 | 2.508 | 2.819 | 3.792 |
| 23 | 1.319 | 1.714 | 2.069 | 2.500 | 2.807 | 3.768 |
| 24 | 1.318 | 1.711 | 2.064 | 2.492 | 2.797 | 3.745 |
| 25 | 1.316 | 1.708 | 2.060 | 2.485 | 2.787 | 3.725 |
| 26 | 1.315 | 1.706 | 2.056 | 2.479 | 2.779 | 3.707 |
| 27 | 1.314 | 1.703 | 2.052 | 2.473 | 2.771 | 3.690 |
| 28 | 1.313 | 1.701 | 2.048 | 2.467 | 2.763 | 3.674 |
| 29 | 1.311 | 1.699 | 2.045 | 2.462 | 2.756 | 3.659 |
| 30 | 1.310 | 1.697 | 2.042 | 2.457 | 2.750 | 3.646 |
| 40 | 1.303 | 1.684 | 2.021 | 2.423 | 2.704 | 3.551 |
| 60 | 1.296 | 1.671 | 2.000 | 2.390 | 2.660 | 3.460 |
| 80 | 1.292 | 1.664 | 1.990 | 2.374 | 2.639 | 3.416 |
| 100 | 1.290 | 1.660 | 1.984 | 2.364 | 2.626 | 3.390 |
| 120 | 1.289 | 1.658 | 1.980 | 2.358 | 2.617 | 3.373 |
| $\infty$ | 1.282 | 1.645 | 1.960 | 2.326 | 2.576 | 3.291 |

*Critical points of t for lower tail areas are found by using the symmetric property.*

# Appendix B

# Answers to Selected Practice Problems

## Chapter 3

**1** (a) Quantitative; (b) quantitative; (c) quantitative; (d) qualitative; (e) quantitative; (f) qualitative; (g) quantitative; (h) quantitative; (i) quantitative; (j) quantitative; (k) quantitative

**3** (a) All individuals who use products of interest imported from China; (b) 500 individuals who were interviewed; (c) ordinal

**5** (b) 29

**9** (b) 3 (c) 14

**11** (b) no outliers

**17** (a) 42% (b) 38% (c) 38%

**21** 0.929

**27** (a) Mean = 120.02, median = 120.10; (b) sd =1.84

**29** (a) Mean = 48.89, median = 47.00, mode = 46.00; (b) right skewed

**31** (a) $Q_1 = 63$, $Q_2 = 69$, $Q_2 = 75$; (b) 21 (c) no outliers

**33** (a) $\overline{X} = 68.42, S = 8.18$, (60.24, 76.60), (52.06, 84.88), (43.88, 93.06), 69%, 97%, 97%, normal

**35** $\overline{X} = 61.33$, Median = 50, Median will be a better measure of central tendency.

**37** One extreme outlier

**39** (a) For SSGB, $\overline{X} = 110.63, S = 3.70$; for non-SGGB, $\overline{X} = 98.25, S = 4.06$. (b) For SSGB, CV = 3.35; for non-SGGB, CV = 4.13; non-SSGB has greater variation.

**41** Normal

## Chapter 4

**1** {2, 3, 4}, {2, 3, 5}, {2, 3, 6}, {2, 4, 5}, {2, 4, 6}, {2, 5, 6}, {3, 4, 5}, {3, 4, 6}, {3, 5, 6}, {4, 5, 6}
$\mu = 4, \sigma^2 = 2, \overline{y}_1 = 3, \overline{y}_2 = 10/3, \overline{y}_3 = 11/3, \overline{y}_4 = 11/3, \overline{y}_5 = 4, \overline{y}_6 = 13/3,$
$\overline{y}_7 = 4, \overline{y}_8 = 13/3, \overline{y}_9 = 14/3, \overline{y}_{10} = 5, E(\overline{y}) = 4, V(\overline{y}) = 0.333$

**3** $\overline{y} = 9.5046$, estimate of margin of error = 0.1314

**5** $\hat{\mu} = 875$ kw, $\hat{T} = 5,250,000$ kw, CI = (464, 1236)

**9** (1083.444 – 1518.956), (6,500,664 – 9,113,736)

**11** $E_\mu = \pm \$0.28634$, $E_T = \pm \$14,317$; ($23.76246, $24.23514); ($1,185,623 – $1,214,257)

**13** $\hat{\mu} = 57360$, $\hat{T} = 286,800,000$

**15** 243

**17** (28,885.45 – 31,344.85); (53,611,055.66 – 58,176,381.14)

*Statistical Quality Control: Using Minitab, R, JMP, and Python*, First Edition. Bhisham C. Gupta.
© 2021 John Wiley & Sons, Inc. Published 2021 by John Wiley & Sons, Inc.
Companion website: www.wiley.com/go/college/gupta/SQC

**19**  58
**21**  135
**23**  $7,200; $6,000,000
**25**  933

# Chapter 5

**7**  If the sample size is increased, then the value of $\beta$ decreases; ARLs = 15, 5, 2, 1, 1
**9**  44, 15, 6, 2, and 1 hr
**11**  $\hat{\sigma} = 5.417$; $LCL = 33.2298, UCL = 47.7702$; $LCL = 0, UCL = 26.6364$
**13**  $\hat{\sigma} = 6.945$; $LCL = 49.8253, UCL = 70.6747$; $LCL = 0, UCL = 32.6326$
**15**  $\hat{\mu} = 20.066, \hat{\sigma} = 0.7021; 17.73\%$
**17**  Same conclusions
**19**  Nonrandom pattern; process not in statistical control
**21**  Process not in statistical control
**23**  Process not in statistical control
**25**  Process in statistical control
**27**  0.60%
**29**  $\hat{\mu} = 10.019, \hat{\sigma} = 0.5817; (8.9767, 11.0611)$
**31**  Process in statistical control; 370
**33**  (a) 0.5% (b) (1.6421, 2.3439)
**35**  Process not in statistical control
**37**  X-bar chart UCL = 16.6886, LCL = 13.3114; S chart UCL = 2.4719, CL = 1.1833, LCL = 0
**39**  Process not in statistical control

# Chapter 6

**1**  A *p chart* plots the fraction of nonconforming units in a sample, and sample size may vary, and *np* chart plots the number of nonconforming units in a sample, and sample size is constant.
**3**  Process is in statistical control; two control charts are identical except for the y-scale
**5**  An appropriate control chart is a *u* chart; process not in statistical control
**7**  UCL = 14.29, CL = 6.76, LCL = 0; process is in statistical control
**9**  *u* chart; process is in statistical control
**11**  (a) n = 142; (b) $\alpha = 0.0013$, $\beta = 0.9734$; (c) $\alpha = 0.0065$, $\beta = 0.9506$
**13**  UCL = 0.08250, CL = 0.04063, LCL = 0; (b) process is in statistical control, so the control chart remains the same
**15**  UCL= 0.1014, CL = 0.0487, LCL = 0; (b) process is in statistical control, so the control chart remains the same
**19**  UCL = 32.55, CL =19.45, LCL = 6.35; (b) process is in statistical control, so the control chart remains the same
**21**  (a) $\hat{p} = 0.0333$; (b) $UCL = 6.17$ $CL = 2$ $LCL = 0$; (c) 0.027
**23**  $UCL = 0.0298$ $CL = 0.0171$ $LCL = 0.0044$
**25**  0.9603, 0.9222
**27**  UCL =27, LCL =12 or UCL = 28, LCL =13; with normal app., UCL = 27.356, LCL = 12.644
**29**  UCL = 13.3541, CL = 10, LCL = 6.6459

# Chapter 7

3  CUSUM control charts plot the cumulative sum of the deviations of sample values (or sample means, if the sample size is greater than one) from the target value.

5  $S_i^+ = Max(0, \overline{Z}_i - k + S_{i-1}^+)$, $S_i^- = Min(0, \overline{Z}_i + k + S_{i-1}^-)$

7  The shift was detected at the 10th sample, whereas in Problem 6, it was detected at the third sample.

9  The conclusions in this problem and in Problem 8 are the same.

11  Process variability has increased.

13  The process is in statistical control where as in problem 6, the CUSUM chart showed that there is a downward shift.

15  The conclusions in Problem 12 and in this problem are the same.

17  In both problems, the conclusions are the same.

19  In this problem, the conclusion is the same as in Problem 18. However, in this problem, the control chart shows a trend.

21  Process is in statistical control. However, the control chart shows a trend.

23  All three control charts show that the process is in statistical control.

25  Process is in statistical control

27  Process is in statistical control

29  Process is not in statistical control. In Problem 28, we had to take eight observations to detect the shift; however, in this problem, we had to take just six observations to detect the shift.

31  Process is not in statistical control

33  Process is not in statistical control

35  Process is in statistical control

37  Process is in statistical control

39  The conclusions in this problem and Problem 37 are the same.

41  Process is in statistical control. However, the control limits and the pattern of the plotted points have changed.

43  Process is in statistical control, and the conclusions in this problem are the same as in Problem 38.

45  $\hat{\mu} = 6.7685, \hat{\sigma} = 0.2894$, and the process is not in statistical control

47  Process is in statistical control. However, the control limits and the pattern of the plotted points have changed.

49  With $m = 2$, the MA control chart shows the process in control; but with $m = 4$, it shows the process is not in control.

51  $\hat{\mu} = 21.7875, \hat{\sigma} = 1.6883$, the process is in statistical control

53  Process is in statistical control. However, both control charts show that the process has a trend.

# Chapter 8

1  (a) 1.026; (b) 0.2%

3  $\hat{C}_{pl} = 1.026, \hat{C}_{pu} = 1.026; \hat{C}_{pl} = 1.282, \hat{C}_{pu} = 0.769$. In Problem 2, the percentage of nonconforming units has increased above the specification limit

7  $\hat{C}_{pk} = 0.708$; percentage of nonconforming units = 1.88%

9  $\hat{c}_p = 2.0, \hat{c}_{pk} = 2.0; \hat{c}_p = 1.67, \hat{c}_{pk} = 1.67; \hat{c}_p = 1.429, \hat{c}_{pk} = 1.429; \hat{c}_p = 1.25, \hat{c}_{pk} = 1.25; \hat{c}_p = 1.10, \hat{c}_{pk} = 1.10$

**11**  2

**13**  1.333, 0.533, 0.537

**15**  2.0, 2.0, 1.0; 1.833, 1.789, 0.7454; 1.67, 1.414, 0.4715; 1.5, 1.109, 0.2773; 1.333, 0.894, 0.1491

**21**  $\hat{C}_{pm} = 1.014$, $\hat{C}_{pmk} = 1.014$, $\hat{C}_{pnst} = 0.014$

**23**  (a) Process not capable; (b) $\hat{C}_{pm} = 0.4375$, $\hat{C}_{pmk} = 0.4375$, $\hat{C}_{pnst} = -0.5625$

**25**  (39,51)

**27**  (a) 1.333; (b) (1.006, 1.660)

**29**  (0.8607, 1.1393)

**31**  yes

**33**  (b) 0.135%, 0.003%

**35**  (a) 1.2026; (b) 2.184; (c) 5.79

**37**  (a) 0.65; (b) 0.0; (c) 0.65; (d) 0.0667

**39**  (a) 31.34; (b) 994; (c) same conclusion

# Chapter 9

Most of the solutions of this chapter are either graphs or large tables. For answers, see the student solution manual, which is available for download via the book's website: www.wiley.com/college.gupta/SQC.

# Bibliography

ANSI/ASQ. (2003a). *ANSI/ASQ Z1.4-2003: Sampling Procedures and Tables for Inspection by Attributes.* Milwaukee, WI: ASQ Quality Press.

ANSI/ASQ. (2003b). *ANSI/ASQ Z1.9-2003: Sampling Procedures and Tables for Inspection by Variables for Percent Nonconforming.* Milwaukee, WI: ASQ Quality Press.

ASQ Statistics Division (2004). *Glossary and Tables for Statistical Quality Control*, 4th ed. Milwaukee, WI: ASQ Quality Press.

Automotive Industry Action Group (AIAG). (2002). *Measurement Systems Analysis (MSA) Reference Manual*, 3rd ed. Dearborn, MI.

Automotive Industry Action Group (AIAG). (2010). *Measurement Systems Analysis (MSA) Reference Manual*, 4th ed. Dearborn, MI.

Barrentine, L. (2003). *Concepts for R&R Studies*, 2nd ed. Milwaukee, WI: ASQ Quality Press.

Bauer, K. (2003). "Who Makes Six Sigma Work?" *DM Review* **13** (8).

Benbow, D., R. Berger, A. Elshennawy, and H.F. Walker. (2007). *The Certified Quality Engineering Handbook*, 2nd ed. Milwaukee, WI: ASQ Quality Press.

Bothe, D.R. (2002). Discussion paper. *Journal of Quality Technology* **34** (1): 32–37.

Boyles, R.A. (1991). "The Taguchi Capability Index." *Journal of Quality Technology* **23** (1): 17–26.

Brown, N. (1966). "Zero Defects the Easy Way with Target Area Control." *Modern Machine Shop Magazine*, 96–100.

Burr. W. (1976). *Statistical Quality Control Methods*. New York, Marcel Dekker.

Champ, C.M., and W.H. Woodall. (1987). "Exact Results for Shewhart Control Charts with Supplementary Run Rules." *Technometrics* **29** (4): 393–399.

Chan, L.K., S.W. Cheng, and F.A. Spring. (1988). "A New Measure of Process Capability: $C_{pm}$." *Journal of Quality Technology* **20**: 162–175.

Chen, S.M., and N.F. Hsu. (1995). "The Asymptotic Distribution of the Process Capability Index $C_{pmk}$." *Communications in Statistical Theory and Method* **24** (5): 1279–1291.

Cochran, W.G. (1977). *Sampling Techniques*, 3rd ed. New York: John Wiley and Sons.

Crosby, P.B. (1979). *Quality Is Free: The Art of Making Quality Certain*. New York: McGraw-Hill.

Crowder, S.V. (1987). "Computation of ARL for Combined Individual Measurements and Moving Range Charts." *Journal of Quality Technology* **19** (1): 98–102.

Crowder, S.V. (1989). "Design of Exponentially Weighted Moving Average Schemes." *Journal of Quality Technology* **21** (2): 155–162.

Goetsch, D.L., and S.B. Davis. (2006). *Quality Management: Introduction to Total Quality Management for Production, Processing, and Services*, 5th ed. Pearson.

Deleryd, M. (1996). "Process Capability Studies in Theory and Practice." Licentiate thesis, Lulea University of Technology, Lulea, Sweden.

*Statistical Quality Control: Using Minitab, R, JMP, and Python*, First Edition. Bhisham C. Gupta.
© 2021 John Wiley & Sons, Inc. Published 2021 by John Wiley & Sons, Inc.
Companion website: www.wiley.com/go/college/gupta/SQC

Deming, W.E. (1950). *Some Theory of Sampling.* New York: John Wiley and Sons.

Deming, W.E. (1951). *Elementary Principles of the Statistical Control of Quality*, 2nd ed. Tokyo: Nippon Kagaku Gijutsu Remmei.

Deming, W.E. (1975). "On Some Statistical Aids toward Economic Production." *Interfaces* **5** (4): 5.

Deming, W.E. (1982). *Quality, Productivity, and Competitive Position.* Cambridge, MA: MIT, Center for Advanced Engineering Study.

Deming, W.E. (1986). *Out of the Crisis.* Boston, MA: MIT, Center for Advanced Engineering Study.

Dodge, H.F., and H.G. Romig (1959). *Sampling Inspection Tables, Single and Double Sampling*, 2nd ed. John Wiley and Sons.

Duncan, A.J. (1986). *Quality Control and Industrial Statistics*, 5th ed. Homewood, IL: Irwin.

Eisenhart, C. (1963). "Realistic Evaluation of Precision of Accuracy of Instrument Calibration System." *Jour. Research and Instrumentation* **67** (2). National Bureau of Standards.

Ford Motor Company. (1984). *Continuing Process Control and Process Capability Improvement.* Dearborn, MI.

Feigenbaum, A.V. (1991). *Total Quality Control*, 3rd ed. McGraw-Hill.

Govindarajulu, Z. (1999). *Elements of Sampling Theory and Methods.* Upper Saddle River, NJ: Prentice Hall.

Gryna, F.M., R.C.H. Chua, and J.A. DeFeo. (2007). *Juran's Quality Planning and Analysis for Enterprise Quality*, 5th ed. New York: McGraw-Hill.

Gupta, B.C. (2005). "A New Process Capability Index $C_{pnst}$." *Journal of Combinatorics, Information & System Sciences* **30**, 67–79.

Gupta, B.C., and C. Peng. (2004). "On the Distributional and Inferential Characterization of the Process Capability Index $C_{pnst}$." *Journal of Statistics and Applications* **1** (1): 63–72.

Gupta, B.C., and Guttman, I. (2013). *Statistics and Probability with Applications for Engineers and Scientists.* John Wiley and Sons.

Gupta, B.C., Guttman, I., and Kalanka, P.J. (2020). *Statistics and Probability with Applications for Engineers and Scientists*, 2nd ed., John Wiley & Sons.

Hawkins, D.M. (1981). "A CUSUM for a Scale Parameter." *Journal of Quality Technology* **13** (4): 228–231.

Hawkins, D.M. (1993). "Cumulative Sum Control Charting, an Underutilized SPC Tool." *Quality Engineering* **5** (3): 463–477.

Hawkins, D.M., and D.H. Olwell. (1998). *Cumulative Sum Charts and Charting for Quality Improvement.* New York: Springer-Verlag.

Hsiang, T.C., and G. Taguchi. (1985). "A Tutorial on Quality Control and Assurance: The Taguchi Methods." ASA annual meeting, Las Vegas, NV.

IBM (International Business Machines Corporation). (1984). *Process Control, Capability and Improvement.* Thornwood, NY: The Quality Institute.

Juran, J., and A.B. Godfrey. (1999). *Juran's Quality Handbook*, 5th ed. New York: McGraw-Hill.

Juran, J.M., F.M. Gryna, and R.S. Bingham Jr. (1974). *Quality Control Handbook.* New York: McGraw-Hill.

Juran, J.M., and F.M. Gryna, Jr. (1980). *Quality Planning and Analysis*, 2nd ed. New York: McGraw-Hill.

Kane, V.E. (1986). "Process Capability Indices." *Journal of Quality Technology* **18**: 41–52.

Kotz, S., and C.R. Lovelace. (1998). *Process Capability Indices in Theory and Practice.* London and New York: Arnold.

Ledolter, J., and A. Swersey. (1997). "An Evaluation of PRE-Control." *Journal of Quality Technology* **29** (2): 163–171.

Lohr, S.L. (1999). *Sampling: Design and Analysis.* Pacific Grove, CA: Duxbury Press.

Lucas, J.M. (1982). "Combined Shewhart-CUSUM Quality Control Schemes." *Journal of Quality Technology* **14** (2): 51–59.

Lucas, J.M., and R.B. Crosier. (1982). "Fast Initial Response for CUSUM Quality Control Schemes." *Technometrics* **24** (3): 199–205.

Lucas, J.M., and M.S. Saccucci. (1990). "Exponentially Weighted Moving Average Control Schemes: Properties and Enhancements." *Technometrics* **32** (1): 1–29.

Montgomery, D.C. (2005). *Design and Analysis of Experiments*, 6th ed. New York: John Wiley and Sons.

Montgomery, D.C. (2013). *Introduction to Statistical Quality Control*, 5th ed. New York: John Wiley and Sons.

Munro, R.A., et al. (2008). *The Certified Six Sigma Green Belt Handbook*. Milwaukee, WI: ASQ Quality Press.

Neubauer, D.V., and S.N. Luko. (2013a). "Comparing Acceptance Sampling Standards." *Quality Engineering* **25** (1): 73–77.

Neubauer, D.V., and S.N. Luko. (2013b). "Comparing Acceptance Sampling Standards." *Quality Engineering* **25** (1): 181–187.

Oakes, D., and R. Wescott, editors. (2001). *The Certified Quality Manager Handbook*, 2nd ed. Milwaukee, WI: ASQ Quality Press.

Ott, E., E. Schilling, and D. Neubauer (2005). *Process Quality Control: Troubleshooting and Interpretation of Data*, 4th ed. Milwaukee, WI: ASQ Quality Press.

Pearn, W.L., and S. Kotz. (1994). "Application Element's Method for Calculating Second and Third Generation Process Capability Indices for Non-normal Pearsonian Populations." *Quality Engineering* **7** (1): 139–145.

Pearn, W.L., S. Kotz, and N.L. Johnson. (1992). "Distributional and Inferential Properties of Process Capability Indices." *Journal of Quality Technology* **24**: 216–231.

Quesenberry, C.P. (1997). *SPC Methods for Quality Improvement*. New York: John Wiley and Sons.

Roberts, S.W. (1959). "Control Chart Tests Based on Geometric Moving Averages." *Technometrics* **1** (3): 239–250.

Ryan, T.P. (2000). *Statistical Methods for Quality Improvement*, 2nd ed. New York: John Wiley and Sons.

Scheaffer, R.L., et al. (2006). *Elementary Survey Sampling*, 6th ed. Belmont, CA: Duxbury Press.

Schilling, E.G. (1982). *Acceptance Sampling in Quality Control*. New York: Marcel Dekker.

Schilling, E.G., and D.V. Neubauer. (2009). *Acceptance Sampling in Quality Control*. CRC Press.

Shainin, D. (1984). "Better Than Good Old X-Bar and R Charts." In *ASQC Quality Congress Transactions*, 302–307. Milwaukee, WI: ASQC Quality Press.

Shainin, D., et. al (1954). *A Simple Effective Process Control Method, Report 54-1*, Rath and Strong, Inc.

Shewhart, W.A. (1931). *Economic Control of Quality of Manufactured Product*. New York: Van Nostrand.

Shewhart, W.A. (1980). Economic Control of Quality of Manufactured Product. ASQ. Published by Quality Press, ISBN 10:0873890760.

Siegmund, D. (1985). *Sequential Analysis: Tests and Confidence Intervals*. New York: Springer-Verlag.

Smith, G. (1991). *Statistical Process Control and Quality Improvement*. New York: Macmillan Publishing.

Spiring, F.A. (1997). "A Unifying Approach to Process Capability Indices." *Journal of Quality Technology* **29** (1): 49–58.

Statistical Research Group, Columbia University. (1945). *Sequential Analysis of Statistical Data: Application*. New York: Columbia University Press.

*Statistics for Engineering and Physical Science*. New York: Springer-Verlag.

Stephens, K.S. (2001). *The Handbook of Applied Acceptance Sampling Plans, Procedures and Principles*. Milwaukee, WI: ASQ Quality Press.

Taguchi, G. (1986). *Introduction to Quality Engineering*. Tokyo: Asian Productivity Organization.

Tague, N.R. (2005). *The Quality Toolbox*, 2nd ed. Milwaukee, WI: ASQ Quality Press.

Traver, R. (1985). "Pre-Control: A Good Alternative to X-Bar–R Charts." *ASQC Quality Progress* (September): 11–14.

Tripathy, P. (2016). "Feigenbaum's Philosophy on Total Quality Management, Leadership and Management." *Quality Progress Magazine*.

Urdhwareshee, H. (2006). "The Power of Pre-Control." www.symphonytech.com/articles/precontrol.htm.

Uselac, S. (1993). *Zen Leadership: The Human Side of Total Quality Team Management*. Loudonville, OH: Mohican.

Van Dobben de Bruyn, C.S. (1968). "Cumulative Sum Tests." In: *Theory and Practice*. London: Griffin.

Vännman, K. (1995). "A Unified Approach to Capability Indices." *Statistica-Sinica* **5**: 805–20

Wald, A. (1973). *Sequential Analysis*. New York: Dover.

Wallgren, E. (1996). "Properties of the Taguchi Capability Index for Markov Dependent Quality Characteristics." Department of Statistics, University of Oreboro, Sweden.

Watson, G.H. (2005). "Feigenbaum's Enduring Influence." *Quality Progress*.

Weeks, J.B (2011). "Is Six Sigma Dead?" *Quality Progress* **44** (10).

Wescott, R. (2007). *The Certified Manager of Quality/Organizational Excellence Handbook*, 3rd ed. Milwaukee, WI: ASQ Quality Press.

Western Electric. (1956). *Statistical Quality Control Handbook*. Indianapolis, IN: Western Electric Corporation.

Wheeler, D., and D. Chambers. (1992). *Understanding Statistical Process Control*, 2nd ed. Knoxville, TN: SPC Press.

Wheeler, D., and R. Lyday. (1989). *Evaluating the Measurement Process*, 2nd ed. Knoxville, TN: SPC Press.

White House. (2011). "White House Announces Steps to Expedite High Impact Infrastructure Projects to Create Jobs." www.whitehouse.gov/the-press-office/2011/08/31/white-house-announces-steps-expedite-high-impact-infrastructure-projects.

## End Notes

[1] http://asq.org/public/six-sigma-training/asqsigma.pdf.

[2] http://www.isixsigma.com/implementation/financial-analysis/six-sigma-costs-and-savings/

[3] US Army. *Stand-To*, April 11, 201. http://www.army.mil/standto/archive/2011/04/11/

[4] The White House. White House Announces Steps to Expedite High Impact Infrastructure Projects to Create Jobs, August 31, 2011. http://www.whitehouse.gov/the-press-office/2011/08/31/white-house-announces-steps-expedite-high-impact-infrastructure-projects

[5] Weeks, J. Bruce (2011). *Is Six Sigma Dead*? Quality Progress, Vol. 44, No. 10.

# Index

## A

Acceptable quality level (AQL) defined   288, 292
Acceptable quality limit (AQL) defined   292
Acceptance sampling by attributes   292–301
Acceptance sampling plan
  intent of   283
  lot-by-lot   283
Acceptance sampling plans   7
Action on the output   133
Action on the process   133
Actions on the system   134
Advantages of Pre-control   260
ANOVA, MSA based on   268–272
ANOVA table with interactions   269
ANOVA table without interactions   270
ANOVA with interaction   269
ANSI/ASQ Z1.4-2003, types of
  sampling standard and plans   305, 309
  levels of inspection   306
ANSI/ASQ Z1.9-2003 plans
  structure and organization of   311
ANSI/ASQ Z1.9-2003, types of
  sampling standard plans   309
  with variable known   309
ANSI/ASQ Z1.9-2003 variable unknown
  range method   313
  standards deviation method   311
Appraiser variation   262
Assignable causes of variation   124, 134
Attribute   173
Average nonconforming   175
Average outgoing quality (AOQ)   286, 288, 292
Average outgoing quality limit (AOQL)
  defined   288, 293

Average run length (ARL)   138, 205
Average sample number (ASN)   301
Average total sampling (ATI)   286, 293

## B

Bar charts   36–41
Basic seven tools   20
Belts, types of   18–19
Binomial distribution   63–65, 175
Bins   32
Box-and-whisker plot   61
Brainstorming   129

## C

Categories   30
Cause-and-effect diagram   126, 129
Center line (CL)   133, 135, 175, 180, 222, 224
Champion or Sponsor   18
Characteristic of measurement system   263
  precision of   263
  accuracy of   263
  stability of   263
  linearity of   263
Check sheet   126
Chi-square distributions   246
Classes   32
Classification of sampling plans   284
Class width   32
Cluster random sampling   91, 92, 112–116
Clusters   112
Combined Shewhart-CUSUM, control
  chart   216–217
Common causes of variation   124
Common ratio   224

*Statistical Quality Control: Using Minitab, R, JMP, and Python*, First Edition. Bhisham C. Gupta.
© 2021 John Wiley & Sons, Inc. Published 2021 by John Wiley & Sons, Inc.
Companion website: www.wiley.com/go/college/gupta/SQC

Index for Chapter 10- Computer Resources to support SOC: MINITAB, R, JMP and Python, that can be downloaded from the book's website  www.wiley.com/college/gupta/SQC.